SSA: Orbital Debris, Space Weather, and Planetary Defense

For a listing of recent titles in the
Artech House Intelligence and Information Operations Library,
turn to the back of this book.

SSA: Orbital Debris, Space Weather, and Planetary Defense

John A. Kennewell

ARTECH
HOUSE
BOSTON | LONDON
artechhouse.com

Library of Congress Cataloging-in-Publication Data
A catalog record for this book is available from the U.S. Library of Congress.

British Library Cataloguing in Publication Data
A catalog record for this book is available from the British Library.

ISBN 13: 978-1-68569-111-0

Cover design by Andy Meaden

Artech House
685 Canton St.
Norwood, MA 02062

10 9 8 7 6 5 4 3 2 1

To my family, who kept me going

Contents

Foreword

The Earth and near-Earth space environments are being challenged by a wide variety of natural and man-made hazards. Nearly 100 spacefaring countries are operating over 10,000 satellites in this limited space but with seemingly limitless external stimuli that are capable of disrupting spacecraft operations. Dr. John Kennewell has taken a holistic approach to his mission of preparing us for the years ahead—space is a dynamic ecosystem that provides equal measures of opportunities and challenges.

John was born before any artificial satellites were orbiting the Earth and he started his career in the early 1960s when fewer than 300 man-made objects orbited the Earth; the vast majority of these objects were already orbital debris. In 1965, he published his first paper, "Determination of Satellite Orbits," when most global citizens were looking skyward for stars and inspiration. He was already building a strong technical foundation of observational techniques, signal processing, scientific models, and empirical experimentation necessary to address emerging space threats and helping mankind prepare to reliably operate in the harsh, unforgiving space environment.

I have been active in the orbital debris community since the mid-1980s when I was the first person to receive a PhD with an emphasis in orbital debris and then coauthored the first book on the topic in 1990 with Mr. Nicholas Johnson. I have worked diligently to contribute to the global orbital debris enterprise by addressing specific weaknesses from my work on breakup modeling through the first impact test of an actual man-made satellite to the establishing of the fragmentation threshold for all current breakup models. I further developed the first ever breakup cause diagnostics, identified highly dangerous clusters of massive intact derelicts, and led the development of the first top internationally selected 50 statistically-most-concerning objects in low Earth orbit. Through this 40-year career, I have emphasized the need for both content and context to solve difficult problems.

In Dr. Kennewell's book *SSA: Orbital Debris, Space Weather, and Planetary Defense*, we get that rare balance of context and content that will make this book valuable for readers for years to come. Orbital debris is a man-made hazard that is evolving at different rates in different ways in different orbits. This is not a simple situation to describe—it requires continual new observations, an ability to challenge norms of thinking, and application of advanced scientific techniques. These three features are brought together seamlessly in this book.

The dimension of space weather as a direct challenge and a catalyst for a series of ripple effects through the space community is nicely laid out in this book. The volatile nature of our Sun is depicted, and its influence is traced to key aspects of satellite reliability due to radiation effects and the direct influence on the atmospheric drag profile for spacecraft in low Earth orbit and spacecraft charging in all orbits. The tracing of these effects to specific orbital regimes and satellites reinforces the adage that "the only generalization that is true is that no generalizations are true." Any serious treatise on space situational awareness must provide the reader with the technical skepticism to reject generalities that oversimplify a complex storyline. John provides practical scientific guardrails throughout the book to assist both the novice and experienced to navigate this domain with a clarity and practicality that is refreshing.

Planetary defense is a welcome topic to broaden the content of the book that also helps to focus the reader on the key signature for all the challenges posed to the global space community. This signature is one of risk identification, assessment, and management. Space is a complicated environment and there are hazards everywhere, however, by focusing on risk in its purest sense—that is to say, risk is probability times consequence—the complex web of risks can be dismantled.

This book is organized in a crafty way with extensive references in every chapter, going back to the basics before covering the advanced topics, and dealing with the life cycle of orbital debris. However, the truly intriguing parts of the book relate to personal experiences, "cheat sheets" on specific focused topics (suitable for pasting on your desktop), and not being afraid to talk about space security. As a matter of fact, this book provides a balanced examination of space safety, space sustainability, and space security—these domains are indeed related, and this book helps to pull that critical intellectual thread.

I applaud the efforts of Dr. Kennewell to create complete coverage of space situational awareness. However, the key aspect of this book is to also be coherent (i.e., material arranged in a logical, self-reinforcing order) and compelling (i.e., continually reminding the reader why and how).

Enjoy reading this book that is half a thorough, well-documented reference suitable to keep at the ready on a shelf near your desk and half a call to arms to start to address risks to space operations in a meaningful manner.

Darren S. McKnight
PhD, University of Colorado, Aerospace Engineering Sciences
Senior Technical Fellow, LeoLabs
November 2025

Preface

There has never been a more pressing time to be aware of the contents and activities in our near-space environment, which is referred to as space situational awareness (SSA). The first widespread introduction to what is going on above our heads was on October 4, 1957, when the Soviets launched *Sputnik 1* into Earth orbit. This event jolted the American military and political establishments and marked the first time that an awareness of the situation in space added to our collective consciousness.

Since then SSA has only grown in importance with the recognition that one who rules the high ground will rule the world. While peaceful activities abound in outer space and state and international legislation has been passed to try and keep it that way, the head of the US Space Force, in a recent address, has acknowledged that US space assets are now routinely under attack daily. At present the attacks have been confined to the cyber domain, but hostile rendezvous and proximity operations (RPO) are increasing and it will only be a matter of time before they become involved with physical property.

It is thus very important that we become acquainted with the many facets of space hazards that exist—those due to both the built environment and the natural environment. I have tried in this text to address many of these issues, from satellites to meteoroids, asteroids and comets, and particularly the electromagnetic environment and the collisional interactions of these bodies.

I would like to thank all those from whom I learned the material that is presented here—teachers, students, colleagues, and the authors of many papers and texts. I would like to thank Artech House and the people there who invited me to write this book and who spent many months helping edit the text and taking me through the production process.

CHAPTER 1
Introduction

1.1 Overview

The overall themes covered in this book are summarized in Figure 1.1. I have tried to concentrate mostly on the principles involved, although there have necessarily been many real-world examples. I have also used the European definition of the term SSA or space situational awareness. If we are going to be aware of the space situation, we need to consider the total space environment, or at least as much of it as affects our activities. Even galactic cosmic radiation, which originates from the massive explosions of stars (supernovae) located around our galaxy provides a limit to our space operations, particularly those that involve humans.

In this introduction I briefly introduce the topic of SSA and follow with an overview (Chapter 2), which includes a section on the major space tracking facilities around the world. This is followed by an overview specifically related to orbital space debris (Chapter 3). We then come to the physical heart of SSA—orbital

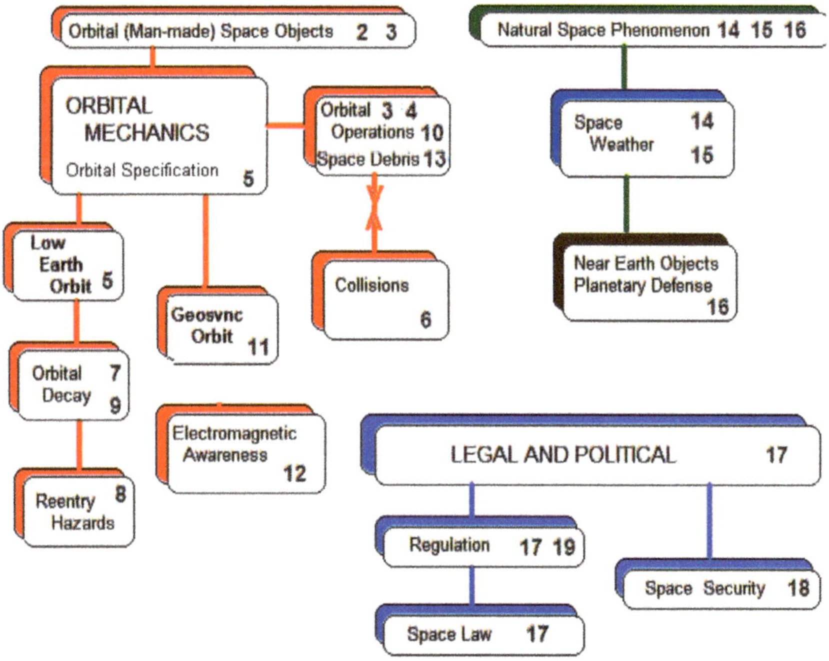

Figure 1.1 Graphical summary of book contents. The numbers refer to chapters in which the material is covered.

mechanics—for it is gravity that is the primary manager of orbital space objects. Newton's laws of mechanics are the basis of all orbital motion. Kepler's three laws can actually be derived from Newton's laws, so the latter are the more fundamental. Of the three laws, it is actually the second that allows us to compute orbits and figure out how much fuel we need to put a satellite into orbit. Many people think of Newton's second law as $\mathbf{F} = m\mathbf{a}$ where $\mathbf{F}$ is the force necessary to give a body of mass m an acceleration of $\mathbf{a}$. However, that is not how Newton stated the law. He gave the more general formulation $\Sigma\mathbf{F} = dp/dt$. This reads that the resultant (or sum) of the impressed forces on a body give rise to a time rate of change of momentum ($\mathbf{p} = m\mathbf{v}$) where $\mathbf{p}$ is the momentum of the body. (Note that bolded letters are vectors with both magnitude and direction.)

It always amazes me that Newton formulated the law in this more general form, because for most terrestrial applications where the mass of a body is constant, the more general form does indeed reduce to $F = ma$. However, when we go into space we use some form of rocket propulsion where mass is pushed out the back and so the rocket becomes progressively lighter. This is where $F = ma$ fails and we must use the more general form, which when expanded becomes $F = m\,dv/dt + v\,dm/dt$.

We should also realize that because the Earth is not a sphere, the gravitational forces acting on a satellite at a fixed height above the surface are not constant. The presence of other celestial bodies (e.g., the Sun and the Moon) also introduce a small perturbation on the gravitational force in orbit, so when we specify the orbit of a satellite or piece of debris we must realize that our orbital specification is only valid for a certain period of time. That period depends on what type of orbit we are dealing with.

Other factors, such as atmospheric drag in low Earth orbit (LEO), and the solar wind and radiation pressure at higher altitudes will also produce a small perturbation to the orbit that is mainly defined by gravity. For low altitude orbits, atmospheric drag will cause the decay (Chapter 7) of the orbit and eventual reentry into the lower atmosphere of an orbital object where it may ablate (burn up) completely (Chapter 9), or pose a hazard (Chapter 8) at the Earth's surface. For the foreseeable future, it appears that reentry hazards are only a minor factor in the overall scheme of space operations.

Orbital operations and accidental collisions (Chapter 6) will produce debris in a diverse spread of orbits, and this debris in turn will increase the probability of such collisions (Chapter 10). However, at the present time it appears that deliberate destruction of large space assets is more likely to result in higher levels of space debris than accidental collisions.

The geosynchronous orbit (GEO) (Chapter 11) is a special orbit with regard to terrestrial applications because satellites in such orbits appear to remain at a fixed point in the sky to an Earth observer. This means that ground stations do not have to track the satellite, a substantial reduction in cost and complexity. However, gravitational and other perturbations mean that the satellite owner has to *station keep* to maintain the satellite in a near geosynchronous orbit. In particular Chapter 11 discusses the observation and tracking of geosynchronous satellites (geosats).

Tracking of and communication with satellites is achieved via the use of electromagnetic energy, whether it be with radio or optical frequencies. An awareness of electromagnetic interactions (Chapter 12) is thus an essential part of SSA.

The other two parts of the SSA triad are then covered: Space weather (Chapters 14 and 15) and near-Earth objects (Chapter 16). Each of these has the potential to cause devastation of manned and unmanned space operations with varying probabilities.

Regulation of space operations is accomplished through space law of individual states (countries), treaties between states (either bilateral or multilateral), and through international space law (Chapter 17). The difficulty of getting international agreement by a diverse range of states with very diverse interests and beliefs has led to guidelines on space operations that may or may not be honored by all parties to such operations. At the present time there seems to be a growing development of space warfare techniques by major parties involved with space operations (Chapter 18).

Because some of the material in this book was first written over many years, it has been decided to relegate some more recent developments to the Epilogue (Chapter 19), rather than disrupt earlier text.

1.2 History

After nearly 70 years into the space age there are already worrying signs that our own actions may be soon come to deny us access to the lower parts of Earth's orbital space.

The space age began in 1957 with the launch of the first man-made satellite. It was also the time at which the first man-made space debris appeared. For every satellite launched into space there were also around 100 pieces of space debris placed into, or generated in, orbit. No one took much notice of this in the early years. Even today, the popular view is that the problem of space debris is due to accidental collisions. So far this has not been true, however, it is likely to be true in the future.

It was not until 1976 that the seminal paper by Donald Kessler and Burt Cours-Palais of the NASA Orbital Debris Program Office gave the prescient warning of a collisional cascade, a chain reaction in which space object collisions would exponentially multiply the amount of space debris in orbit. This will cause the whole of low Earth orbit to become unusable. At this time a satellite launched into orbit would suffer a collision within months if not days of its placement into orbit, and render it yet another useless piece of debris. This became known as the Kessler syndrome.

This, however, was not the situation in the early years. With a low Earth orbit space volume of around one trillion cubic kilometers, it appeared to be an almost infinite resource. The early satellites were placed in low Earth as it was believed that the expense of launch was a major factor in the commercial use of higher orbits. It was not until the successful launch of Syncom 2 into geosynchronous orbit that it was realized that this orbit offered much better commercial, government, and military opportunities. Gradually LEO was relegated to the military and scientific research.

It was not until later that there was a gradual resurgence of interest in LEO. This was mainly due to two areas of space commercialization: communication and high-resolution imaging of the Earth, useful for a multitude of applications. For the latter, LEO was required to achieve submeter image resolution. Even though GEO provided a very wide overview of the Earth below, it was not possible to achieve the high resolution the commercial world desired. Similarly, although GEO provided cheap high-bandwidth transcontinental communications, the one-half-second delay over a two-way space circuit was annoying at best, and unusable at worst for high speed communication.

In fact, for long distance (e.g., England to Australia) telephony a geosynchronous satellite was normally paired with an underwater fiber to reduce the two-way delay to a more manageable one-quarter second. It was not only the pressure of high-speed internet but also the development of affordable phased array antennae that led to the reconsideration of LEO for such communication.

The development of phased array antennae and associated electronics allowed for totally electronic tracking (no moving parts) of LEO communication satellites (commsats) by a user terminal. Networking technology, where one satellite is seamlessly handed off to another in a LEO megasatellite constellation, also made a large contribution to this development. Such technology had of course been highly developed for ground systems over several decades.

Meanwhile, even before this most recent development in LEO, initiated by the entrepreneur multibillionaire and space enthusiast Elon Musk, the collection of satellites in GEO had been growing. The two main areas of GEO space applications, communication and remote sensing, were well represented by both civilian and military interests. Civilian applications were divided into government and commercial interests. An interesting development, particularly in the United States, was that commercial interest, because of the money involved, began to outpace the others, and started providing extensive services from space at such a low price that governments preferred to outsource to these companies rather than develop their own use of the space sector.

The third main use of space, the provision of global navigation and timing services, was established and has grown up in middle Earth orbit (MEO). The four orbit regimes (LEO, MEO, GEO, and HEO) are shown in Figure 1.2. HEO is generally assumed to extend out to at least 100,000 km.

The geosynchronous orbit, where satellites appear to an Earth observer to be stationary in the sky, is a special orbit with major significance to satellite operators and users. The concepts of satellite and debris density and management are quite different than for low Earth orbit. This is treated in Chapter 11.

The desire to try and regulate space activities came to fruition between the mid-1960s and the early 1970s with the publication by the United Nations of five legal principles pertaining to operations in outer space, and the consequent passage of four outer space treaties. A fifth treaty was tabled but was never ratified by a sufficient number of countries for it to be enacted by the General Assembly. This fifth treaty, called the Moon treaty, attempted to reserve all of space for "the good of mankind" and sought to prevent individual countries (nation states in international legal parlance) from ever claiming to own any piece of territory anywhere in the universe.

The unrealistic wide scope of this treaty may have stemmed from a failure of the legal teams to realize the extent of the universe, and how much the treaty goes against historical precedent and the deep inbuilt desire of mankind to explore and own property and territory. There is also an idealistic standpoint, which sought to prevent the exploitation of celestial bodies by the more advanced nations to the detriment of the less developed ones. The treaty was of course based on a similar terrestrial treaty about Antarctica, but this only passed because (1) there was no adjacent territory, and (2) the Antarctica environment makes it extremely difficult

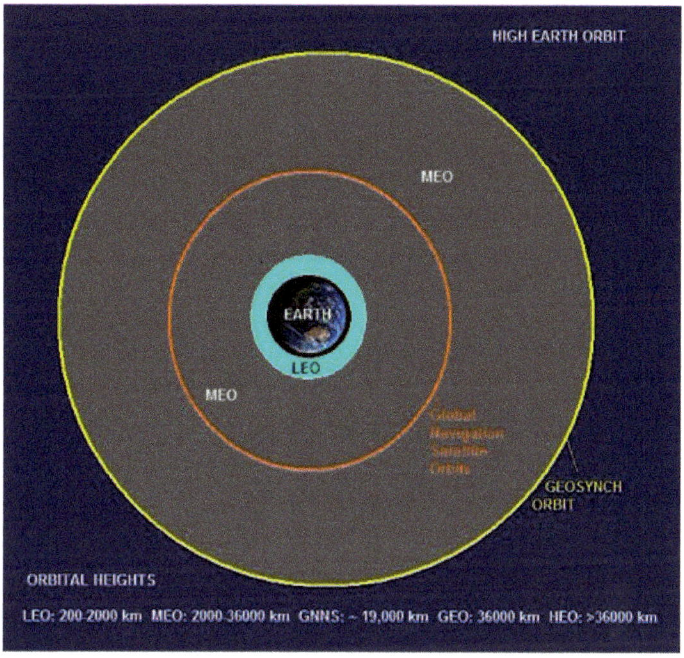

Figure 1.2 Earth space orbital regions.

to build, occupy, and maintain significant infrastructure. This has not prevented countries from claiming title, Australia once claimed one-third of the Antarctic land mass.

There is also a very strong interest among space faring nations to mine celestial bodies, particularly asteroids, for the water and precious metals they contain. The Australian Centre for Space Engineering Research (ACSER) hosted by the University of New South Wales, was the first in the world to run a continuing series of Off-Earth Mining forums. This is totally ironic given that Australia was the only developed country to ratify the Moon Treaty. This was pointed out in humor at the OEMF in 2013.

The failure of the Moon Treaty and a subsequent later attempt to pass a treaty on space debris does not bode well for any subsequent attempts to use international law to control and remediate this growing problem. It would appear that state law, and bilateral and multilateral treaties, are going to be the only way that space operations are controlled. If state law diverges between different nations or blocks of nations, we can only hope that it does not escalate into military action. But with the increasing number of state actors (around 50 as compared to 2 major players in the 1960s) and their increasing belligerence, the prospect of conflict is problematic.

Some authors consider that space situational awareness started with the Cold War, and virtually with the start of the space age (or space race as it sometimes called), simply because that is when the first space tracking systems started to be put in place. However, the use of the term space situational awareness, and the implications it is associated with, certainly come from a much later era. In fact I have been unable to find a reference to the first use of the term, but would tentatively ascribe it to a date after the turn of the century/millennium. Before that time papers or books on the main current concern of SSA would have titles with the words *space debris* in the heading. The first book devoted to this topic was *Artificial Space Debris* by Nick Johnson and Darren McKnight. Both of these authors, along with many similar pioneering people, were products of the US Air Force. Johnson went on to head the NASA Orbital Debris Program Office for many years but has since passed away. McKnight, by contrast, is now the senior technical fellow for the commercial firm LEOlabs, which runs a global S-band space tracking network and provides SSA products to various companies and agencies. Other companies provide similar services.

The last two decades have certainly seen a rapid growth in the terms related to SSA, and the number of groups with an interest in it. However, there is no internationally agreed definition of the term. It appears to have evolved differently in Europe than it did in the United States where a typical definition states that "space situational awareness is the knowledge and characterization of space objects and their operational environment," with the implication that the objects under discussion are man-made, resulting from human activity in the Earth orbital space. The European definition is usually much broader and gives acknowledgment to the fact that space, including Earth orbital space, is filled with a variety of particles and energy that affect man-made objects in that space. This definition thus includes the areas of space weather and what is now called planetary defense, in acknowledgment that the Earth has and will be subject to catastrophic collisions with asteroids and comets.

Space weather has led to several disasters and near disasters for space assets. In 1967 a missile-tracking system in North America experienced jamming of their radar, used in searching for enemy missiles traversing orbital space on their way to a nuclear detonation over the United States. But for the action of an astute officer versed in the basics of space weather, who pointed out that the jamming was due to solar radio emissions, a catastrophic event may have occurred. Space weather has also been responsible for the loss of quite a few military satellites. Surely we need to ask if there is a difference between the loss of a space asset due to a collision with a small piece of man-made debris, or due to a space weather event, or to the impact of a piece of natural space debris broken off from a comet or asteroid. We need to be aware of the possibility of all three situations.

The related term, SDA or space domain awareness, is also used differently by various stakeholders. The term appears to have originated in the US military in the latter part of the first decade of the twenty-first century, and in the United States is often defined (by comparison with terrestrial and marine domains) as "the ability of decision makers to understand, as completely as necessary, their current and predicted operational space environment." Then again, many parties use the terms SSA and SDA interchangeably (or to confuse and confound the enemy).

The extent of SSA is also another area that is rarely directly mentioned. Most of human activity in space is confined to Earth orbits below a height of 100,000 km and that might seem a logical limit in which to confine our interest. However, in the last decade we have seen a great interest in human-directed space activity in and around lunar orbit. We need to keep at least a minimal awareness of this activity lest it take us unaware.

There are many papers appearing on SSA topics and quite a number of volumes have appeared on its legal aspects. Because the scope of SSA is so great it is usually necessary to concentrate on specific areas; however, this book aims to give an overview of the topic but also to concentrate on the space physics underlying the subject.

There will be a discussion of orbits and how they are reported. This is an area fundamental to SSA, that of producing the most comprehensive catalog of resident space objects (RSO), either satellites or artificial debris, which orbit the Earth. We will consider collisions between RSOs and the debris that results. We will also consider the evolution of the space environment. Some simple models will be introduced to allow more detailed examination of these concepts. The models will be introduced through mathematical formula and/or algorithms and other general purpose language. The simplicity of a model sometimes belies its effectiveness, especially in cases where the space environment cannot be specified with precision.

The term space traffic management (STM) is often discussed as the natural outcome of SSA, or what we do with our *awareness* of the space domain. It is a concept developed mainly in the legal and political arena, although it is stated to have originated in 1982 by the Czech astronomer Lubos Perek (a polymath who worked on a diversity of problems from distant galaxies to the management of orbital space). I imagine he coined the term STM with a degree of ironic humor for it is indeed a misleading term to many people, especially those without a grounding in the physical laws that govern orbital motion. One does not *manage* space *traffic* in the sense that we manage traffic on the Earth. It is gravity that manages orbital motion, and smaller environmental perturbations that modify this motion. It is true that many satellites now have the capability to be directed to maneuver by their ground operators, and this applies to the large constellation of satellites that are now being launched into space, such as the SpaceX Starlink constellation. The International Space Station has had this capability since its birth, and in fact undergoes several collision avoidance maneuvers (COLA) every year. However, STM can only be applied to spacccraft that have such maneuvering capability, and which is activated when a detailed conjunction assessment (CONASS) indicates the possibility of a collision. Some military organizations like to try and conceal their satellites and so collisions with such satellites will not show up in a CONASS undertaken by commercial organizations. Similarly, impending impacts with small bodies under 10 cm in size may not reach a general awareness because of sensor limitation for bodies this size, and smaller. However, a hypervelocity impact is much more akin to an explosion than a low velocity collision, and destroys both the larger and smaller bodies involved. And it should be noted that nearly all collisions in low Earth orbit are hypervelocity collisions, a term we will define later.

The term STM is also, like SSA and SDA, defined differently by different people and organizations. However, it is a term that has caught on and is now used as an umbrella term for many different aspects of SSA responses, and over a diverse industry with a large annual expenditure (which is sure to perpetuate its use).

In conclusion, we should mention two recent developments. For several decades there has been a clear divide between the functions of satellites in low Earth orbit and those in geosynchronous orbit. Nearly all broadcast and communication satellites have been placed in GEO. In this orbit the satellite appears stationary from a ground station and the user can thus employ a simple fixed low-cost antenna that does not need to track the satellite it is using.

However, the disadvantage is that the distance to GEO introduces a non-negligible time delay that limits data rate in a two-way communication. This delay is around one-quarter of a second. For a two-way conversation this doubles to half a second, which is quite noticeable and can be annoying to some people. This is why communication companies usually only employ a GEO link on one side of the circuit in a transatlantic or transpacific circuit, with the other side carried by undersea cable. For high-speed two-way data links that require acknowledgment of packets sent, this GEO delay is simply intolerable.

High-speed communication links using LEO sats do not have the same delay, but require antenna tracking and hand-off technology. Until recently this has not possible for low-cost systems such as mobile phones. However, the development of phased array antenna steering and the use of cellular technology to switch between many satellites in a constellation has made LEO satellite communication more desirable now than GEO satellite systems. However, we still need to remember that the incredible effort by many engineers and financiers to bring this about in practice is a large part of the equation.

We also need to address the impact that these communication megaconstellations will have on the LEO environment with the increased risk of collision, and the need to develop and devote many more resources to the task of ensuring that this risk is kept under control.

The second development we need to consider in orbital object SSA is the push to establish permanent settlements on the Moon. This poses a whole new series of problems for spacecraft tracking and STM. Due to the distance to the Moon, only the largest ground telescopes will be able to even see a one-meter object, and only if the light from the Moon does not wash out the image. Passive radar (where the spacecraft does not carry a transponder) is totally out of the question. Thus even a large defunct satellite, which may pose a collision hazard, will not be detectable. Thus, it may be necessary to establish a series of SSA tracking sensors on or around the moon to accurately monitor the increasing lunar orbital object population.

CHAPTER 2
Overview of Space Situational Awareness

Space situational awareness means different things to different people, but in its broadest view refers to a knowledge of our near-space environment. This includes both a natural and man-made component. SSA is a term that has become prominent due to several collisions of orbiting space objects. This chapter describes the components of SSA and gives an overview of their significance.

2.1 Definitions

In the broadest sense, SSA may be defined as a knowledge of the energy and particle fluxes in near-Earth space, natural and artificial objects passing through or orbiting within this space, including the past, present, and future state of these components (Figure 2.1). The realm of near-Earth space may be left rather vague at this stage. It is definitely within cislunar space, but extends to an Earth radius of at least 100,000 km to include nearly all man-made objects currently in orbit.

Not everyone agrees with this definition. Some reserve the term only for macroscopic objects in near-Earth space. The space foundation [1] states that "Space situational awareness refers to the ability to view, understand, and predict the physical location of natural and man-made objects in orbit around the Earth, with the

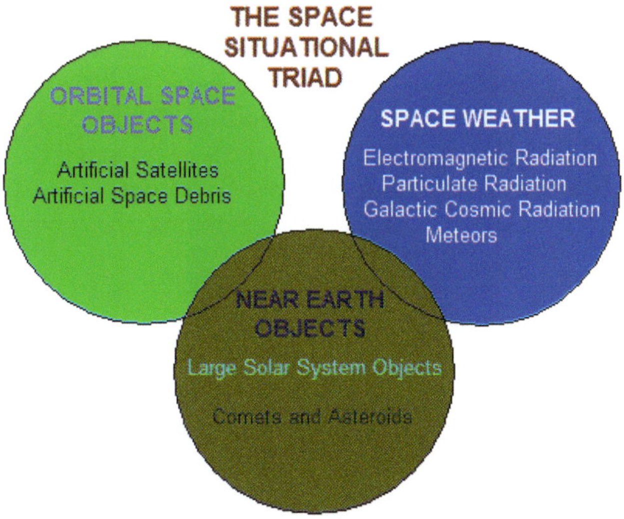

Figure 2.1 The three components of comprehensive SSA.

objective of avoiding collisions." Not only is this very restrictive, but it is not very useful, as most natural objects do not orbit the Earth, but rather transit through near-Earth space.

The European Space Agency [2], however, uses the fuller definition, and specifically lists three segments of knowledge in SSA: (1) SST: Space surveillance and tracking of objects in Earth orbit (watching for active and inactive satellites, discarded launch stages, and fragmentation debris that orbit the Earth). (2) SWE: Space weather (monitoring conditions at the Sun and in the solar wind, and in Earth's magnetosphere, ionosphere, and thermosphere, that can affect space-borne and ground-based infrastructure or endanger human life or health). (3) NEO: Near-earth objects (detecting natural objects that can potentially impact Earth and cause damage).

The US Strategic Command [3], which possesses the largest SST assets on the planet, defines SSA as "the requisite current and predictive knowledge of space events, threats, activities, conditions and space system (space, ground, link) status capabilities, constraints, and employment—to current and future, friendly and hostile—to enable commanders, decision makers, planners, and operators to gain and maintain space superiority across the spectrum of conflict." This includes not only the knowledge of the space segment, but also the ground-based capabilities that enable the knowledge, as well as specifying the reason that SSA is important to this organization.

A good SSA is invariably linked to threats and hazards, but SSA can also provide opportunities to both mitigate or reduce the hazards, and even to benefit from the potential resources present in both man-made and natural debris.

2.2 SSA Awareness

The previously mentioned Chinese ASAT test on January 10, 2007, produced the single largest cloud of space debris in our space-faring history–debris that will remain in orbit for hundreds of years and debris that routinely threatens the International Space Station several times a year. There have been several other collisional incidents before and since (Figure 3.4 documents many of these incidents).

However, in February 2009, a collision occurred between an active Iridium communication satellite and a defunct Russian satellite. This was at an altitude that will ensure a long life for the resultant debris cloud. It was really this totally unexpected accidental collision that catapulted the concept of SSA to the fore.

2.3 Three Components of SSA

Figure 2.2 indicates one arrangement of components in SSA.

These components are a knowledge of space weather, the electromagnetic and subatomic particle fluxes incident on the Earth, and which are mainly derived from solar activity; a knowledge of the current man-made orbital space object population and the means to propagate this state into the future; and a knowledge of the flux of natural space debris that transits through and/or impacts upon the near-space environment, including the Earth.

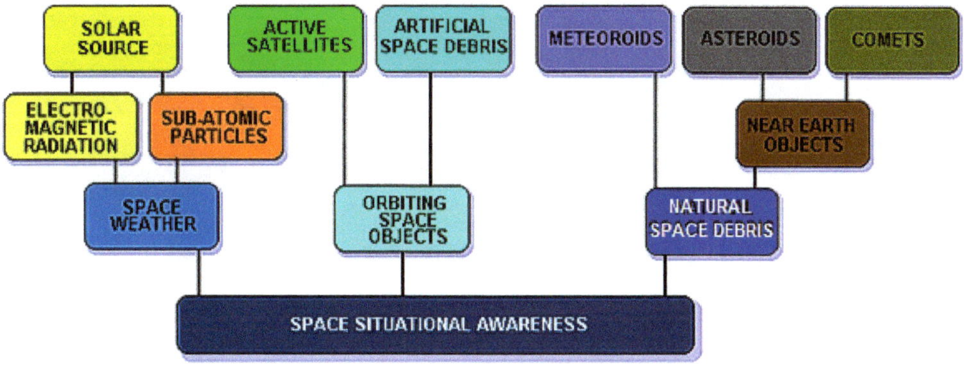

Figure 2.2 Components of SSA.

2.4 Space Weather

The source of most of our space weather is the Sun. We can separate this into a background and a transient component. This division and the classification of the subcomponents is illustrated in Figure 2.3.

The transient components of space weather can impact the orbital population of active satellites and space debris. Figures 2.3 and 2.4 indicate some of these and other effects of severe space weather. Increased and variable atmospheric drag can impact the ability to propagate satellite orbits to future times. Single-event upsets can influence the ability to control a satellite's attitude, and in extreme cases, the ability to station-keep in geosynchronous orbit. Radiation damage to solar cells can reduce the useful life of a satellite, and thus the income such an asset produces commercially.

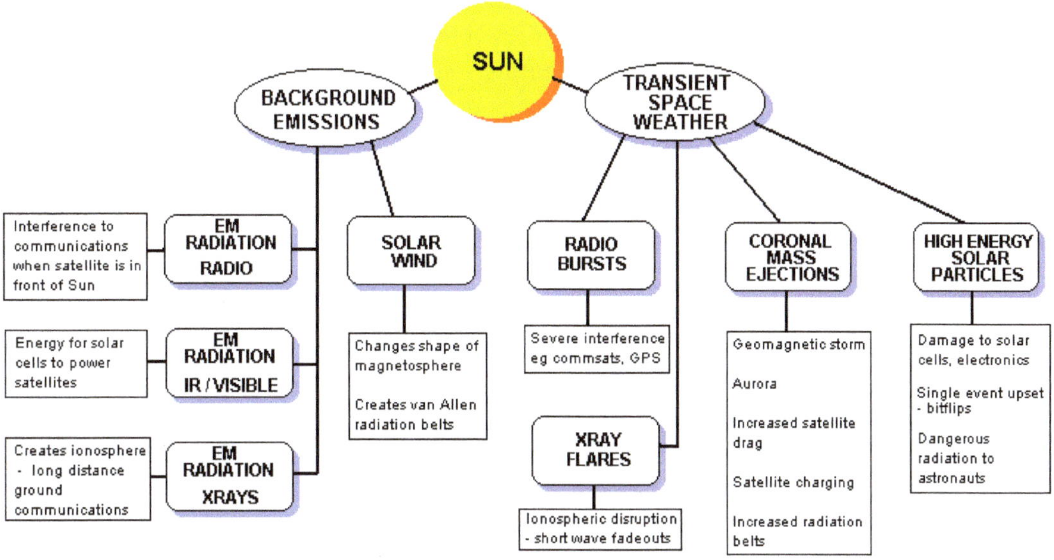

Figure 2.3 Components of space weather.

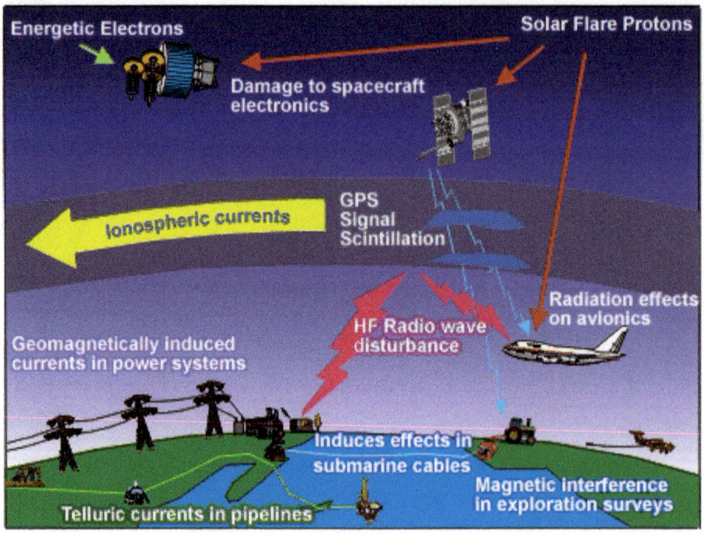

Figure 2.4 A graphical view of significant space weather effects. (*From:* Louis Lanzerotti of Bell Labs.)

Global ground-based and space sensors continually monitor space weather. This data and forecasts derived from it are made available by regional space weather centers. In the United States this space weather information is available from the Space Weather Prediction Center [4], in Australia from the Space Weather Service [5], and in Europe from the Solar Influences and Data Analysis Center of the Royal Observatory of Belgium [6].

2.5 Natural Space Debris

At the beginning of the space age, there was considerable concern about the natural meteoroid flux and its possible impact (literally) on satellites that were placed in Earth's orbit. Some of the first scientific sensors on early satellites were meteoroid detectors. Fortunately the natural space-debris flux transiting orbit is small enough not to cause serious problems, although a few space shuttle windows were replaced due to impacts from small pieces of natural debris (most window replacements however, were due to artificial space debris impacts).

The flux of natural debris through LEO was exceeded by the artificial debris flux in 1995 for objects around one centimeter in diameter. However, this does not tell the whole story as natural space debris (meteoroids) pass through LEO in less than 100 seconds whereas artificial debris remains in orbit for months to centuries so the exposure time is vastly greater. The velocities of each population is also quite different. Artificial space objects in LEO orbit around the Earth at a velocity of around 7 km/s, whereas natural space debris has velocities ranging from 11 to 72 km/s with respect to the Earth.

It normally requires an object of at least one centimeter to cause major spacecraft damage. Despite the fact that meteoroids drop around 10,000 kg of material on the Earth each year (most of which floats slowly down from its demise through ablation into dust at around 80 to 100 km and ends up on the ocean floor), the exposure time for collision is so much smaller than for orbital space debris (OSD) that it may be ignored. However, for meteoroid collisions that do occur, the damage is likely to be a lot greater because the collision energy is equal to the square of the collision velocity.

The transition from meteoroid to asteroid is ill-defined but is often given as around 10 meters in diameter. The size range of comet nuclei is even less well-known. As an object's size increases, the number of known objects decreases. It is estimated that there exists around 1,000 asteroids with diameters of one kilometer or greater that could pose a hazard, not to satellites, but to the Earth itself (or www .minorplanetcenter.org/iau/mpc.html rather human society). Most of these objects have been detected and have reasonably well-known orbits. Objects above 100m in size are believed to be much more plentiful (~10,000) but only a fraction of this population has been detected. Objects less than 100m in size are virtually undetectable with current technology, unless the body approaches close to the Earth. Due to their hypervelocity, such pieces of rock have kinetic energies measured in multiple megatons of TNT, and are quite capable of destroying a city if a direct impact occurred. On February 15, 2013, a large meteoroid (or small asteroid) of around 15m in size with a mass of 7,000 tons, deposited 300 kT TNT energy over a Russian city in the Ural mountains, shattering windows and injuring around 1,000 people. On the same day a larger (about 45m) asteroid passed below geosynchronous orbit with an energy of around 2.5 Mt TNT. These and other asteroid impacts or near misses are recorded in the HORIZONS database [26].

Many observatories, including many amateur astronomers, are devoted to the detection of NEOs, and routinely provide positional data to the Minor Planet Center [7]. Orbital dynamics and predictions of NEOs are computed and made available via the CNEOS NASA/JPL site [8] and the University of Pisa NEODyS site [9].

2.6 Orbiting Space Objects: Man-Made Space Debris

2.6.1 Overview

Orbiting space objects (OSO), also known as resident space objects (RSO), consist of active satellites and space debris. The latter are a mixture of defunct satellites, launch vehicles, and pieces of space hardware produced by operational activities, deterioration, and fragmentation.

A large part of SSA for OSO consists of the tracking of these objects with the aim of creating a catalog of such objects. The catalog then forms the basis of orbital evolution and conjunction assessment; the latter with the aim of protecting active assets from collision. In the future this will also extend to the minimization of space debris production through collisions.

SSA also extends to knowledge of the space capability of each space-faring state. Maneuver analysis is often necessary in attempts to determine spacecraft functionality. Some states use catalog information to determine overpass times for intelligence spacecraft in order to hide certain ground-based activities from enemy eyes. Human intelligence (HUMINT) can often provide information that is just as valuable as that provided by technical means, especially in the area of SSA prediction (e.g., knowledge of a proposed space nuclear explosion that has the capacity to disable a substantial fraction of satellites in low Earth orbit).

Figure 2.5 gives an overview of the areas associated with SSA for OSO.

Modeling collisional probabilities and outcomes is very important for the prediction of the evolution of the orbital space debris population. Reentries of large orbital space objects pose a small hazard to humans on the ground. Antisatellite (ASAT) activities pose a problem not only to the targeted spacecraft, but also to the entire orbital population through the generation of a large amount of space debris. High-altitude nuclear explosions (HANE) have the potential to cripple large numbers of active satellites through the electromagnetic pulse (EMP) produced and the increased hazard of trapped radiation (which may last for months). Although space-faring states are unlikely to resort to HANE in any war, rogue states and possibly large terrorist organizations (particularly those sponsored by state organizations) may view this as a way to "equalize" asymmetric warfare capabilities. MIJI is a US-military term that stands for meaconing, intrusion, jamming, and interference. Ground- and space-based MIJI has already been encountered by some space operators. The US military has set up a system to rapidly determine and locate interference to space assets. Weaponization of space (threatening ground and space assets) is a problem of concern to some organizations, particularly with respect to the enforcement of treaties regarding the militarization of outer space.

2.6.2 Orbital Space Debris

Orbital space debris (also referred to as artificial or man-made space debris) is basically any space object or hardware fragment that no longer has a useful

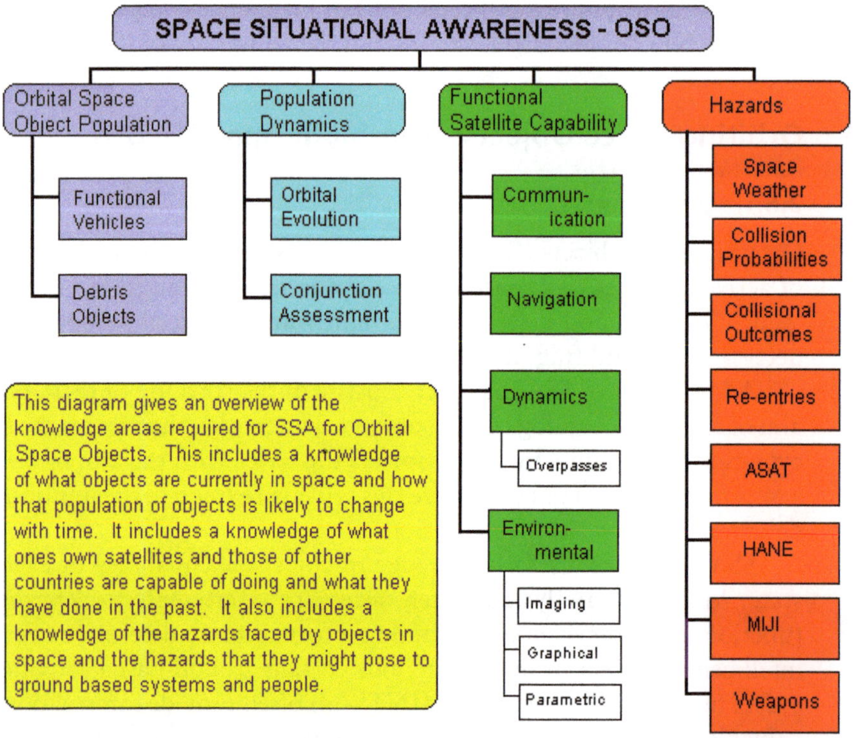

Figure 2.5 Knowledge areas for artificial space objects in SSA.

function. (Note that some satellites are launched and put in orbital storage until required. These are obviously not pieces of space debris, even though they are currently nonfunctional.)

The production of space debris started with the launch of the first artificial satellite, *Sputnik-1*, in October 1957. Initially, every space launch produced around 100 pieces of debris, from launch vehicles that remained in orbit, to discarded shrouds, to smaller fragments produced by pyrotechnic devices used to separate the satellite from the launch vehicle. This number has gradually been reduced, particularly with the transition from pyrotechnic to mechanical means of satellite separation and dispersion [27]. This is also discussed in NASA's *Handbook for Limiting Orbital Debris* [28]. Mandates by various governments (specifically the United States and EU) regarding the decommissioning of inactive spacecraft have also helped in this regard.

The orbital space debris population is a balance between debris sources and debris sinks. Since the initial creation of space debris, the number of debris sources has generally exceeded the number of sinks and so the orbital space debris population has kept increasing. Governments, in particular the United States, have enacted regulations to reduce the extraneous debris produced during launch and other mission operations. There is now an onus on launch operators to remove the final stage of the launch vehicle from orbit after mission insertion has been accomplished

Debris sources include new launches, operations in space, satellite deterioration, and fragmentation. At present most fragmentation debris is produced by explosions that result from stored energy in launch vehicles and defunct satellites. This includes unspent fuel and batteries. Six accidental collisions have been reported up to the end of 2023, the last being the March 18, 2021 collision between the Chinese *Yunhai-1* Earth observation satellite and most probably debris from the Zenit-2 rocket body that launched the Russian *Kosmos 2333* satellite. The size of the debris that caused this collision was probably under 10 cm and essentially untrackable. Such accidental collisions are likely to become more common in the future as the smaller sized untrackable debris population increases.

Debris sinks include orbital decay (due to atmospheric drag), intentional deorbiting (by fuel burns), and retrieval. Orbital decay accounts by far for the largest debris sink in the lower parts of low Earth orbit.

The concept of collisional cascade and the exponential growth of space debris is discussed in Section 3.8. Because a collision between even a small debris particle and an active satellite can cause fatal damage to the latter, space debris mitigation is a high-priority concern for SSA.

2.6.3 The Orbital Space Debris Population

Figure 2.6 is a graph from the NASA Orbital Debris Program Office (ODPO) [11] showing the increase in the orbital space object population since 1957.

In 2023, the US Space Surveillance Network (SSN) tracked over almost 25,000 orbital space objects above 10 cm in size. Of these, only a few thousand are active satellites. The step-like increase in early 2007 is due to the deliberate collision of a Chinese ASAT missile with a defunct weather satellite. A much smaller increase in 2008 is due to the US destruction of a classified reconnaissance satellite (*USA–193*). The debris generated in this deliberate collision quickly disappeared from orbit due

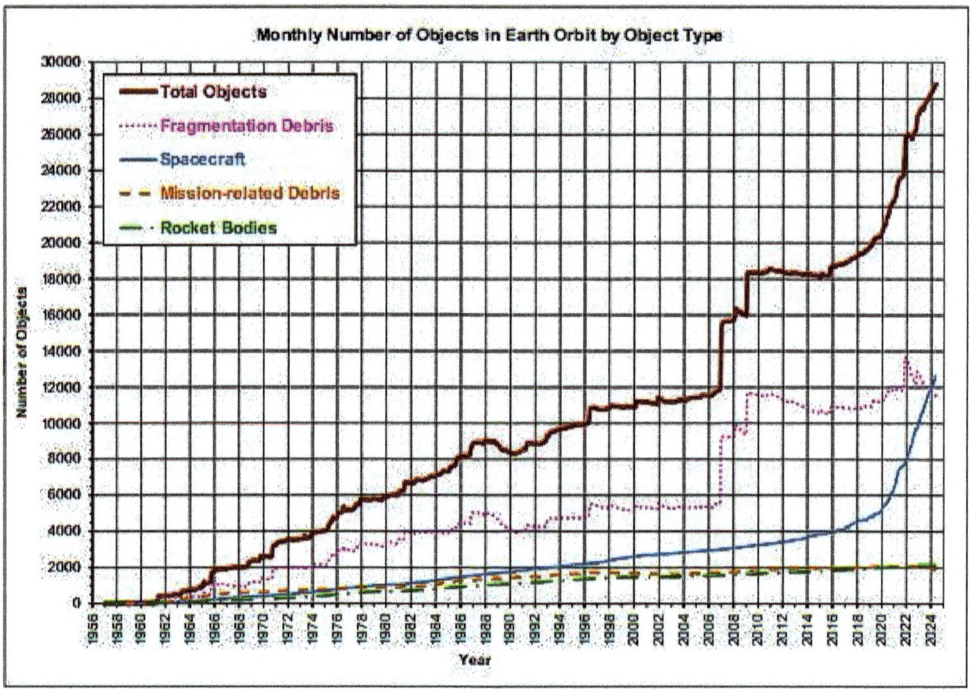

Figure 2.6 Orbital space objects by type. (*Source:* NASA/ODQN.)

to the low altitude of the event and the subsequent atmospheric decay of the fragments. The third step increase in 2009 was due to an accidental collision of an active Iridium communications satellite with a defunct Russian communications satellite. The events in 2007 and 2009 occurred at such an altitude that normal decay of the collisional debris will take hundreds of years to occur.

The NASA Orbital Debris Program Office produces a quarterly newsletter [12] with updates on the tracked orbital space object population, together with news of the latest research findings and meetings related to orbital space debris.

2.7 Tracking Orbital Space Objects

The United Nations Office of Outer Space Affairs (UNOOSA) maintains an index [13] of objects launched into space. This is compiled from information supplied by member states.

However, the population of orbital space objects (also known as resident space objects) consists of a lot more than the objects launched into orbit. Operations in space and environmental degradation, including collisions, produce a much larger population than the number of active satellites in orbit. Keeping track of these is vital to an awareness of the orbital space regime and forms the basis of a database that allows modification of this environment, most specifically, the remediation of space debris, a fundamental component of what some refer to as space traffic management.

2.7.1 United States

The United States has by far the largest network of sensors that track orbital space objects, with some 20 sensors deployed globally. These include radar and optical systems. The first dedicated system to be used to track orbiting space objects was the US Navy's NAVSPASUR, a multistatic radar consisting of 9 sites spread across the southern United States, from Southern California to Georgia (Figures 2.7, 2.8, and 2.9). There were 3 transmitter sites and 6 receive sites. The system was operated by the US Navy from 1960 until 2004 when it was turned over to the US Air Force,

Figure 2.7 The main space fence transmitter at Lake Kickapoo near Wichita, Texas. US Navy image of the VHF 3-km long transmitter array (now decommissioned). (*Source:* US Navy.)

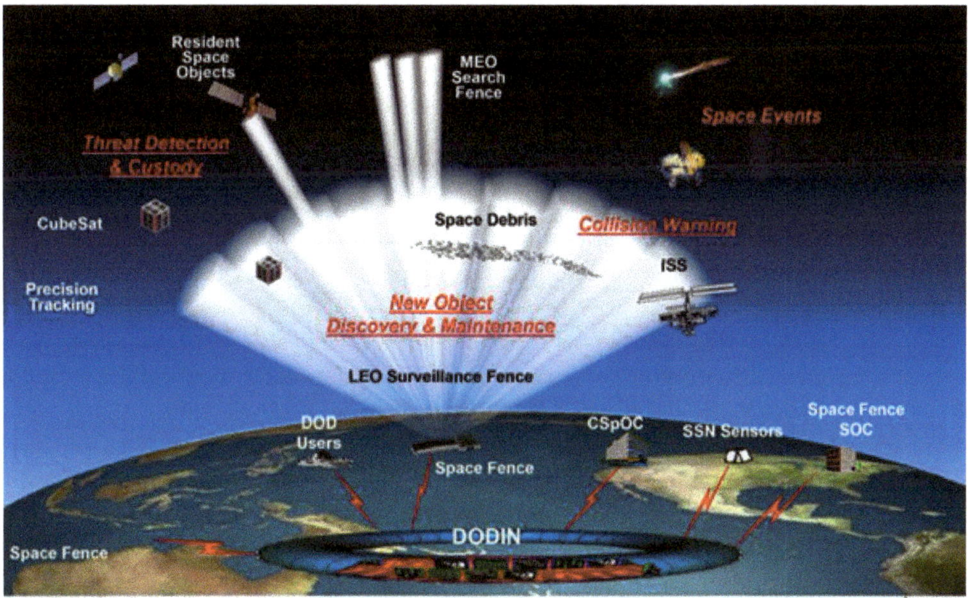

Legend

CSpOC - Combined Space Operations Center
DOD - Department of Defense
DODIN - Department of Defense Information Network
ISS - International Space Station
LEO - Low Earth Orbit
MEO - Medium Earth Orbit
SOC - Space Fence Operation Center
SSN - Space Surveillance Network

Figure 2.8 The new S-band space surveillance fence. (*Source:* US DoD.)

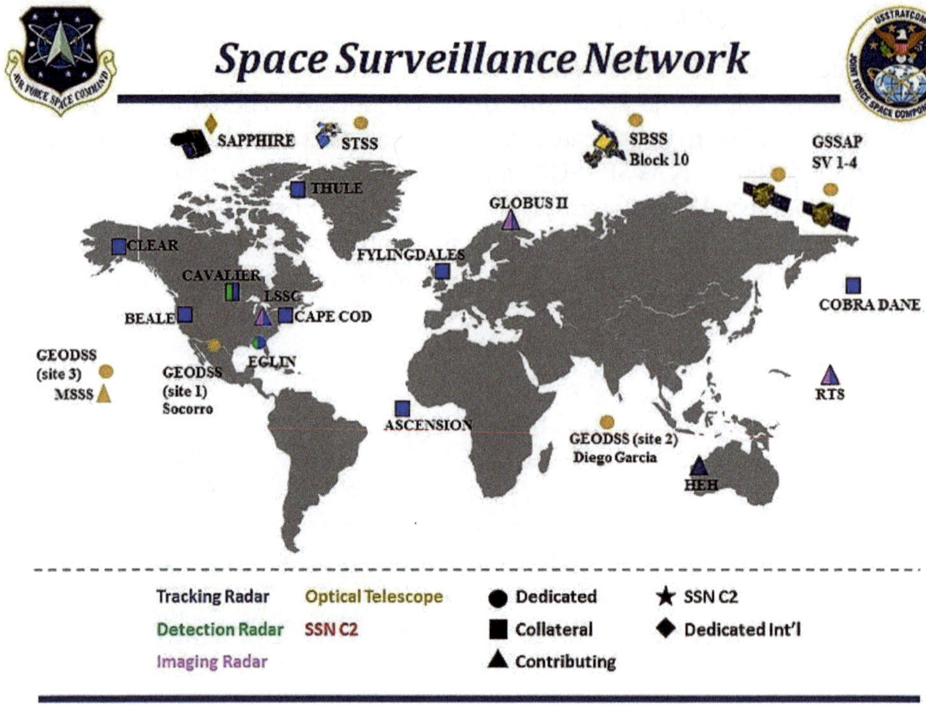

Figure 2.9 United States space surveillance sites (*Source:* US DoD.)

where it became known as the Air Force Space Surveillance System (AFSSS—system designation AN/FPS-133). The operational frequency was 217 MHz (wavelength 1.4m). The system remained in operation until 2013.

The radar signal returned by an object depends critically on the object size. For objects less than the radar wavelength, the signal decreases as the fourth-power of the size (see Section 12.8 and in particular Figure 12.50). Because of this, enormous power was required to allow the radar to detect objects down to 10 cm in size. At almost 7 GW effective radiated power, this was probably the highest power transmitter in the world (and a strong beacon for any extraterrestrial beings who might be listening!).

Colloquially known as the *space fence*, the system radiated power in a narrow fan beam that could detect any space objects passing over it. The stated detection limit was a 75-cm object at 30,000 km. The system was nontracking and thus able to provide uncued detection of objects transiting its beam.

This array created an east-west fan beam with an effective radiated power of 6 GW. Space objects passing through this beam reflected energy back to receiving stations on the ground that recorded time, intensity, and Doppler shift of the echoed signal. This system processed over 10,000 observations per day. Because it was nontracking, it was able to start the generation of a space object catalog with no a priori information. In contrast to this, other sensors, most of which track, require approximate positional data to provide more accurate orbital information.

In the 1960s and 1970s, the space fence was used together with Baker-Nunn cameras, which provided optical detection of orbiting space objects. It was initially run by the US Navy but was turned over to the US Air Force in 2004.

Its final configuration consisted of an array of 2,556 dipoles with a total length of 3.2 km and a gain of 40 dBi. The beamwidth of the antenna was 0.02 degrees in the north-south direction and approximately 120 degrees east-west. Each dipole was fed by a 300W power amplifier for a total radio frequency (RF) power of 767 kW. The effective radiated power was 98 dBW or 6.3 billion watts. This massive radiated power was necessary to compensate for the Rayleigh-scattered reflection of small debris objects at the VHF frequency used, where the signal returned is inversely proportional to the fourth-power of the object size.

This system was decommissioned in 2013, and has been replaced by the more efficient and smaller S-band space fence on the Kwajalein atoll in the Marshall Islands in the western Pacific Ocean (Figure 2.8). This became operational in 2020 by the US Space Force and feeds data to the Joint Space Operations Center at Vandenberg Air Force Base in California.

Today, many countries maintain some form of space object tracking (SOT) capability.

Only the United States, Russia, and China have well-developed networks to track orbital space objects. The US Space Surveillance Network. Europe is also developing its own SOT network as part of a broad SSA effort.

The goal of space object tracking in the context of space situational awareness is the creation of a catalog of all orbiting space objects down to the limiting magnitude of the sensors involved. Ideally this should extend to objects as small as 1 cm, this being the size at which significant damage will result if a collision should occur with such an object. There are two components to catalog generation. The first requires sensors, optical and radio, that can provide object positions and velocities at specified times. The second requires orbital prediction programs that can ingest these object *states* and produce an orbital ephemeris for each orbital object. This second component also includes identification of each object. This is not always possible, and catalogs contain substantial numbers of uncorrelated objects.

To generate a catalog or partial catalog from scratch requires a wide beam, nontracking sensor that can provide first look data (low accuracy) for all objects, so that higher accuracy (narrow beam) sensors can then be tasked to track each object in turn.

The SSN (Figure 2.9) until recently, was the most comprehensive SOT network in the world. It is land and space based and distributed throughout many countries in the world. Independent sensors also feed into this network, and some commercial companies operate their own SOT and provide services directly to customers for a fee.

The US SSN uses data from over 30 land-based sensors as well as space-based assets. The primary optical system is based on the three Ground Based Electro-Optical Deep Space Surveillance (GEODSS) sites at Socorro, New Mexico; Maui, Hawaiian Islands; and Diego Garcia, Indian Ocean. These can track objects in all orbits (e.g., LEO and GEO).

The US radar sensors are a mix of single dish and phased array radars. Phased array radars use coplanar antenna patches that can be electronically steered to the desired position on the sky. Electronic steering is done by feeding the many patches with signals of different phases.

The US SSN also uses collaborating sensors that provide SOT information on a time-limited basis. These include the Haystack and Haystack Auxiliary radars run by the Massachusetts Institute of Technology, and the Air Force Research Laboratory (AFRL) sensors based at the Air Force Maui Optical Site (AMOS), which includes a 3m infrared telescope.

There are now also space-based sensors in the US SSN system. These include two Geosynchronouus Space Situational Awareness Program (GSSAP) satellites.

As of 2007, control and data processing for the US Space Surveillance Network was moved from Cheyenne Mountain in Colorado, to Vandenberg Air Force Base in California. (Cheyenne Mountain remains as a backup site.) The data from the diverse systems of the SSN feed into this Joint Space Operations Center (JSpOC) in California which compiles the US Department of Defense (DoD) space object catalog (SATCAT or SATC). Data for all tracked objects is available in a low-resolution format (Keplerian orbital elements). A higher precision catalog is also compiled for a limited set of objects, and this is in state vector format (position and velocity at a specified time). Both of these formats must be propagated forward in time using a particular model that includes perturbational forces due to the Earth's nonsphericity, atmospheric drag, radiation, and lunar and solar gravity. The low-resolution catalog is made available publicly [14] for all but US (and some other) military satellites. A subset of the SATCAT is freely available via celestrak.org.

In the last ten years there has been an upgrading of US SOT sensors around the world. The original space fence has been replaced by an S-band version. This is located on Kwajalein Atoll in the Marshall Islands (Western Pacific). The use of the higher S-band frequency has allowed the detection size limit to be pushed lower with the use of much lower power than the original VHF AFSPASUR (Air Force space surveillance). The United States has also placed newer optical and infrared SOT sensors into space.

2.7.2 Russia

While the US SSN and its SATC are by far the largest source of information on orbital space objects, they are not the only ones who track OSO. Russia maintains its own catalog. Most of the Russian sensors lie in the Asian continent, but this is supplemented by mobile surveillance vessels. The Russian sensors consist of radar, optical, and lidar stations.

The breakup of the Soviet Union in 1991 into 15 separate republics caused Russia to lose substantial space-tracking capability. This was particularly so around the Baikonur Cosmodrome in Kazakhstan, although this is still the most used Russian launch facility.

Among the various Russian radar sites it is not easy to separate the tasks of space tracking from those of missile warning and signals intelligence, and to separate the old systems (e.g., Krona) from newer constructions or upgrades.

Because land-based systems in Russia cannot detect low-inclination LEO satellites, Russia maintains a maritime mobile space-tracking system.

2.7.3 China

China is known to have extensive SOT facilities in its own country, but releases few details about its network, which includes stations in other countries. China is thus able to cover regions of space not possible from its own territory. China maintains and operates up to six Yuanwang ships as maritime mobile space-tracking platforms. An optical tracking facility is believed to be located at the Purple Mountain Observatory.

China also has at least two extraterritorial space track sensors. In return for launching the *Nigsat*, Nigeria has allowed China to build a large space track facility on its territory. Pakistan also hosts a Chinese space track facility.

2.7.4 Europe

European governments have expressed frustration with data delays and denials from the US SSN, and European Space Agency (ESA) has proposed a European tracking system for space situational awareness. Initially, they used two radars and a dedicated optical telescope at Tenerife in Spain to observe high-interest items and produce a very limited local catalog. However, about a decade ago an initiative was started to develop an indigenous SOT network as soon as possible.

The German FGAN TIRA radar uses 1 MW at L-band for tracking and 13 kW at K-band for imaging. Normally used in monostatic mode, it can be used with a large radiotelescope at Effelsberg, Germany, in a bistatic mode. The antenna is 34m diameter.

The French operate the GRAVES bistatic radar and a maritime C-band space track radar. The United Kingdom defense establishment hosts two radars with space track facility. European space surveillance facilities feed into the European space operations center (ESOC) at Darmstadt, Germany for processing.

2.7.5 Japan

The Japanese Space Forum operates two space-tracking facilities from Okayama prefecture, the Bisei Spaceguard optical site, and the Kamisaibara radar site.

2.7.6 Australia

Australia has hosted many facilities related to satellite tracking and communications over the years. These have been mostly for the United States, but an ESA radio site at New Norcia, northeast of Perth, is now part of the ESA Deep Space Network.

More recently, several US space object tracking facilities have been established on North West Cape, Western Australia. These include the Extended High Accuracy Network Determination System (HANDS), an FPQ-14 C-band radar, and the 3m Space Surveillance Telescope. Electro-Optical Systems (EOS), a commercial firm based in Canberra, has also installed a space lidar optical facility [18].

The Curtin University (Perth, Western Australia) Desert Fireball Network has joined with Lockheed Martin (Australia) to produce the FireOpal optical SOT network spread across the Western Australia and South Australia desert regions. This is a nontracking very wide field-of-view imaging network that can see objects down to 13th magnitude.

The Curtin Institute of Radio Astronomy, which runs the Murchison Widefield Array [19] radio telescope (a large phased array VHF, 70–300 MHz radio telescope) has experimented with using the Murchison Widefield Array (MWA) as a passive SOT sensor, where FM and TV signals provide the transmitted signal.

The University of Western Australia runs the Zadko Observatory at the Gingin Gravity Precinct. It operates the 1m Zadko telescope that has been used by CNES (the French Space Agency) for SOT. It was also part of the French TAROT network for SOT, and also hosts multiple SOT telescopes for government and commercial operators around the world.

Following the establishment of the Australian Space Agency in 2018 there was a rush by many commercial entities to place SOT sensors in Australia. These have included ExoAnalytics (optical), LeoLabs (phased array radars), and many others.

2.7.7 Commercial and Other

Because the US SSN does not make available data on US (and some other) military satellites, a group of around 100 dedicated amateur space observers routinely make observations of the larger space objects in low Earth orbit and produce a catalog [20] of about 100 military satellites that do not appear in the US SATC.

Satellite operators/owners frequently know the location of their assets more accurately than does the US SSN, either through ranging data (geosats) or the inclusion of a GPS unit (leosats) in their satellite. The US JSpOC has thus set up a Commercial and Foreign Entities (CFE) program to exchange and ingest this data.

The new, large array radio telescopes that are coming on line in the next few years may also be a source of space-track data, using OSO reflections from existing TV and FM radio transmitters on the ground. However, the lower frequencies of these systems limit the size of OSO that can be tracked.

Space object tracking is currently a growing commercial field with an expanding array of ground-based sensors being deployed by commercial companies. In particular, LeoLabs, with an increasing contingent of radar tracking stations around the world, is the only company that maintains an entire LEO public catalog and provides more frequent updates than the US Space Force 18th Space Defense Squadron. They currently (early 2025) average six state vector updates a day for the entire catalog compared to an average of four for the USSF catalog. The use of radar rather than optical sensors enables LeoLabs to maintain this enviable position and provide comprehensive conjunction assessment to other companies.

2.7.8 Catalogs, Accuracy, and Processing

The SATC currently does not have the accuracy (spatial position) to perform adequate and routine conjunction assessment on the large number of OSO. It also does not contain information on debris down to the critical size of one centimeter. These limitations are basically due to sensor constraints. The US DoD does have plans to upgrade the SSN with equipment that will provide more accurate positional information on all objects down to one centimeter. One example of this is the S-band space fence operating at 2500 MHz that allows detectable returns from the smaller sized objects. Problems with the US SSN are discussed by Brian Weeden [22, 23]. Dr. Weeden is a technical advisor for the Secure World Foundation [24].

Possible duplication of the newer space fence in Western Australia will also provide a greater number of observations with which to compile a more accurate and complete catalog.

Laser tracking of space objects, currently under investigation by the Australian firm Electro-Optical Systems [18] holds great promise to increase the accuracy of data available to generate OSO catalogs.

One of the problems with such an increase in sensor capability is the vastly increased computing power, transfer bandwidth, storage, and efficient processing software to deal with this increase in data.

Space situational awareness operators will also require new techniques of data fusion and display to be able to rapidly monitor and interpret changes in the environment of interest, with data incoming from multiple diverse sensor systems spread around the globe, and in space.

Some software for SSA is available commercially [25], and there is an effort underway to try to develop open source software for this field.

2.8 Summary

The number of active satellites is increasing slowly, but the number of pieces of artificial space debris is rapidly increasing and may soon reach a critical population size, where accidental collisions will cause a runaway exponential increase in the number of orbital space objects. Monitoring of the near-space environment is becoming more critical to enable conjunction assessment with subsequent collision avoidance in cases of high-value space assets. A new generation of space situational sensors, together with an increasing number of states who wish to perform independent SSA analysis is projected to give rise to a large and diverse array of measurements and analysis techniques that will require data fusion and innovative display techniques to provide a useful tool for planning and operational purposes. This will need to be combined with space weather data, and data on near-Earth objects, to provide a truly integrated and meaningful approach to space situational awareness.

References

[1] www.spacefoundation.org/programs/public-policy-and-government-affairs/introduction-space-activities/space-situational.

[2] www.esa.int/Our_Activities/Operations/Space_Situational_Awareness2.

[3] www.onorbitwatch.com/feature/space-situational-awareness-topic-deserves-discussion.

[4] US Space Weather Prediction Center, www.swpc.noaa.gov.

[5} Australian Space Weather Service, sws.bom.gov.au.

[6] Solar Influences and Data Analysis Center of the Royal Observatory of Belgium, space weather.eu.

[7] Minor Planet Center, minorplanetcenter.org/iau/mpc.html.

[8] JPL Center for NEO Studies, cneos.jpl.nasa.gov.

[9] NEO DYnamic Site, newton.spacedys.com, University of Pisa, Italy.

[10] Donald J Kessler & Burton G Cour-Palais, "Collision Frequency of Artificial Satellites: The Creation of a Debris Belt," *Journal of Geophysical Research: Space Physics*, Vol. 83, Issue A6, pp. 2637–2646.

[11] NASA ODPO (Orbital Debris Program Office), orbitaldebris.jsc.nasa.gov.

[12] NASA ODQN (Orbital Debris Quarterly News), orbitaldebris.jsc.nasa.gov/quarterly-news/.
[13] UNOOSA Space Object Index, oosa.unvienna.org/oosa/osoindex.html.
[14] SATCAT, space-track.org.
[15] Global Space Tracking Systems, globalsecurity.org/space/world/index.html.
[16] Vallado, globalsecurity.org/space/world/china/space-surveillance.htm.
[17] French GRAVES radar upgrade, SDC7-paper297.pdf.
[18] Electro-Optical Systems, eos-aus.com/space/optical-tracking-systems/.
[19] Murchison Widefield Array, https://astronomy.curtin.edu.au/research/murchison-widefield-array/.
[20] Amateur SSA, satobs.org.
[21] Amateur SSA, satobs.org/seesat/seesatindex.html.
[22] Weeden, Brian, "The Numbers Game," thespacereview.com/article/1417/1.
[23] Weeden, Brian, "Billiards in Space," thespacereview.com/article/1314/1.
[24] Secure World Foundation, swfound.org.
[25] SSA software, agi.com/solutions/main.aspx/id/space-situational-awareness.
[26] NASA Center for NEO Studies, https://cneos.jpl.nasa.gov.
[27] MacCaire, Alexis, "Space Debris Management of Launch Operations," Space Debris Conference 7, paper 419 SDC7-paper419.pdf.
[28] NASA Handbook for Limiting Orbital Debris, https://standards.nasa.gov/standard/nasa/nasa-hdbk-871914.

Selected Bibliography

Every year the US Air Force hosts the AMOStech conference on the island of Maui. Many technical papers are presented each year relevant to various aspects of SSA. These papers are downloadable (pdf) and may be accessed at amostech.com/amos-technical-papers/.

CHAPTER 3
Orbital Space Debris

3.1 Classification of Space Debris

Figure 3.1 gives a graphic classification of the two fundamental types of space debris that either exist in or pass through the orbital regime associated with the planet Earth.

Space debris exists throughout all regions of space that we have so far been able to observe, and it has done so for a very long time. We refer to this as natural space debris. It is present in the solar system, in other stellar systems, in the space between stars, and in the space between galaxies. Some of this debris hinders our direct observation of various regions. An example of this is the dust that obscures much of the Milky Way, including the galactic center, from our vision. This is seen as dark clouds in the night sky. One prominent southern hemisphere dust cloud is the Coal Sack, just to the left of the Southern Cross.

What is new since the start of the space age is the presence of artificial space debris in orbit around the Earth. This is referred to as OSD and is one component of the RSO in Earth's orbit. It comes from human activities in space, and it is starting to cause us problems. This chapter will concentrate on discussing artificial orbital space debris, although comparison will occasionally be made with the natural space debris near the Earth.

3.2 Generating Space Debris

The first satellite launched into space was the Soviet *Sputnik-1*. This was placed into low Earth orbit on October 4, 1957 [1].

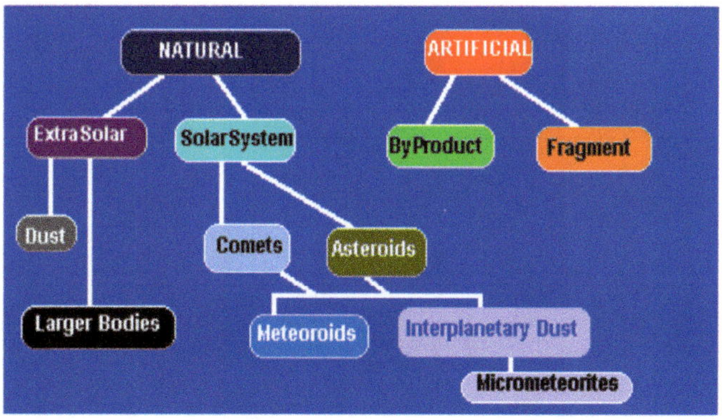

Figure 3.1 Classification of space debris. (*Source:* National Academies Press.)

As well as the 58-cm diameter sphere (mass 84 kg) which was *Sputnik-1*, the final stage of the launcher, with a length of 30m and a mass of 4,000 kg, also made it to orbit. Also in orbit was the protective shroud enclosing the satellite during the ascent through the atmosphere, as well as the mating adaptor from which the satellite was released, and most likely smaller debris resulting from the pyrotechnic release of the shroud.

And so was born the first orbital artificial space debris.

The generation of artificial space debris has continued throughout the space age. Figure 3.2, from the NASA Orbital Debris Program Office, shows the number of space objects tracked and cataloged by the US Space Surveillance Network.

Significant changes, both increases and decreases in the orbital space object population, have been indicated by comments on the graph.

In general all objects shown here will be 10 cm or greater in size, as that is the limit below which current tracking is not reliable. Also not shown on the graph are several thousand space objects that are being tracked but that have not been cataloged. These are referred to as uncorrelated objects.

3.3 Natural Space Debris and Space Travel

As well as artificial space debris there is also a population of natural space debris (meteoroids) that, while they don't orbit the Earth, do pass through all orbital altitudes. In fact prior to the space age, some scientists predicted that the hazard from

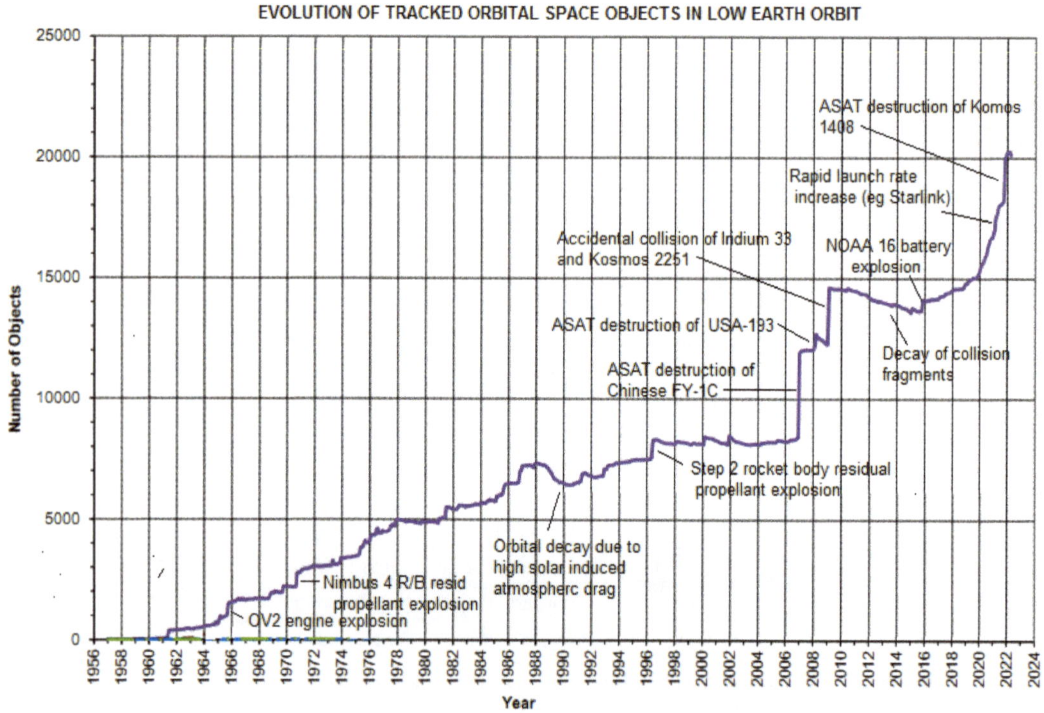

Figure 3.2 Labeled artificial space population growth in LEO. (*Source:* NASA ODQN.)

natural debris might be so great as to make space travel very dangerous. In fact NASA spent considerable effort in trying to evaluate this hazard. Ground visual and radar observations were examined, and most of the early satellites carried meteoroid detectors.

The early satellite detectors were often microphones to record the sound of a meteoroid impact. A large number of impacts were recorded by these detectors and this appeared to reinforce the idea of a large, natural orbiting debris cloud. This was referred to as an anomalous debris cloud because it indicated a population much larger than ground observations (mainly radar) could account for. Continued investigation found that microphone-type detectors were registering impacts that were in fact expansion and contraction noises of satellite panels.

And thus it was found that the meteoroid collision hazard is usually low but not negligible (Figure 3.3).

3.4 Components of Orbital Space Debris

We might define orbital space debris as essentially any object in Earth's orbit that does not have a useful purpose. Figure 3.4 is one classification of Earth-orbiting space objects [2]. This shows the wide range of sources that contribute to, and make up, orbital space debris.

3.5 Debris Measurements and Models

The debris population is measured with the use of both ground and space sensors. And models have been developed to interpolate in size regions where sensors do not cover, and to extrapolate to objects not accessible to sensors. Models are also

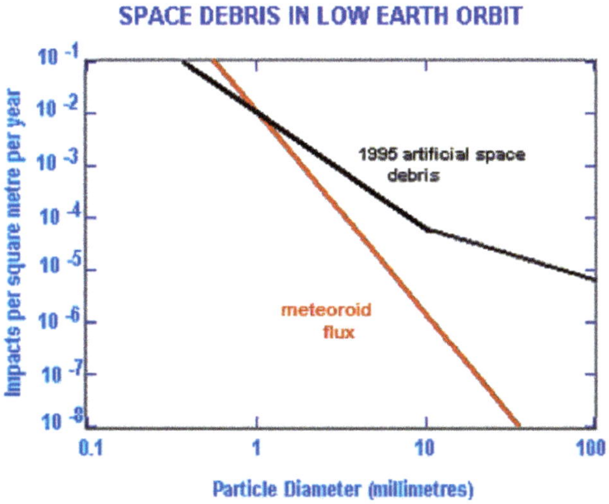

Figure 3.3 In 1995, artificial space debris numbers exceeded the natural space debris flux in low Earth orbit for all particle sizes above 1 mm.

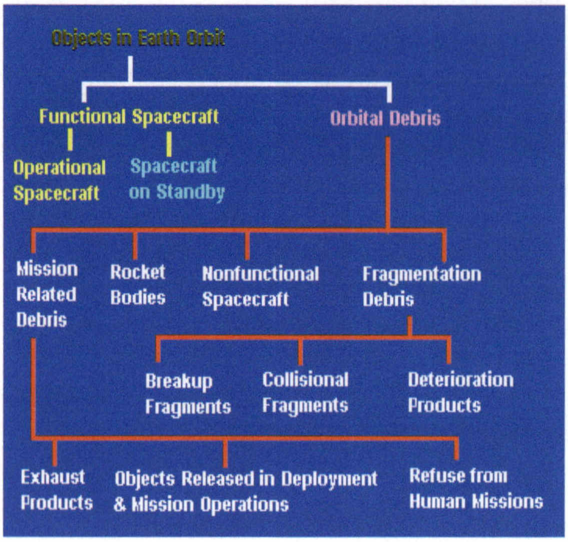

Figure 3.4 Classification of orbital space debris. (*After:* Orbital Debris, US National Research Council.)

used to propagate debris populations forward in time to examine the future consequences of various scenarios.

Natural space debris flux can be monitored through its interaction with the Earth's atmosphere:

- Low frequency meteor radar for continuous surveillance. These normally operate in the frequency range from 20 to 50 MHz. A single radar can observe a collection area of around 1,000 square kilometers. The newer meteor radars can provide directional and velocity information.
- Visual and/or video observations are sometimes used to augment the radar patrols for better characterization of the meteoroid that produces the meteor phenomenon.

Artificial space debris requires considerably more resources to monitor because of the constantly changing environment, and the need to assign specific space objects to specific space events (e.g., producing an object catalog). Both optical and radar ground-based sensors are used.

- Optical observations for larger objects, and in general for geosynchronous objects (because of radar range limitations);
- UHF radar for continuous monitoring;
- SHF radar for smaller objects and precision data.

Space-based sensors are also starting to be used for space object surveillance. Special spacecraft with collection panels and other passive monitors have been left in low Earth orbit and then retrieved. Examination of the impacts on these satellites gives very useful information about the smaller sizes of debris objects—those which are much too small to be detected by other sensors.

Another view of the overall orbital space debris population, both measured and extrapolated down to very small sizes, is shown in Figure 3.5 [3]. This also indicates the different means used to monitor this population.

The smallest debris populations are estimated from microscopic examination of spacecraft panels that have been exposed to this flux in orbit. Much of this data has come from the NASA Long Duration Exposure Facility (LDEF), shown together with a close-up of one panel that shows a multitude of small impacts (Figure 3.6).

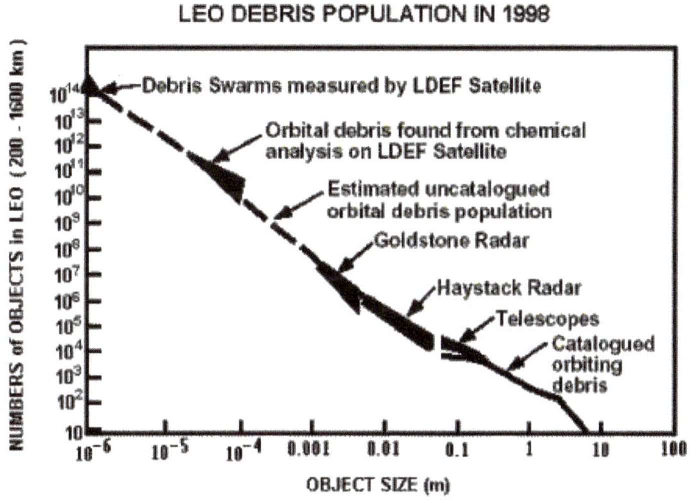

Figure 3.5 Space debris size spectrum. (*Source:* NASA ODPO report 1998.)

(a) (b)

Figure 3.6 (a) LDEF satellite, and (b) LDEF panel showing impacts. (*Source:* NASA ODPO.)

3.6 Data Analysis and Availability

Data from all SSN sensors was initially sent to the Space Control Center underneath Cheyenne Mountain near Colorado Springs. This function has now been transferred to the Joint Space Operations Center in California.

Data is processed and produces the SATCAT or satellite catalog of orbital objects, either as a state vector or Keplerian orbital elements.

Data is made available to registered users via the NASA Orbital Information Group (OIG). Since 2002, only approved parties are issued registration. This includes most US entities and entities of friendly states. Data does not include orbital elements of US military satellites.

A subset of the SATCAT is freely available on the web [4]. While this is very useful for tracking most active satellites, it is of little use for CONASS and COLA purposes.

Many governments (particularly European) have expressed frustration with data delays and denials, and perceived data degradation, and ESA has proposed a European tracking system for space situational awareness. At present they use two radars and a dedicated optical telescope at Tenerife to observe high-interest items and produce very limited local catalogs. Other telescopes are being enlisted to help in this task.

A group of dedicated amateur satellite observers track and produce orbital elements for many US military satellites; these can be accessed at [5].

3.7 Problems with Orbital Space Debris

Orbital space debris can give rise to a range of problems in four main areas:

1. Collisions
   - With operational spacecraft—minor damage to total destruction;
   - With other debris—generation of more debris.
2. Space Science
   - Debris effects may be mistaken for other phenomena;
   - Debris effects may mask other phenomena.
3. Astronomy
   - Debris trails are increasingly prevalent on astronomical images;
   - Produce transient radio astronomical events due to reflection of ground and space transmitters;
   - Over time debris will increase the background sky glow.
4. Other
   - Reentry hazard from large debris objects;
   - Radiation hazard from some debris.

These problems are now discussed in more detail.

3.7.1 Hypervelocity Collisions

Orbital velocity in low Earth orbit is around 7 to 8 km/sec. Obviously, collisional relative velocities can vary from near zero up to nearly 16 km/sec. Because of orbital dynamics the average relative velocity between two colliding objects is around 10 km/sec.

A hypervelocity collision is defined as one wherein the kinetic energy of the impactor is greater than the explosive energy released by the same mass of high explosive such as TNT.

For an explosion of TNT, the energy released is 4.2×10^6 joule per kilogram of material. Equating this to a kinetic energy per kilogram:

$$\frac{1}{2}v^2 = 4.2 \times 10^6$$

which gives

$$v_h \sim 3 \text{ km/s}$$

Thus 3 km/s is the lower limit for a hypervelocity impact.

A hypervelocity collision thus displays more properties of an explosion than a typical low-velocity collision. Because the average relative collisional velocity in low Earth orbit is around 10 km/s, practically all space debris collisions in LEO will be hypervelocity collisions. The impactor is usually totally destroyed.

A particle size of one centimeter is typically regarded as the transitional size above which severe damage to a satellite or spacecraft is produced. The current dedicated sensors of the US Space Surveillance Network are not able to track or provide orbital information on particles this small. Collision with a 10-cm-sized object will generally result in total destruction of an active space asset.

3.7.2 Reentry Hazard

Most reentering space debris ablates totally as it passes through increasingly dense atmosphere in a fiery conflagration. Only the largest objects (usually >10 tons initial mass) partially survive the reentry process. Fragments normally fall to ground at their terminal velocity, which may range from almost zero to ~ 200 m/s (Figures 3.7 and 3.8).

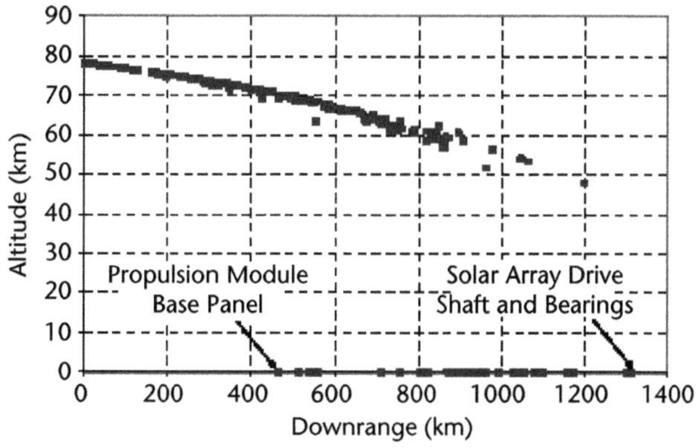

Figure 3.7 Component shedding during reentry. (*From:* NASA ODPO.)

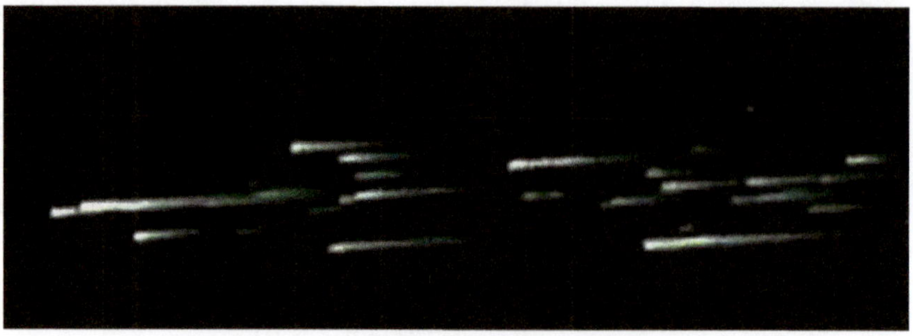

(a)

(b)

Figure 3.8 (a) Trails of a large reentry in the night sky (*Source:* NASA), and (b) a titanium pressure tank from a FOTON rocket that survived the reentry process and landed intact in outback Western Australia (*Source:* John A. Kennewell).

It now appears that a few people have suffered minor injuries from debris reentries, but as far as we know there have been no deaths. NASA now routinely conducts a survivability analysis on all large satellites.

3.7.3 Radiation Problems

In the early 1970s US gamma ray detectors on some scientific satellites detected strange events. The US military declassified some of this data in 1973.

It turned out that Soviet RORSAT (radar ocean reconnaissance) satellites were carrying nuclear reactors to provide enough energy to run the high-power radars. One of these satellites (*Cosmos 954*) survived reentry over Canada on January 24, 1978, and scattered radioactive material. The Canadian government presented the Soviets with a $6 million dollar cleanup bill, of which the Soviets paid half.

To avoid similar problems, further satellites ejected the reactor cores to higher altitudes at the end of the satellite's useful life. However, this process has resulted in the release of highly radioactive sodium-potassium (NaK) coolant particles. These droplets represent a unique space debris class.

3.7.4 Space Science and Debris

Abrasion of optical surfaces on satellite sensors by prolific micron-sized debris particles can reduce sensor life.

Mistaken identity: early space particle collection experiments (e.g., by *Skylab*) found titanium in space. It was initially thought that this must have come from the Sun. Later on it was found to be from paint that had flaked off various orbital rocket bodies and some satellites.

On May 23, 1985, a Greek astronomer captured on film a flash from the part of the Moon's surface not illuminated by the Sun. Such transient lunar phenomenon, or TLP, have always been a much debated topic, but this appears to settle the matter once and for all. Could it be a meteoroid impact, a volcanic eruption/outgassing, or maybe some kind of ionization phenomenon? It turned out to be none of these. Several years later satellite investigators showed that it was in fact a glint off a derelict US military weather satellite. Several other claims about new astronomical phenomenon have later been revealed to be due to space debris.

3.7.5 Astronomy: Image Debris Trails

Space object trails showing up on astronomical images is becoming an increasing problem. A recent astronomical imagery book by a famous astrophotographer containing about 200 spectacular astronomical images had satellite/debris trails across a dozen of them. A trail through a beautiful galactic image that has taken several hours to expose can be very frustrating, but in photometric research, where many hundreds of images are often processed automatically without human oversight, a trail through the target object could result in false data.

When photographing near the celestial equator the problem becomes a lot worse because of the number of large satellites and debris objects that are in or near geosynchronous orbit.

3.7.6 Radio Astronomy: Transient Events

Large pieces of space debris, as well as operational spacecraft, will reflect radio signals from ground transmitters back to the ground. Meteors do the same thing, but usually at frequencies less than 200 MHz. Metallic debris will echo microwave frequencies. Even radio quiet zones cannot legislate against this type of transient interference.

3.8 Collisional Cascade

Each collision in LEO produces more debris, and this debris has the potential to create even more debris, and so on.

In 1978, two NASA employees wrote a paper that was published in the *Journal of Geophysical Research* titled *Collision Frequency of Artificial Satellites: The Creation of a Debris Belt*, by Donald Kessler and Burt Cours-Palais [6]. This was written before any accidental collision had happened. The article pointed out that when the debris numbers in low Earth orbit become great enough, a chain reaction could occur that would rapidly increase the debris population to a point where normal satellite operations would no longer be possible.

Today this scenario is termed the Kessler syndrome.

Eleven years after this article was published, the science fiction writer Frederick Pohl wrote a story titled *Homegoing* [7] about this scenario. The story is set in a future Earth surrounded by garbage belts containing 90,000 trackable pieces of space debris. An astronaut in the story describes a final attempt to launch a spaceship into orbit. The ship was destroyed and all her colleagues were killed. She states that previous generations "shot us out of space forever!" The narrative notes that "the thing that keeps the human race trapped on the surface of the Earth is its own previous activities in space. Just as has happened often before in human history, the human race has been defeated by its own success."

In 2006, Kessler's successor in the NASA Orbital Debris Program Office wrote a paper saying that the orbital debris population is now sufficient for the Kessler syndrome to proceed in future years without any further injection of debris into low Earth orbit [8] (Figure 3.9).

3.9 Debris Visual Magnitude

Orbital space objects are visible only by the light they reflect/scatter from the Sun. Objects in low Earth orbit are visible only during the twilight hours; morning twilight

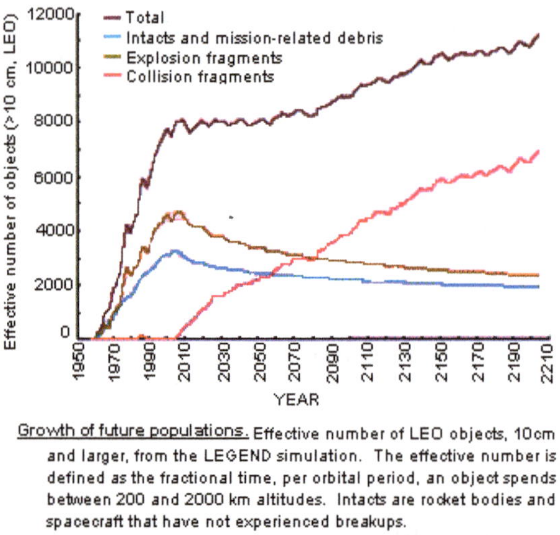

Figure 3.9 In a paper published in the January 2006 issue of *Science*, Kessler's successor stated that the debris population is now sufficient, and that without any further launches to LEO, collisional cascading will exponentially increase the space debris population, it becoming the dominant production mechanism in 50 years. With the launching of megaconstellations to LEO, the approach of rapid collisional cascade is drawing much closer. (*Source: Science* NASA ODPO.)

and evening twilight. In the daytime, satellites are not normally visible because the observer is swamped by sunlight. In the nighttime, the Earth blocks the sunlight from reading the object. Only around twilight, when the Sun is below the observer's horizon but the satellite is still illuminated by the Sun will the satellite be visible. The situation is shown in Figure 3.10.

This figure shows the Earth in cross section (despite surface topography being shown) with the center at point C, the observer at O, and the solar terminator at T. The terminator is the surface boundary between daytime and nighttime. The radius of the Earth is R. A satellite is shown at point S in an orbit around the Earth at an altitude h above the surface (at least at the point S, the satellite does not have to be in a circular orbit). This satellite, shown at the observer's zenith, must have a minimum height h to be illuminated by the Sun.

If the satellite was closer to the terminator T, it would be visible at a height less than h, as long as it was in the sunlit area (shown in yellow). If the observer was further away from the terminator (i.e., later in the night), it would need to have an altitude greater than h to be visible.

The time that a satellite in LEO is visible will be from one to two hours after sunset. Close to sunset, the satellite must be brighter than the twilight sky. Further into the darker hours, the satellite must be high enough to catch the sunlight.

Techniques are currently being developed to allow observations in the daylight hours. One of these techniques is to use narrow band filters to reduce the background sky brightness.

Visual measurements of orbital space objects use astronomical magnitude to give a precise value for this quantity. This is a scale that was devised by the Greek

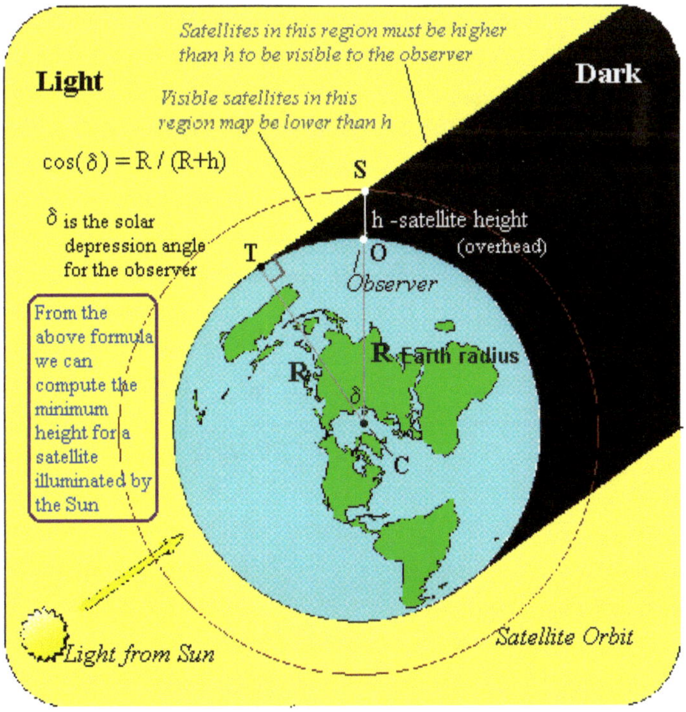

Figure 3.10 Diagram to illustrate LEO satellite visibility.

astronomer Hipparchus who placed the brightest visible stars at magnitude 1 and the faintest visible stars at magnitude 6 (i.e., a five-point scale). More accurate measurements have revealed that the brighter stars actually span a greater range than thought and the brightest star, Sirius, is now set at magnitude −1.6.

As the eye responds logarithmically to light intensity or brightness, the mathematical formula to compute magnitude difference from light intensity L is given by:

$$m_0 - m_1 = 2.5\log\left(\frac{L_1}{L_0}\right)$$

In astronomy the zero point of this scale is given by the star Vega ($m = 0$). Any star fainter than Vega will have a positive magnitude, and any brighter star will have a negative magnitude. If $L_1/L_0 = 100$ then $\Delta m = m_0 - m_1 = 5$, so 5 magnitudes corresponds to a luminosity difference of 100. Magnitude values for typical celestial objects are shown in Figure 3.11.

It is not easy to predict the visibility, from the ground, of a space object. Many factors influence the actual brightness, but we can make a rough estimate by considering a spherical object that is half illuminated by the Sun when directly overhead an observer (Figure 3.12). The albedo (reflection coefficient) of the object is assumed to be around 0.1, a value not atypical for an object in low Earth orbit.

Figure 3.12 shows sunlight normally incident on a spherical orbiting space object. The flux of this visible radiation is $S \approx 1370$ watts per square meter (the same that it is for the Earth). The portion of this power intercepted is $P = SA$ watts

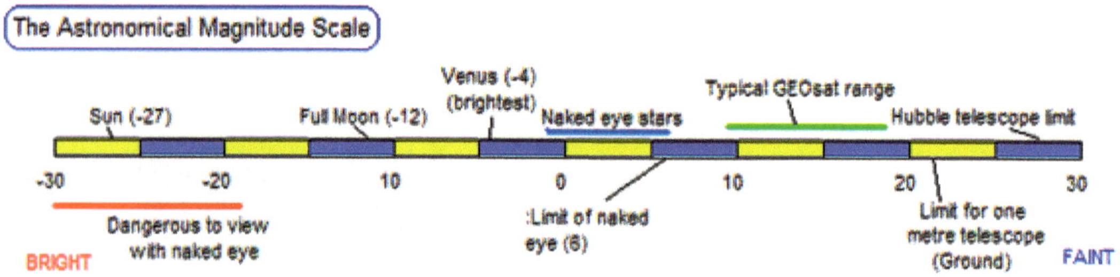

Figure 3.11 The astronomical magnitude scale over 60 magnitudes.

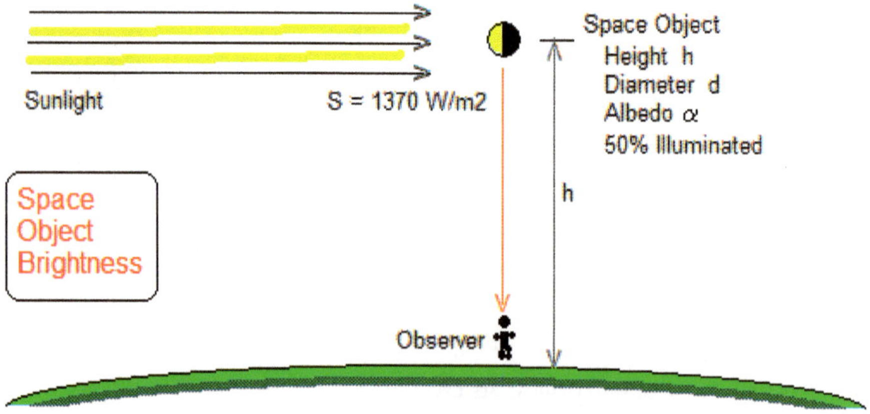

Figure 3.12 Geometry for calculation of space object brightness magnitude.

where A is the area of the 50% illuminated part of the object, and $A = \pi(d/2)^2/2 = \pi d^2/8$. Only a fraction a of this power is reradiated. The flux at the Earth's surface, a distance d away is $S_e = \alpha SA/4\pi h^2$.

Converting flux ratio to magnitude:

$$m_{\text{obj}} - m_{\text{sun}} = 2.5\log\left(\frac{1{,}370}{S_e}\right)$$

and thus

$$m_{\text{obj}} = 2.5\log\left[\left(\frac{32}{\alpha}\right)\left(\frac{h}{d}\right)^2\right] - 27$$

where −27 is the visual magnitude of the Sun.

This formula is valid for satellites or space debris observed near the zenith with 50% solar illumination, where m is the standard astronomical magnitude, α is the satellite albedo, h is the height of the assumed spherical object with a diameter d. The output of this model is plotted in Figure 3.13. A typical albedo of 0.1 gives a very approximate visual magnitude for various objects in low Earth orbit as a function of their size and altitude.

3.10 Current Issues In Space Debris

There are basically five broad areas in space debris research and operations:

1. Measurements;
2. Modeling;

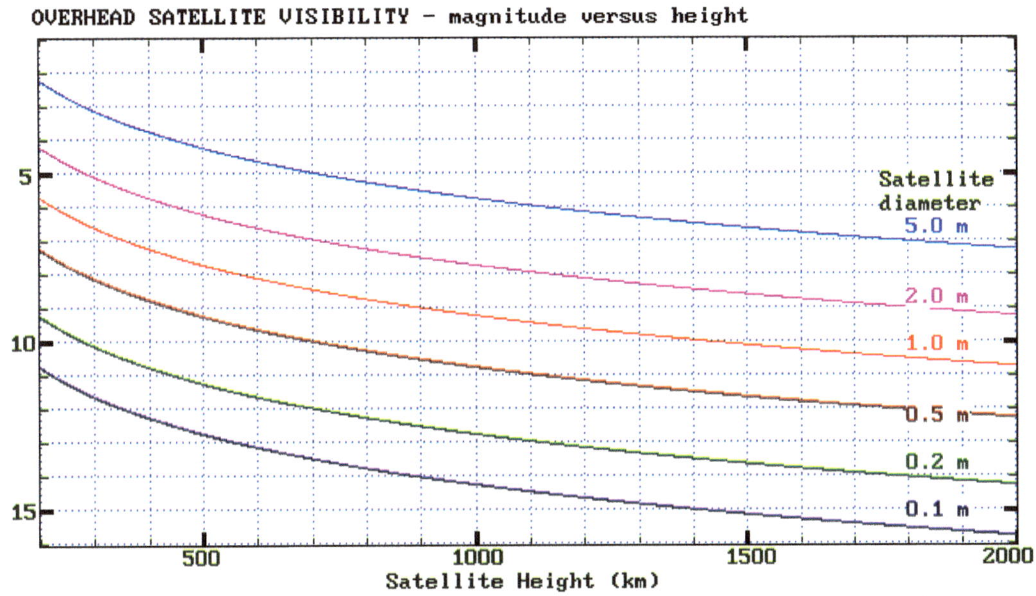

Figure 3.13 Graph of typical satellite or space debris magnitudes.

3. Impact phenomenology;
4. Mitigation;
5. Reentry.

The terms *space situational awareness* and *space domain awareness* are currently used to indicate a knowledge of what is going on in the space domain, particularly with reference to man-made orbital space objects. With the current interest in going back to the moon and the establishment of lunar bases (particularly with the presence again of international rivalry), the domain of space interest is expanding from high Earth orbit to cis-lunar space. This means all the volume of space centered about the Earth that extends to lunar orbit. A knowledge of the space object population within this volume is a large part of SSA.

- Issue 1: The ability of space surveillance networks to be able to track objects down to 1-cm size and cope with an exponentially increasing number of objects above this size. The US Space Surveillance Network is being upgraded and private companies are joining in the effort to track orbital space objects. Time will tell whether these efforts are adequate.
- Issue 2: The European space community is becoming increasingly concerned about reliance on the United States for space situational awareness and is slowly developing an indigenous capability. It is yet to be seen whether their network can stand alone as a wholly autonomous system.
- Issue 3: Finally, the most pressing issue for space debris at the current moment is the development of a sound scientific method for active removal of large amounts of space debris, both large and small sizes, from low Earth orbit. Not only must it be able to be engineered and be financially possible, it must also be politically acceptable.

3.11 Early Space Debris People

Joseph Loftus (1930–2005)

Joseph Loftus was one of the earliest proponents of the importance of space debris. He gained an international reputation in this field, and was known as the godfather of the NASA Orbital Debris Program Office. His efforts led to the funding and establishment of that office in 1979.

Donald Kessler

Donald Kessler was the first chief scientist and program manager of the NASA Orbital Debris Program Office. He is widely remembered for the 1978 paper on the possibility of collisional cascading leading to a severe orbital debris crisis. This is now referred to as the Kessler syndrome. He was the first to predict the presence of large amounts of small uncataloged space debris. He retired from NASA in 1996 to become an independent space debris consultant.

Nicholas Johnson (1950–2021)

Nicholas Johnson was the chief scientist and program manager for the NASA Orbital Debris Program Office. Prior to this appointment he worked in private industry on space debris problems and as an expert on the Soviet space program. He was the coauthor of the first book solely devoted to orbital space debris.

John Africano (1951–2006)

John Africano was an astronomer with diverse interests. In the latter part of his life he concentrated on measurements of space debris, working in both the civilian and military communities. He was an avid communicator and educator, and co-organized the annual AMOS Technical Conference on Maui.

Walter Flury

Walter Flury, now retired, was a leading European scientist on space debris issues, bringing these to the attention of a wide audience in the 1980s. He was the first coordinator of space activities within ESA and launched the first European Conference on Space Debris in 1993.

Heiner Klinkrad

Heiner Klinkrad was for many years the head of the European Space Agency Space Debris Office. He assumed this position in 2006. He led the ESA delegation to the Inter-Agency Space Debris Coordination Committee. He has written a book on space debris.

3.12 Space Debris Places and Organizations

NASA Orbital Debris Program Office, www.orbitaldebris.jsc.nasa.gov.

European Space Research and Technology Centre, www.esa.int/SPECIALS/Space_debris/index.html.

Inter-Agency Space Debris Coordination Committee, www.iadc-online.org/index.cgi.

United Nations Office for Outer Space Affairs, http://www.unoosa.org/.

International Academy of Astronautics, http://www.iaaweb.org.

Center for Orbital and Reentry Debris Studies (a division of The Aerospace Corporation), www.aero.org/cords.

The International Institute of Space Law, www.iislweb.org/.

Center for Space Standards and Innovation (CSSI is a research arm of Analytical Graphics, Inc. (AGI)), www.centerforspace.com.

References

[1] https://nssdc.gsfc.nasa.gov/nmc/spacecraft/display.action?id=1957-001B

[2] *Orbital Debris: A Technical Assessment*, (US) National Research Council, published by National Academy Press, Washington, DC, 1995, Figure 1, page 12.

[3] "Orbital Debris—A Chronology," in *NASA/TP-1999-208856*, 1999, Fig. 2, p. 3.

[4] Celestrak, celestrak.org

[5] Satellite Observations, satobs.org

[6] Kessler, Donald J., and Burton G. Cour-Palais, "Collision Frequency of Artificial Satellites: The Creation of a Debris Belt," *Journal of Geophysical Research: Space Physics*, Vol. 83, No. A6, pp. 2637–2646.

[7] Pohl, Frederick, *Homegoing*, New York: Del Ray Books, 1979.

[8] Liou, J-C., and N. L. Johnson, "Risks in Space from Orbiting Debris," *Science*, Vol. 311, 2006, pp. 340–341.

Selected Bibliography

Books

Johnson, N. L., and D. S. McKnight, *Artificial Space Debris*, Malabar, FL: Orbit Book Company, 1987. This was the first published book completely devoted to orbital space debris.

Orbital Debris: A Technical Assessment, (US) National Research Council, published by National Academy Press, Washington DC, 1995, www.nap.edu/books/0309051258/html/index.html.

Journals

Advances in Space Research, www.sciencedirect.com/science/journal/02731177.
Space Debris, www.springer.com/engineering/mechanical+eng/journal/11193/.

Online

Australian Space Academy, spaceacademy.net.au/watch.htm/sd.

ESA, www.esa.int/Space_Safety/Space_Debris.

NASA Orbital Debris Program Office, www.orbitaldebris.jsc.nasa.gov.

CHAPTER 4

Fast Facts on Space Debris

4.1 What is Space Debris?

Space debris comes in two types: natural and artificial.

- Natural space debris consists of small pieces of cometary and asteroidal material called meteoroids. We see these as meteors when they travel through the Earth's atmosphere.
- Artificial space debris is any nonfunctional man-made object in space (usually orbiting the Earth).

4.2 Where Does Artificial Space Debris Come From?

- Satellites that have reached the end of their life;
- Satellites and spacecraft that have failed;
- Rocket stages that have launched satellites into space;
- Nose cones, payload covers, shrouds, bolts, and other launch hardware;
- Solid propellant slag;
- Space activity castaways (accidental or deliberate), such as wrenches and human waste;
- Deterioration fragments (e.g., peeling paint);
- Fragments from exploding batteries, fuel tanks (not totally empty), and other flammables;
- Fragments from collisions, both accidental and deliberate.

4.3 When Was the First Piece of Artificial Space Debris Created?

At the start of the space age, on October 4th, 1957, when the last stage of the rocket that launched *Sputnik-1*, remained in orbit.

4.4 Is Space Debris a Problem?

The main worry about space debris is possible collision with active or functioning satellites or spacecraft.

4.5 How Likely Is a Collision with Space Debris?

Prior to 1957, many astronomers warned that natural space debris (i.e., meteoroids) would pose such a high collision risk that man's use of space might be severely compromised. Fortunately, this turned out to be not so. The collision probability from meteoroids is very low, although not negligible. The larger the spacecraft and the longer it remains in space, the greater the chance that it will suffer a collision.

At the present time, only a handful of dangerous collisions with artificial space debris are known to have occurred. However, the production of this type of debris is increasing at such a rate that it gives concern for the future.

4.6 What Other Problems Are Due to Space Debris?

- Small pieces of space debris (less than 1/10 mm) are prolific enough to cause erosion of optical surfaces. This is like sandblasting, and can ruin telescope mirrors, and decrease the efficiency of solar cells.
- Particles such as paint flakes (under 1 mm) can cause small craters in walls and windows. Almost 100 Space Shuttle windscreens have had to be replaced (as of 2008) due to pits caused by such impacts.

4.7 What Is the Minimum Size for Dangerous Damage?

It is believed that any fragment of space debris larger than 1 cm will penetrate the walls of existing satellites and spacecraft.

4.8 Why Are Space Debris Impacts So Dangerous?

It does seem slightly incredible that a paint flake can cause a crater in a windscreen. However, the reason is the velocity of space debris impacts. A typical impact occurs at a closing velocity of 10 km/s or 36,000 kilometers per hour!

Such impacts are called hypervelocity impacts to indicate their extreme nature. At these velocities a piece of debris has more kinetic energy (energy due to its motion) than an equal mass of high explosive. An impact with a 1-kilogram object will thus cause more damage than the explosion of 1 kilogram of TNT.

4.9 Can You Express a Space Debris Collision in Everyday Terms?

- A 2-mm space debris fragment colliding at 10 km/s is like being hit with a cricket ball at 100 km/h.
- A 10-mm fragment at the same speed is like being hit by a large motorbike at 100 km/h.

4.10 Why Are Collision Velocities So High?

Velocities of objects in space are determined by the laws of physics and the gravitational field of the body around which the objects orbit.

For the Earth, an object needs to be accelerated to a velocity of around 7 kilometers per second to stay in low Earth orbit (this is why it takes so much rocket fuel to launch a satellite).

Objects are in many different orbits around the Earth, some traveling in the opposite direction to others. With these high orbital velocities it is only natural that collisional velocities will be correspondingly high.

In the case of natural space objects (meteoroids) that orbit around the Sun, these travel at even greater velocities. To remain in its orbit around the Sun, the Earth has to move at 30 km/sec. Other objects that come near the Earth may have relative velocities ranging anywhere from 11 to 72 km/sec. The average collisional velocity between a meteoroid and a satellite is about 20 km/sec.

4.11 How Much Debris Is in Space Now?

Table 4.1 shows, as of 2022, the estimated number of pieces of artificial space debris in orbit around the Earth.

4.12 Is Anything Being Done About Space Debris?

In June 2007 the United Nations General Assembly adopted a set of 7 orbital debris mitigation guidelines for member states (countries) to follow. However, these are legally nonbinding under international law.

Most space-faring countries realize that space debris is a problem and have their own programs to try and reduce the creation of more space debris in future space activities. NASA has the most proactive program in this area.

The problem is that these programs do not reduce the amount of debris that is currently in orbit.

4.13 Can Satellites Be Protected from Space Debris?

There are two ways that satellites and spacecraft may be protected from orbital space debris impacts:

Table 4.1 Fragment Size Distribution

Size Range	*Number of Fragments*
1 to 10 mm	50 million
10 to 100 mm	300,000
> 100 mm	25,000

- Computer programs can search for possible collisions between large space debris objects and high-value spacecraft. When they detect the likelihood of such a collision, the spacecraft is maneuvered (by small thruster rockets) out of harm's way. Such maneuvers are now being carried out for large spacecraft such as the International Space Station and the Hubble Space Telescope among others. However, these operations are expensive and can disturb delicate experiments. Also, not every satellite has the ability to maneuver.

  Space-tracking networks can only track space objects larger than about 100 mm. Since even a 10-mm object can severely damage a satellite, it is obvious that collision avoidance will never be 100% effective.
- Debris shields can be designed to provide additional protection for a spacecraft. One obvious way is simply to increase the thickness of the vehicle walls. However, this adds a lot of mass to the craft and makes it a more expensive to launch into space.

  Specially designed shields take advantage of the fact that two thin walls separated by a space are more resistant to debris penetration than a single thicker wall. This type of design is called a Whipple shield after the astronomer Fred Whipple who came up with the idea in the 1950s. The outer wall absorbs a lot of the debris energy so that the inner wall is not punctured. This type of shield and modifications to this design are currently installed on the International Space Station. However, once again this does not offer 100% protection.

4.14 What About Debris Already in Orbit?

Although the probability of collision with a piece of orbital space debris is currently very low, it will not remain so forever, even if no more debris is put into orbit.

The reason for this is that collisions between objects in space will slowly but surely increase the number of dangerous debris fragments. These in turn will produce more debris and so on in a chain reaction. In a few hundred years, the amount of debris will be so great that space operations will be severely limited.

In 1989 science fiction writer Frederick Pohl predicted a similar situation in his novel *Homegoing*. Set many years in the future, mankind had generated so much space debris that he was confined to the surface of the Earth.

4.15 What Can Be Done to Remove Debris from Space?

A number of solutions have been proposed to this problem:

- Objects in low altitude orbits (below about 500 km) are affected by atmospheric drag. This lowers their orbit until they reenter the atmosphere and are thus naturally removed from orbit. The lower the orbit the faster it decays.
- Space tugs could be employed to catch large space debris objects and either lower their altitudes for natural decay, or bring them back to Earth.

- Giant sponge-like objects could be deployed to catch or soak up small debris pieces. After a time, the sponge would be removed from orbit.
- Attach tails or tethers to large pieces of space debris to increase the drag they experience and lower their orbits.
- Use large ground-based lasers to push small pieces of debris into lower orbits.

All of these schemes will be very expensive and use technology that is still to be developed. The laser scheme looks as if it might be the most promising. However, political problems might outweigh technical problems.

4.16 Is Anyone Actively Involved in Debris Reduction?

There are currently many organizations involved in the investigation of how space debris might be removed from orbit, and many interesting ideas have been proposed. However, as of 2024 there are no agencies involved in active removal of existing space debris from orbit. Natural decay due to atmospheric drag is significant for orbiting objects below an altitude of around 500 km.

It is the opinion of this author that laser removal of space debris may be the best option for objects smaller than 10 cm in size. If intact derelict satellites or rocket bodies still have functioning rocket systems and sufficient fuel, they can be made to deorbit themselves. If these objects are large enough to be expected to at least partially survive reentry, the preferred location is the southern South Pacific Ocean due to relative lack of land masses. The removal of single large orbiting objects is likely to be so expensive that few organizations will consider it.

4.17 What Happens to Space Debris Reentering Earth's Atmosphere?

Most pieces of space debris will burn up as they enter the Earth's atmosphere. This ablation process starts around a height of 100 km and is usually complete by the time the object has descended to about 20 km.

Very heavy or refractory pieces (and occasionally very light pieces) may not burn up completely, and some part of the object may make it down to the ground. However, even these objects have lost most of their space velocity and hit the ground at no more than 100 ms per second.

4.18 Are Returning Objects Hazardous?

It has been estimated that one piece of space debris may make it down to the Earth's surface almost every day. Most of these are very small and most will fall into the ocean or an unpopulated region of the Earth's surface. Very few are ever recovered, much to the disappointment of reentry scientists.

The probability of being hit by a piece of space debris is very low and a lot less than the probability of your being hit by a car while crossing a road. There have been a few reports of people being hit by a piece of reentering debris, but because of the low velocities none of these has been serious. Reports of property damage have been made.

Occasionally a large reentering derelict satellite may contain hazardous cargo. Two cases of radioactive debris have actually made it to the ground (these were from radioactive power sources). Such reentries are known well in advance (although the exact point of landfall can never be known precisely), and trained teams can be sent to secure the site and ensure no danger to any people in the area. Australia has a contingency plan set up to deal with such an event.

4.19 Can Anything be Done to Reduce Reentry Risk?

If a satellite has maneuvering capability and still has remaining fuel at the end of its life, it can be set up so that it reenters over a large area of ocean. NASA carries out a reentry survivability analysis for all large spacecraft, and if this indicates that a significant fraction of the object may survive the reentry process, they will attempt to control the reentry of such objects into a safe area. These controlled reentries are generally directed to impact in the southern Pacific Ocean where the population density is very close to zero.

If a large object is noncontrollable, it may be possible to destroy it through collision with an antisatellite missile launched from the ground. Such a collision produces many thousands of pieces of space debris, each of which will burn up when it reenters the atmosphere. However, as long as this is done at a sufficiently low altitude (below 300 km), the additional space debris produced will all decay from orbit (through atmospheric drag) within a month or two. However, the cost of such an operation is very expensive, and it could be argued whether this approach is worthwhile, considering the very low probability of damage or injury.

4.20 How Long Does It Take for Space Debris to Decay?

By "decay" we mean a reduction in orbital height due to atmospheric drag.

The decay lifetime of a space object depends on its altitude, the level of solar activity, and its mass to cross-sectional area.

Objects with a large mass-to-area ratio will remain in orbit longer as they are less affected by drag.

High solar activity increases the density of the atmosphere in low Earth orbits and reduces satellites' decay lifetimes.

On the average, a satellite in an initial 300-km high orbit will have a decay lifetime of only a few months. One in a 500-km orbit has a lifetime of around 10 years, and one at 1000-km altitude will stay in orbit for thousands of years.

4.21 Are There Any Other Future Problems with Space Debris?

Some astronomers have been worried about orbital space debris proliferation because of two concerns:

- Increased numbers of medium- to large-scale space debris will cause light trails across astronomical images, decreasing their scientific and aesthetic value. The trails also may confuse automatic computer analysis of large numbers of images.
- Extremely large numbers of very small objects (fractions of a millimeter) are expected to increase the background or ambient nighttime sky glow, limiting the extent to which astronomers can see faint objects (e.g., very distant galaxies).

CHAPTER 5

Orbits: Parameters And Specifications

5.1 Low Earth Circular Orbits

A circular orbit (Figure 5.1) is mathematically the simplest orbit, although it can be difficult to achieve exactly in practice. Only one parameter is needed to specify the size of the orbit, and that is the radius *r*, the constant distance from the center of the Earth to the satellite.

If the radius of the Earth is denoted by *Re* and the satellite radius by *r*, then the altitude or height of the orbit (and the satellite) above the Earth's surface is $h = r - Re$.

Unfortunately for our simplicity, the radius of the Earth is not a constant, varying from 6,357 km at the poles to 6,378 km at the equator. If our satellite has a low inclination (i.e., it spends most of its time near the equator), then we might use the equatorial radius. Otherwise we could use the mean radius of 6,371 km. Of course, for high-accuracy applications, we need to consider the varying shape of the Earth at each point beneath the satellite's orbit.

If we know the period of the satellite (the time it takes to make one complete orbit around the Earth) we can compute its height, and vice versa. We can also

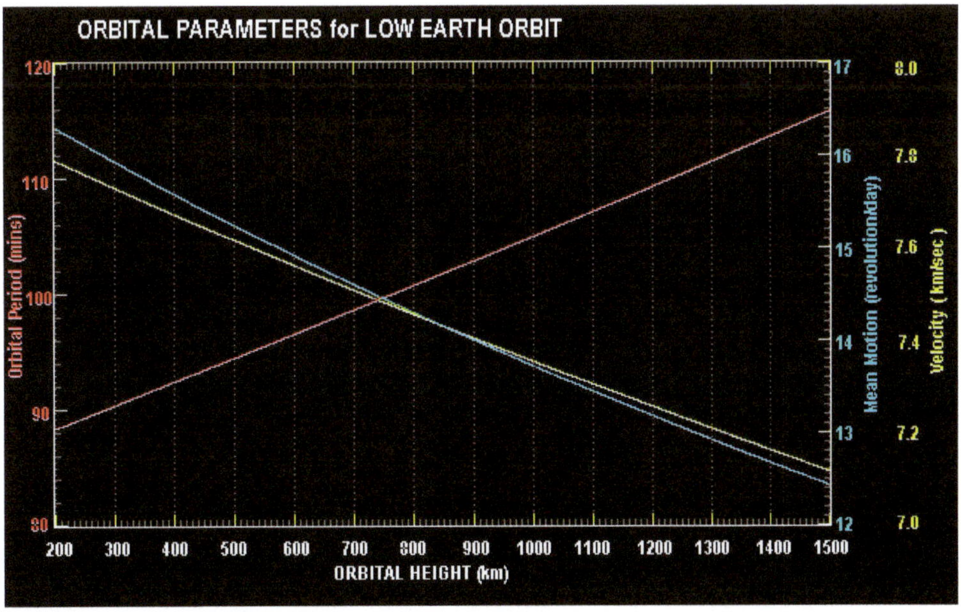

Figure 5.1 LEO orbital parameters in graphical form.

compute its orbital velocity (v) and a quantity known as its mean motion, the number of orbits it completes in one day.

Isaac Newton provided the formulae we need for these calculations. Starting with the famous second law (N2) of motion $F = ma$ where F is the force on the satellite, m is its mass, and a is its acceleration.

Newton's law of gravitation $F = GMm/r^2$ is the required force that the Earth exerts on the satellite.

A body moving in a circle is actually being accelerated toward the center of the circle (by the Earth's gravitational force). The acceleration is referred to as the centripetal (directed toward the center) acceleration, and the force that produces the acceleration is known as the centripetal force.

The magnitude of the acceleration is given by $a = v^2/r$.

If we substitute for the force on the left-hand side of N2 and the acceleration on the right-hand side we end up with:

$$\frac{GMm}{r^2} = \frac{mv^2}{r}$$

Where G is the universal constant of gravitation, and M is the mass of the Earth.

We now note that the mass of the satellite (m) disappears from the equation, and we are left with an expression relating the orbital velocity to the orbital radius:

$$v = \sqrt{\frac{GM}{r}}$$

The orbital period (T) may be computed by dividing the distance traveled in one orbit by the velocity. The former quantity is the circumference of the circle.

$$T = \frac{2\pi r}{v}$$

which after substitution for v becomes

$$T = 2\pi\sqrt{\frac{r^3}{GM}}$$

Classically the mean motion of the satellite is then defined as

$$N = \frac{2\pi}{T}$$

This has the units of radians per second (in SI units) and is a very small number. Because of this, it is more usual to define the mean motion (n) of the satellite as the number of orbits traversed in one complete day which is 86,400 seconds long, and this is given by:

$$n = \frac{86{,}400}{T}$$

Note that with the exception of n, all quantities are strict SI units. Distances are in meters, velocities in meters/seconds, and times in seconds. The exception, the mean motion, is in revolutions per day.

In this system of units, the values for the gravitational constant and the mass of the Earth are given as:

$$G = 6.67259 \times 10^{-11}\ \text{m}^3\text{kg}^{-1}\text{s}^{-2}$$
$$M = 5.9736 \times 10^{24}\ \text{kg}$$

It is interesting to note that the uncertainty in the measured value for G is larger than the uncertainty in the measurement of the product of G times M, which is given as:

$$k2 = GM = 3.9860 \times 10^{14}\ \text{m}^3\text{s}^{-2}$$

Note that strict SI units use meters for the distance unit. However, because this is a small unit with regard to distances and velocities in space, it is more usual to refer to distances in kilometers and orbital velocities in km/s.

Figure 5.1 displays various LEO parameters in graphical form.

The red curve shows the orbital period in minutes, and its corresponding scale is on the left-hand side of the graph. The right-hand side scale is used for both the orbital velocity (yellow) which ranges in value from 7 to 8 km/s, and the mean motion (blue) which varies from 12 to 17 orbits per day.

5.2 Geosynchronous Satellites

Among all the orbits, the geostationary or geosynchronous orbit is very special. It exists at only one height, that height at which the orbital period is exactly equal to the Earth's sidereal rotation period (23 hours, 56 minutes, and 4 seconds). Because of the applications that such an orbit presents, we have devoted a whole chapter to this special orbit and will only consider the locational specification of satellites in this orbit.

5.2.1 Introduction

Geosynchronous satellites (Figure 5.2) are those that appear to remain nearly stationary in the sky as observed from a point on the Earth's surface. They do so because their period of revolution about the Earth is identical to the period of rotation of the Earth about its polar axis.

A true geostationary satellite will remain absolutely fixed at the same point in the sky as seen by a ground observer. However, perturbing forces such as radiation pressure, Earth nonsphericity, and lunar and solar gravitational forces, all conspire to remove a satellite from a true geostationary orbit. Ground-directed maneuvers are required to counteract these forces and maintain the satellite in what is really a geosynchronous orbit—an orbit that is close to but not exactly geostationary. These corrections are termed *station keeping*, and are carried out to keep the satellite in a

Figure 5.2 Australia's first geosynchronous satellite, *AUSSAT-A1*, being deployed from the payload bay of Space Shuttle mission STS-51I. The satellite was deployed into low Earth orbit, and was later boosted into GEO by the attached booster rocket. (*Source:* NASA.)

virtual box around the true geostationary point. This box usually has dimensions of less than one degree of arc, as viewed from the ground.

Station keeping has several objectives. If the satellite is used to broadcast TV or data, it means that small receiving antennae do not need to be moved to track the satellite. It also helps minimize interference between adjacent satellites using the same frequency band, and is one step toward minimizing collisions between active satellites.

Sometimes, if the station keeping fuel on board a geosat becomes low, the inclination of the satellite (the angle that its orbital plane makes with the Earth's equator) is allowed to increase from zero. Such satellites show a figure eight motion in the sky. The satellite sweeps around the figure eight within a period of 24 hours.

Thus a ground observer is likely to see a combination of secular drift, periodic motions (e.g., figure eight), and abrupt correctional changes (due to thruster firings) in the motion of a geosynchronous satellite. However, these are usually small, and for many locational purposes it is adequate to compute look angles assuming the simplicity of geostationarity.

5.2.2 Specification of a Geostationary Orbit

A geostationary orbit has specific orbital parameters. It is first of all a circular orbit. This means that it has an eccentricity of zero—its distance from the center of the Earth is always the same (42,165 km). It furthermore has an orbital inclination of zero—its orbital plane is identical to the Earth's equatorial plane. Lastly, it has a period (related to its orbital radius) of 86,164 seconds (about 23 hours, 56 minutes, 4 seconds), which is the time it takes the Earth to rotate through 360 degrees.

The location of a satellite is usually computed using an ephemeris program and a set of orbital elements. However, for a stationary geosat, this is unnecessary and its position may be computed using only geometry. The only parameter needed to specify a geosat is its longitude. This is the same as the longitude of a point on the Earth's equator directly below the satellite itself.

Geosat longitude may be specified from 0 to 360 degrees, measured eastward around the equator from the zero degree meridian passing through Greenwich, England. Alternatively, it might be specified from 0 to +180 degrees (east) and 0 to −180 degrees (west). Note that +180 and −180 are the same point.

5.2.3 Geosat Geometry

We need both plane trigonometry and spherical trigonometry to derive the equations needed to compute geostationary satellite look angles.

We start with spherical trigonometry. This will give us the azimuth angle we require and also specify the plane in which we use plane trigonometry to find the satellite elevation.

Figure 5.3 shows points on the Earth: O is the observer, C is the center of the Earth, P is a point on the equator with the same longitude as the observer, and T is the point directly below the satellite. All curved lines on the diagram are parts of great circles.

We are interested in the spherical triangle OPT. The distance t is the latitude of the observer, q is the longitudinal difference between the observer and the satellite

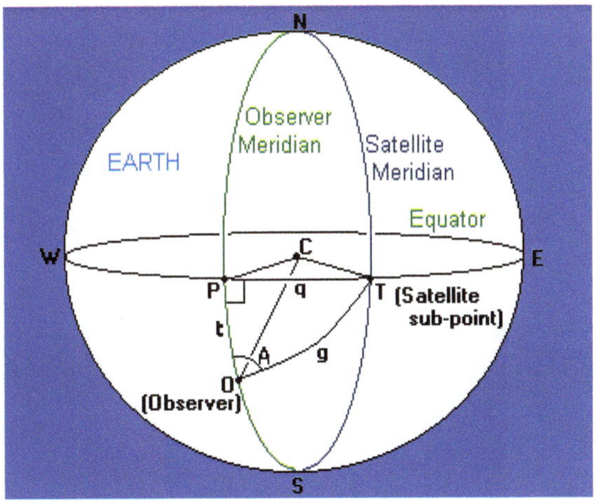

Figure 5.3 Spherical geometry for geosat location.

(or subsatellite point), and g is the great circle distance from the observer to the subsatellite point. The azimuth angle A is the angle measured from north of the satellite, and this is the angle we are trying to calculate.

The sides of the spherical triangle are related by:

$$\cos(g) = \cos(q)\cos(t)$$

Since we know the angles t and q, we can calculate the angle g, then we can calculate the azimuth from the equation:

$$\sin(A) = \frac{\sin(q)}{\sin(g)}$$

We then move to consider a plane triangle in the plane OTC, which contains both the observer and the satellite (Figure 5.4).

In the triangle COS, the satellite elevation is given by:

$$\tan(e) = \frac{[(R+h)\cos(g) - R]}{[(R+h)\sin(g)]}$$

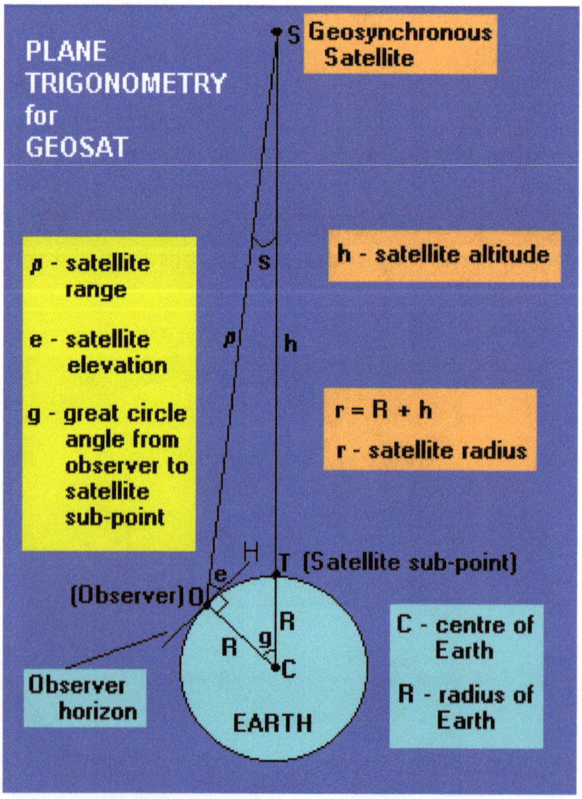

Figure 5.4 Plane geometry for geosat location.

and the slant range to the satellite can be found from:

$$\rho = \sqrt{\left[R^2 + (R+h)^2 - 2R(R+h)\cos(g)\right]}$$

To complete our calculations we need only to know that R = 6,378 km and r = (Rah) = 42,165 km, which are the equatorial radius of the Earth, and the satellite orbital radius, respectively.

These formulae assume a perfectly spherical Earth, whereas it is in fact an oblate spheroid with the polar radius less than the above-quoted equatorial radius. Calculations with an oblate Earth model show that the errors incurred from the spherical approximation amount to less than 0.1 degree in azimuth and elevation, and 10 km in slant range, at midlatitudes. This error is less than many station-keeping regimes.

5.2.4 Geosat Look Angle Charts

The above formula can be used to plot look angle charts. These show both azimuth and elevation angles as a function of observer latitude and longitudinal difference from the satellite.

Figure 5.5 is the chart to use at midlatitudes.

The azimuth contours are shown in white or red and are plotted every two degrees. The elevation contours are also plotted every two degrees in blue or cyan. The chart can thus be read to an accuracy of one degree with ease.

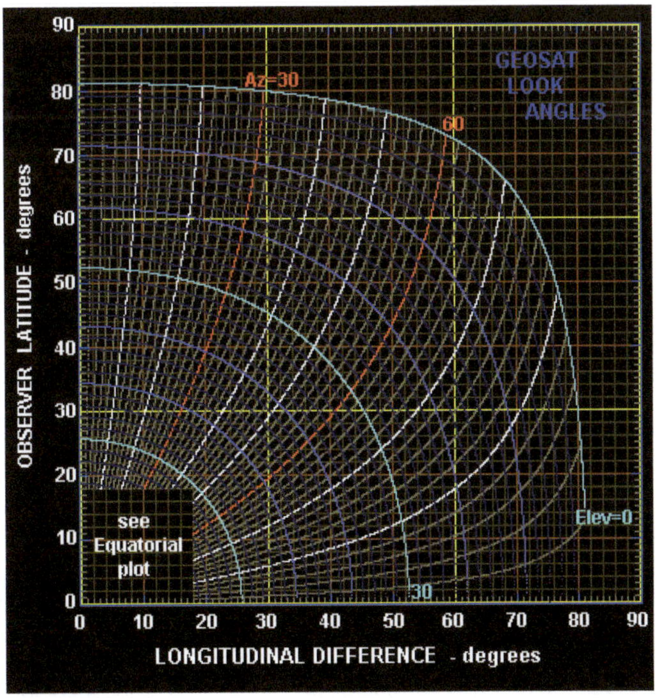

Figure 5.5 Geosat location chart to use for midlatitudes.

The azimuth angles shown in Figure 5.5 are in the range from 0 to 90 degrees. The true azimuth is computed from these values according to the quadrant in which the observer lies with respect to the satellite. The modification formulae are shown in the quadrant diagram of Figure 5.6.

As an example, consider an observer at a latitude of 40-degrees north and 32 degrees to the east of the desired satellite. From the look angle chart we find an elevation of 34 degrees and an azimuth of 44 degrees. However, as the observer is in the quadrant to the northeast of the satellite, the true azimuth is 180 + 44 or 224 degrees.

For observers in the tropics the look angle chart becomes very congested, and an equatorial supplement is provided in Figure 5.7.

Note that a geosynchronous satellite is below the horizon when the great circle distance to its subpoint exceeds 81 degrees. This means no geosynchronous satellite coverage in the Polar regions. Geosat communication is possible around the coastline of Antarctica but necessitates very large ground antennae (e.g., 20-meter class dishes).

5.2.5 Equatorial/Polar Coordinates

For astronomical and some other purposes, it is more convenient to have the coordinates of a geosynchronous satellite in terms of declination (δ) and hour angle (HA). These may be computed from elevation and azimuth according to the formula:

$$\sin(\delta) = \cos(A)\cos(t)\cos(e) + \sin(t)\sin(e)$$

$$\cos(HA) = \frac{\sin(e) - \sin(t)\sin(\delta)}{\cos(t)\cos(\delta)}$$

Note that δ is positive to the north of the equatorial plane and negative to the south. The HA is negative in the east and positive in the west.

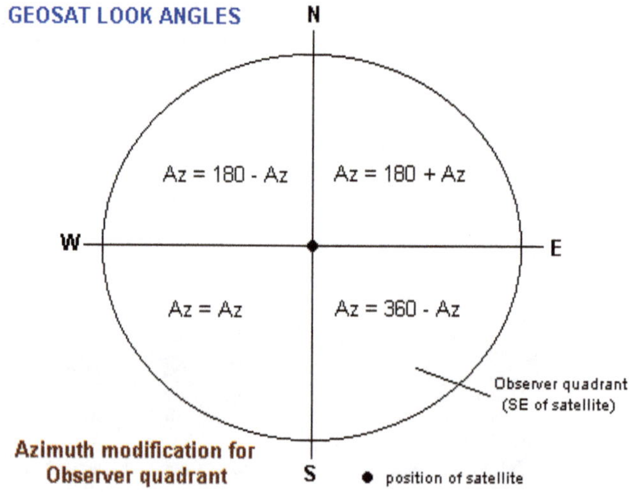

Figure 5.6 Quadrant diagram to resolve the correct azimuth.

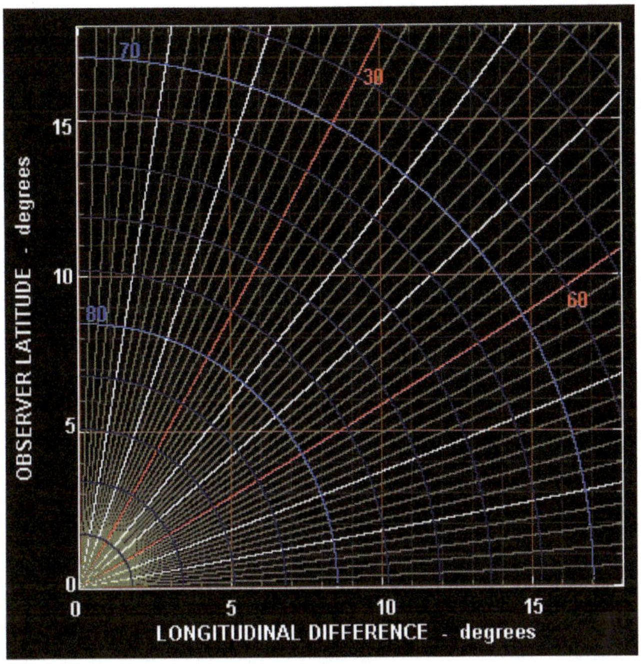

Figure 5.7 Geosat look angles for near equatorial observers.

The above two equatorial coordinates are constant because the satellite is fixed in the sky. However, the stars are not fixed and accurate positions are often derived from stellar position comparison. For this purpose it is necessary to change the hour angle coordinate into right ascension. This is done using the formula:

$$RA = LST - HA$$

where LST is the local sidereal time at the observing location.

Note that RA, LST, and HA must all be in the same units, which is usually hours, but may also be degrees. Note also that the right ascension of a geosat is continuously changing, and if this coordinate is specified it is vital to specify the time of measurement.

Figure 5.8 shows how the declination of a satellite changes with longitudinal difference.

The declination of a geosat will always lie between + and −9 degrees. For mid-latitudes the declination is around −5 degrees in the Northern Hemisphere and +5 degrees in the Southern Hemisphere. For a given location, the declination of any visible geosat never varies by more than about half a degree.

The band in the sky across which geostationary satellites are found is sometimes referred to as the Clarke Belt, a tribute to the writer Arthur C. Clarke who first suggested the use of this orbit in 1945.

5.2.6 Algorithm Output

Coding the above formulas into a computer program is relatively straightforward. The main difficulty usually arises in putting things in their correct quadrant.

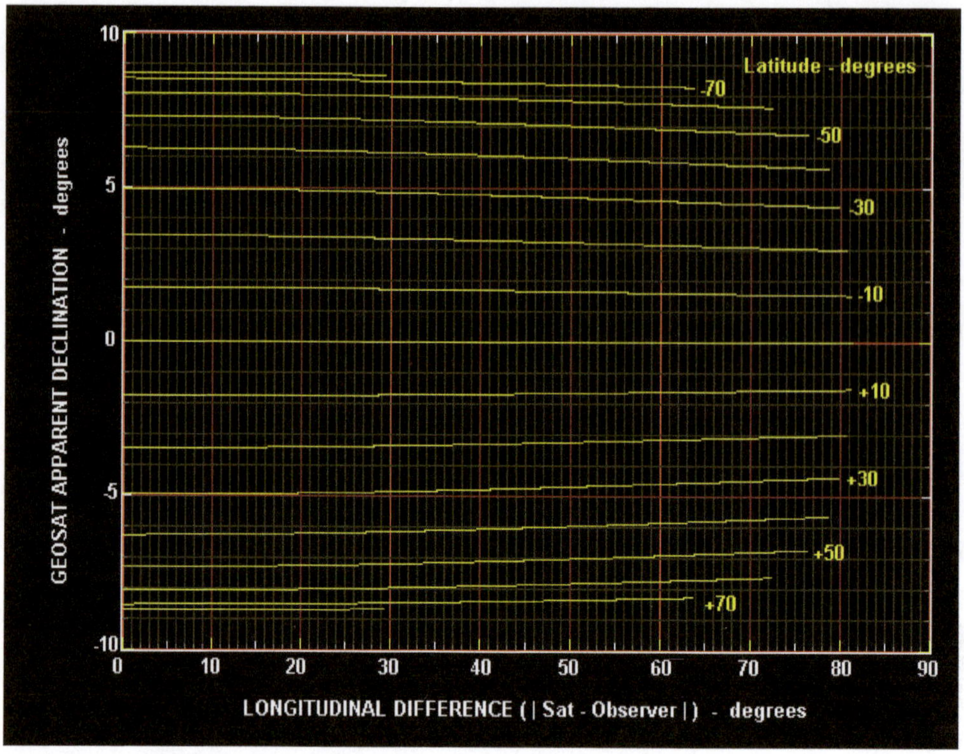

Figure 5.8 The apparent declination of a geostationary satellite.

A typical output is displayed in Table 5.1. This can be used as a check for anyone who wishes to develop their own program.

The last column in Table 5.1 is the great circle distance from the observer to the satellite subpoint on the Earth's surface.

5.3 Specifying Satellite Orbits in General

5.3.1 Introduction

Satellites are launched into many different types of orbit around the Earth for many different reasons. All of these orbits, however, share one characteristic: they follow an elliptical path with the center of the Earth at one focus of the ellipse. This ellipse can be specified as a set of parameters. However, this does not completely define the orbit, because an ellipse lies in a plane and that plane can be oriented in many different ways relative to the Earth's geometry.

It turns out that six numbers suffice to totally specify the parameters of a satellite in orbit about the Earth (or a body in orbit about the Sun). If we know these numbers we can then calculate where a satellite will be at a future time.

This section is concerned with what those numbers are and how they are specified.

5.3.2 Orbital Specification

As stated above, there are two basic aspects to the motion of an object in orbit:

Table 5.1 Geosat Look Angles for an Observer in Western Australia

Geostationary Satellite Look Angles
Observer LAT = –32.0 LON = 117.0

SAT-LONG (deg)	*ELEVANGLE (deg)*	*AZANGLE (deg)*	*HA (hrs)*	*DEC (deg)*	*SLANT (km)*	*GND (km)*
42.0	4.0	278.1	5.487	4.70	41236	8607
52.0	12.5	283.9	4.801	4.81	40320	7681
62.0	21.0	290.4	4.098	4.91	39457	6779
72.0	29.3	297.9	3.380	5.01	38678	5917
82.0	37.1	307.1	2.647	5.10	38011	5120
92.0	44.0	318.7	1.901	5.17	37485	4427
102.0	49.3	333.2	1.145	5.22	37120	3896
112.0	52.4	350.6	0.382	5.25	36934	3601
122.0	52.4	9.4	–0.382	5.25	36934	3601
132.0	49.3	26.8	–1.145	5.22	37120	3896
142.0	44.0	41.3	–1.901	5.17	37485	4427
152.0	37.1	52.9	–2.647	5.10	38011	5120
162.0	29.3	62.1	–3.380	5.01	38678	5917
172.0	21.0	69.6	–4.098	4.91	39457	6779
–178.0	12.5	76.1	–4.801	4.81	40320	7681
–168.0	4.0	81.9	–5.487	4.70	41236	8607

1. The shape and size of the orbit;
2. The orientation of the orbit with respect to the Earth (or other primary body).

The first aspect determines the period of the object and the second determines its ground coverage.

5.3.3 Orbital Shape and Size

Figure 5.9 represents a satellite in an elliptical orbit around the Earth. The long or major axis of the ellipse is the length AP, and the short or minor axis of the ellipse is the length $m1 - m2$. The ellipse can be specified by two quantities, the semimajor axis a, and the semiminor axis b. The two points, f1 and f2, are called the foci (plural of focus) of the ellipse, and the Earth is always at one focus of the ellipse, in this case f1. The distance of the foci to the center of the ellipse is c, and $c^2 = a^2 - b^2$.

The shape of the ellipse is defined by the relative values of a and b, and is specified by another quantity e, the eccentricity of the ellipse, where

$$e^2 = 1 - \left(\frac{b}{a}\right)^2$$

When $b = a$, the eccentricity is zero, and the ellipse becomes a circle. The eccentricity can have values from 0 up to nearly 1. At a value of $e = 1$, the geometry

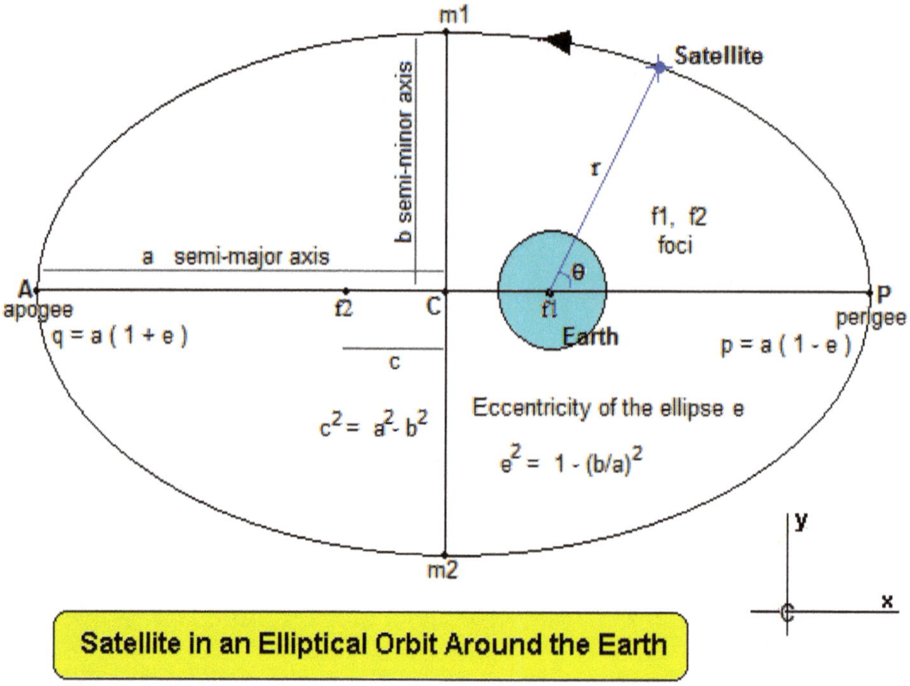

Figure 5.9 Satellite parameters in the orbital plane.

becomes a parabola, which is no longer an ellipse. When this occurs the satellite would no longer be in orbit around the planet. Values of e close to 1 result in a very elongated orbit.

The point P is called the perigee of the orbit and this is where the satellite is at its closest point to the Earth. The point A is called the apogee and this is where the satellite is at its furthest distance from the planet. Perigee and apogee distances can be calculated using

$$q = a(1+e)\ldots\text{apogee}$$
$$p = a(1-e)\ldots\text{perigee}$$

The height of the satellite above the Earth's surface is equal to

$$h = r - R$$

where R is radius of the Earth. Note that this varies with latitude of the observer. At the equator $R = 6{,}380$ km and using this will give an approximate value for the height. At apogee and perigee the heights are given by

$$h_p = p - R \text{ and } h_q = q - R$$

The position of a satellite in the orbit can be specified by the quantities r and θ. However, in specifying the orbital elements of a satellite it is more usual to specify the time at which the satellite passes the perigee (t_p or t_0), and then to use orbital mechanics (based on Newton's law of gravitation) to compute the actual position around the orbit.

It is customary to specify the shape and size of the orbit by the two parameters:

a—semimajor axis (in km);

e—the orbital eccentricity (a dimensionless number).

The time it takes the satellite to move from perigee around the orbit and back to perigee again is called the orbital period (P) of the satellite.

The orbital period depends only on (and can be calculated from) the semimajor axis and the mass of the Earth (M):

$$P = \sqrt{\left[\left(4\pi^2 a^3\right)/(GM)\right]}$$

where G is Newton's constant of gravitation ($G = 6.6725985 \times 10^{-11}$ in SI units) and $M = 5.9736 \times 10^{24}$ kg. This gives a value for the product of G and M of 3.9860×10^{14}. The semimajor axis must be in meters and this will give a period in seconds.

5.3.4 Orbital Orientation

In Figure 5.10 the celestial equator is the projection of the Earth's equator onto the sky, and forms the reference plane with which to describe the orientation of the satellite orbit. The reference plane is also the Earth's equatorial plane. The points at which the satellite orbit crosses this reference plane are called the nodes of the orbit (the ascending node is where the satellite is moving from below the reference plane to above it, the descending node is where it moves from above to below).

Three angles are needed to describe the orientation of the orbit with respect to the Earth:

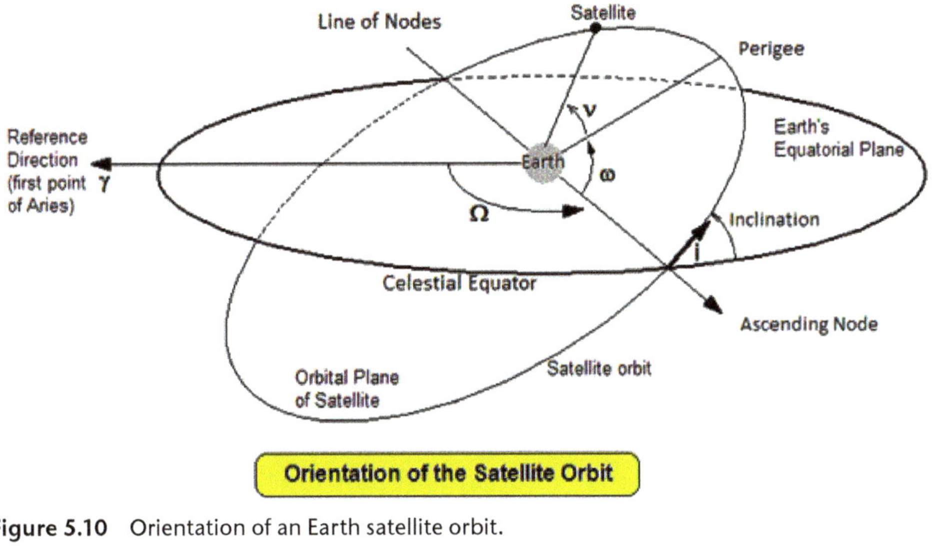

Figure 5.10 Orientation of an Earth satellite orbit.

1. i is the inclination of the orbit, is the angle between the orbital plane and the reference plane (i.e., the celestial equator).
2. Ω is the angle between the line of nodes and a reference line in the reference plane. This reference line points to an astronomical point in the sky known as the first point in Aries (a constellation). It is designated by γ.
3. ω is the angle between the line of nodes and the semimajor axis of the ellipse in the direction of perigee. This angle is measured in the plane of the orbit.

The angle v is the angle from perigee to the satellite. This is called the *true anomaly* of the body. Many orbital element specifications also refer to a *mean anomaly*, which refers to an angle that goes out to a sphere circumscribed around the ellipse. One can be calculated from the other using a formula called *Kepler's equation*. The position of the satellite is computed at any specified time t by knowing both the time of perigee passage, and computing the mean anomaly M from the time t, and the satellite's period P: $M = 2\pi\, t/T$ radians. The true anomaly is then calculated from the mean anomaly by Kepler's equation:

$$M = v - e\sin v$$

This equation is solved by iteration. In the days before computers much attention was paid to the most efficient methods to solve this equation. Even today, when many orbits must be computed for conjunction assessment, attention should still be paid to fast solutions of this equation.

5.3.5 Orbital Elements

The position of an orbital space object can thus be specified by six parameters:

1. a, the semimajor axis specifies the size of the orbit;
2. e, the eccentricity specifies the shape of the orbit;
3. i, the inclination specifies one orientation of the orbit;
4. ω, the argument of perigee specifies a second orientation of the orbit;
5. Ω, the argument of the ascending node specifies the third orientation;
6. t_p, the time of perigee passage is used to calculate the object location in the orbit.

These parameters are sometimes referred to as the Keplerian elements of the orbit.

However, they are not the only way to specify the position of an object, and in particular the classic Keplerian elements have a problem in specifying object position in circular orbits where there is no perigee (inclination = 0). This is also the case for truly geostationary orbits. It is thus necessary to use a modified version of these elements for real-world applications.

The US Strategic Command, which operates the US space tracking network issues space object orbital information in a form called TLE (two line elements) because the specified parameters are printed over two lines. The actual six parameters used are a modification of the six elements we have used previously.

1. n, the mean motion of the satellite in revolutions per day (this gives a);
2. e, the orbital eccentricity (identical);
3. i, the orbital inclination (identical);
4. RAAN, the right ascension of the ascending node (this gives Ω);
5. ω, the argument of perigee (identical);
6. M, the mean anomaly at the time specified in the TLE and substitutes for the t_p parameter;
7. D/T, date and time of the TLE.

There is also a decay value included, which refers to orbital change due to atmospheric perturbation of the elements.

To calculate the satellite mean anomaly you subtract the TLE time from the satellite prediction time and multiply the result by the mean motion n.

5.3.6 TLE Format

A typical TLE has the format:

```
2014-037E
1 40073U 14037E   14195.06978695  .00000547  00000-0  79836-4 0    207
2 40073  98.4035 246.9202 0004647 235.4646 124.6133 14.80036563    782
```

The first or header line is the name of the satellite or object in question. The next two lines contain the actual elements.

This TLE thus specifies the orbital elements for a space object referred to as *2014-037E*, one of a group of satellites launched by the Russians in 2014. The E means that it was the 5th object put into orbit by the one-launch vehicle.

The format of the following two lines is as follows:

Line 1

Group 1	Line number (= 1)	column 1
Group 2	NORAD catalogue number follower by	columns 3 to 7
	classification (U=unclassified)	column 8
Group 3	International designation =	
	Last two digits of year +	columns 10 & 11
	Launch number of the year +	columns 12 to 14
	Object number deployed (alphabetic)	columns 15 to 17
Group 4	TLE epoch date (yyddd.dddddddd)	columns 19 to 32
Group 5	Rate of change of mean motion	columns 34 to 43
Group 6	Second derivative of mean motion	columns 45 to 52
Group 7	Atmospheric drag term	columns 54 to 61
Group 8	Ephemeris type (=0)	column 63
Group 9	Element set number	columns 65 to 68
Group 10	Checksum (to check data integrity)	column 69

Line 2

Group 1	Line number (=2)	column 1
Group 2	NORAD catalogue number	columns 3 to 7
Group 3	Orbital inclination (degrees)	columns 9 to 16
Group 4	RAAN (degrees)	columns 18 to 25
Group 5	Orbital eccentricity (leading decimal point is assumed)	columns 27 to 33
Group 6	Argument of perigee (degrees)	columns 35 to 42
Group 7	Mean anomaly (degrees)	columns 44 to 51
Group 8	Mean motion (revolutions per day)	columns 53 to 63
Group 9	Revolution number (orbit number)	columns 64 to 68
Group 10	Checksum (modulo 10)	column 69

Thus for the above TLE, the satellite has a NORAD designator of *40073*, and a COSPAR (international) designator of *14037E*. The latter means that it was the 5th object deployed from the 37th launch of the year in 2014.

The date or epoch of the TLE is day 195 of year 2014. The fractional part of the date (0.06978695) gives a time of 01:40:29.7 Universal Time (UT).

The mean motion of the object is 14.8003 revolutions per day and the change in this motion is 0.00000547 rev/day/day. Groups 6 and 7 are usually not very accurate and are often not used in orbit prediction programs. (The last two characters of these two groups are the exponent of the number.)

The orbital inclination is 98.4 degrees. Any value over 90 degrees indicates a retrograde orbit, which indicates it moves over the Earth's surface from east to west, whereas a prograde (normal) orbit moves from west to east.

The right ascension of the ascending node is 246.0 degrees, the argument of perigee is 235.4 degrees, and the mean anomaly is 124.6 degrees.

The eccentricity of the orbit is 0.0004647, which being so close to zero indicates a near-circular orbit. The mean motion of the orbit is 14.800 revolutions per day, which indicates an orbital period of 97.3 minutes.

The numbers at the end of each line indicate that this is the 20th set of elements (TLE) issued for this satellite and that the satellite was on its 78th orbit at this time.

The following TLE is for practice. After decoding the elements, use the previous information to deduce the approximate altitude of the satellite, which was the first Australian communications satellite.

```
OPTUS A1 (AUSSAT 1)
1 15993U 85076B   18165.48109566 -.00000294 +00000-0 +00000-0 0  9999
2 15993 015.3011 003.5014 0006102 312.6931 228.2493 00.99759997090415
```

5.3.7 TLE Availability

Two line elements are available from a number of websites. Dr. Tom Kelso makes available TLEs for a range of different satellites at celestrak.com.

US Strategic Command tracks tens of thousands of objects in Earth orbit and access to the TLEs for these objects is available at www.sat-track.org to approved customers. Application for approval is made via the same website. This database does not include US military satellites.

5.3.8 Decoding the TLEs

All satellite prediction programs include preprocessors that decode the TLE format into orbital elements for subsequent processing. However, it is often useful to have a small stand-alone TLE decoder that allows a quick look at the orbital status of a particular satellite. The following TLE for the International Space Station and its plain language decode can be used to check such a TLE decoder (Table 5.2).

5.3.9 The Real World

Unfortunately life is not quite as simple as we would like. The Keplerian elements are only truly correct for the time of issue. They gradually change over time due to perturbing forces acting on the satellite. These forces change depending upon the orbital altitude of the object. Perturbing forces can be divided into two types:

1. Conservative perturbative forces:
   Gravitational forces due to the Earth, Sun, Moon, and so forth. Due to the oblate shape of the Earth, gravitational forces can be divided into the primary gravitational force (which produces the instantaneous orbit) and harmonics, which perturb that orbit. For practical purposes it is only necessary to consider the first two harmonics, J2 and J4, and for many purposes only J2 needs to be considered.

Table 5.2 Satellite TLE and Decoded Output

```
Satellite ID : ISS
1 25544U 98067A   11190.42490741  .00001183  00000-0 -80029-5 0  6797
2 25544  51.6526  48.6986 0010535 312.2137  50.7032 15.59819830724381

    Epoch time = year 2011 day 190 hour 10 minute 11 UT

    Orbit Size and Shape
        Semimajor axis =   6766 km
        Eccentricity   = 0.0011
    Orbit Orientation
        Inclination    =  51.65 deg
        RAAN           =  48.70 deg
        Perigee arg.   = 312.21 deg
    Satellite Information
        Period         =   92.3 minutes
        Perigee height =    381 km
        Apogee height  =    395 km
        Epoch anomaly  =  50.70 deg
    Orbit number     =  72438
    TLE set number =     679
```

2. Dissipative perturbative forces:
   - Atmospheric drag;
   - Solar radiation pressure;
   - Earth albedo radiation pressure;
   - Solar wind pressure.

In low Earth orbit atmospheric drag and J2 are the primary perturbing forces.

Conservative forces can be calculated and do not change the total energy of the space object. Dissipative forces change the total energy of the object they act upon and are in general more difficult to deal with than conservative forces.

The end result of perturbations is that the TLEs have a limited life for orbital prediction and new TLEs need to be acquired at regular intervals. For objects in very low orbits (e.g., below 300 km) they become unusable in a day or so. Generally, for LEO objects the TLE needs to be updated every few weeks, sometimes weekly.

For space objects in higher orbits, such as geosynchronous orbit, TLE lifetime is generally longer, maybe as much as a few months.

5.3.10 State Vectors

The accuracy of TLEs issued by US Strategic Command will rarely give orbital positions to better than 100 meters and typically only to a kilometer. In cases where higher accuracy is required orbital positions are specified not by orbital elements but by quantities called *state vectors*. These specify the three-dimensional position and velocity of an object at a given time. There are thus three position coordinate numbers and three velocity numbers. Thus we see that six quantities are still required to specify the orbit.

Although state vectors can be used by orbital propagation programs they don't give humans any idea of the orbital parameters. It is thus useful to be able to convert from state vectors to orbital elements. The algorithm to do this in SI units is as follows:

Constants:

For the Earth, the gravitational parameter is $\mu = GM_e = 3.9860 \times 10^{14}\ \text{m}^3\ \text{s}^{-2}$

where G is the universal gravitational constant and M_e is the mass of the Earth.

Inputs:

Position vector $r = Rx\, i + Ry\, j + Rz\, \mathbf{k}$ and $r = |\mathbf{r}|$

Velocity vector $\mathbf{v} = Vx\, \mathbf{i} + Vy\, \mathbf{j} + Vz\, \mathbf{k}$ and $v = |\mathbf{v}|$

where **i**, **j**, and **k** are unit vectors in the x, y, and z directions, respectively.

Note that the Rs must be in meters and the Vs in meters per second.

Compute:

Angular momentum vector $\mathbf{L} = \mathbf{r} \times \mathbf{v}$ (per unit mass—$\text{m}^2\ \text{s}^{-1}$) and $L = |\mathbf{L}|$

Eccentricity vector $\mathbf{e} = (v^2 - \mu/r)\mathbf{r} - (\mathbf{r} \cdot \mathbf{v})\mathbf{v}$ and $e = |\mathbf{e}|$ (e is the orbital eccentricity)

[check that $e < 0.999$ for an elliptical orbit – if not stop]

Nodal vector $\mathbf{n} = \mathbf{k} \times \mathbf{L}$ and $n = |\mathbf{n}|$

Total energy $E = v^2/2 - \mu/r$ (per unit mass—J/kg) (this is kinetic plus potential energy).

Then:

Semimajor axis $a = -\mu/(2E)$

MM (mean motion) $MM = \sqrt{(\mu/a^3)}$ (radians per second)

Inclination inc $= a\cos(L_k/L)$ where L_k is the component of L in the z direction

Ω (RAAN) $\Omega = a\cos(n_i/n)$ and if $n_j < 0$ then $\Omega = 2\pi - \Omega$

$\omega = a\cos(\mathbf{n} \cdot \mathbf{e}/(ne))$ and if $e_k < 0$ then $\omega = 2\pi - \omega$

True anomaly $\nu = a\cos(\mathbf{e} \cdot \mathbf{r}/(er))$ and if $\mathbf{r} \cdot \mathbf{v} < 0$ then $\nu = 2\pi - \nu$ [at time of $\mathbf{r}$ & $\mathbf{v}$ vectors]

Refer to Vallado [5] for more details and a worked example. Note however that Vallado uses canonical units where μ is set to unity ($\mu = 1$) and the distance unit is taken as the radius of the Earth (1 DU = 1 ER = 6,378.14 km) which gives a time unit TU of 1 TU = 13.446852 minutes.

5.4 Orbital Ephemeris Programs

The generation of an orbital ephemeris is orbital prediction (Figure 5.11).

Orbital elements are input to the prediction program along with the desired date and time range required. The orbital elements may be in TLE format or directly as

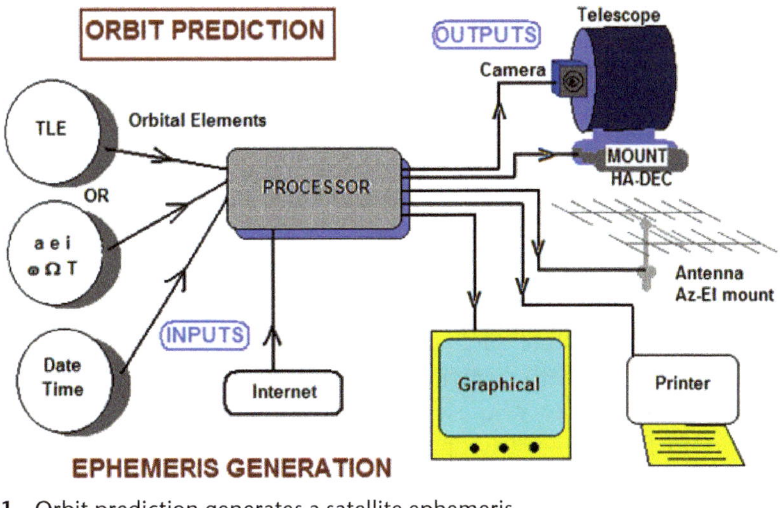

Figure 5.11 Orbit prediction generates a satellite ephemeris.

orbital quantities. The processor runs the program through the time specified and produces an output that may be viewed as text or in graphical format. It may also be used to control optical, radio, or radar sensors, both in their pointing and/or with regard to data acquisition. There are many orbital prediction programs available from internet websites for many different purposes. Some predictions are computed online at the remote website. The only input is the object name or designation. The output may be either text or graphic. Some programs may be downloaded and run on a user's computer.

5.4.1 Online Orbital Prediction

5.4.1.1 Heavens Above

Chris Peat provides an online orbital prediction service at https://www.heavens-above.com/. Your observing location can be specified and sky views requested. Ten-day predictions are generated for specified satellites and many output formats are available.

5.4.1.2 NY2O

Live real-time satellite tracking and predictions are available at https://www.n2yo.com/.

Select any satellite orbiting the Earth and check where it is located now, or check satellite passes over your location and receive alerts on your phone.

5.4.1.3 AMSAT

The Amateur Satellite organization AMSAT provides online satellite pass predictions at https://www.amsat.org/track/. This site concentrates, although not exclusively, on amateur satellites.

5.4.1.4 ESA

The European Space Agency provides an online orbit prediction tool at https://evdc.esa.int/orbit/.

5.4.1.5 OTHER

Other online satellite tracking services are provided at https://www.orbtrack.org/, https://uphere.space/, https://www.satflare.com/home.asp, and elsewhere.

5.4.2 Satellite Tracking and Prediction Software

CelesTrak, a satellite data site maintained by Tom Kelso, provides a quite comprehensive list of satellite tracking software at https://celestrak.org/software/satellite/sat-trak.php. A few of these are listed below.

GPREDICT—http://gpredict.oz9aec.net/ is free software licensed under the GNU General Public License. Gpredict is a real-time satellite tracking and orbit prediction application. It can track a large number of satellites and display their position and other data in lists, tables, maps, and polar plots (radar view). Gpredict can also predict the time of future passes for a satellite, and provide you with detailed information about each pass.

Another useful software package is GMAT. This is a NASA open source package from the Goddard Space Flight Center, https://opensource.gsfc.nasa.gov/projects/GMAT/index.php. Much more than orbit prediction, the General Mission Analysis Tool (GMAT) is the world's only enterprise, multimission, open-source software system for space mission design, optimization, and navigation. The system supports missions in flight regimes ranging from low Earth orbit to lunar, libration point, and deep space missions.

STK (Systems Tool Kit) is available from Analytic Graphics, Inc., http://www.agi.com/products/stk/, and has an enormous range of features and modeling capability.

The Sky Professional is available from Software Bisque, https://www.bisque.com/product/theskyx-pro/. This is a program for astronomers and has the capability to control many telescope mounts to track satellites. Together with Maxim DL available from Diffraction Limited, https://diffractionlimited.com/product/maxim-dl/, to control cameras attached to the telescope, these two programs provide a base for satellite optical tracking and imaging. Maxim DL operates scientific-grade cameras, DSLR cameras, low-cost imagers, webcams, and other video sources. It also controls filter wheels, focusers, autoguiders, telescopes, focal plane rotators, and observatory domes. A wide variety of hardware is supported through ASCOM drivers.

SATPC32 is software for calculating and tracking Earth orbiting satellites by Erich Eichmann, call sign DK1TB. It works with various radios and antennas, and provides Doppler correction on amateur radio transceivers as well as frequency control, and rotor control. It is available for electronic purchase at the AMSAT website https://www.amsat.org/product-category/software/.

If a reader wishes to produce their own orbital prediction program, the text by Bate, Mueller, and White [2], contains all the relevant algorithms.

5.5 Orbit Determination

Orbit prediction is the operation where orbital elements are ingested by a program to produce an ephemeris that will allow the object to be located in the sky. Orbit determination is the inverse of this where observations of an orbital space object are made, either optically or by radar means, and these observations are used to generate orbital elements of that object. Orbit determination is a more complex process than orbital prediction and is generally only done by central agencies that receive continuous observations from a multitude of sources and use these to regularly update and publish orbital elements for a large number of orbital objects.

From optical observations, two angular coordinates can be derived for each time, either azimuth and elevation, or hour angle and declination. These can generally be determined to high precision with an accuracy down to a few seconds of arc. Radar observations can determine range and range rate from time and frequency (Doppler) measurements, with high accuracy (a few tens of meters for range), and a few tens of meters per second for velocity (range rate). Both these measurements are in the line of sight of the radar to the orbital object. Radar can also determine azimuth,

azimuth rate, elevation, and elevation rate to low accuracy. A combination of radar and optical measurements can thus provide the highest accuracy.

Crude orbital determination methods started in the early 17th century when Johannes Kepler enunciated his three laws of planetary motion. This set the basis for Isaac Newton and the universal law of gravitation. However, not until Laplace and Gauss in the 19th century did the development of orbital determination techniques occur that are still used today. Because the problems solved were related to astronomical bodies in the solar system well beyond the Earth, the subject became known as celestial mechanics. Not until it was applied to artificial satellites orbiting the Earth did it become known as orbital dynamics. This occurred in the 1950s.

Because there was initially no way to determine distances, the first techniques used angles only, although it may seem counterintuitive at first that measurement of only angles could lead to a determination of distance. However, the resolution comes about because a priori we accept the universal law of gravitation that includes a distance. Because of this constraint the state of a body in orbit only needs 6 quantities to specify it completely, and from which its future position and motion may be calculated, instead of the nine quantities required to specify the generalized state of a body in free space. These six quantities could be three positional components and three velocity components at a specified time, and they could be the six orbital elements specified earlier in this section.

For angles-only orbit determination, azimuth and elevation, or hour angle and declination of the orbital object must be specified at three different points in time. From these and an appropriate method (e.g., that of Gauss) a preliminary orbit (i.e., six orbital elements) may be computed. This orbit may be improved with further observations through iterative techniques. All this discussion has skipped over a multitude of corrections that must be performed. One of the most important is that the observations (i.e., Az-El angles were measured in a topocentric coordinate system that rotates with Earth, whereas the satellite orbit is specified in a geocentric coordinate system. Now for a perfectly spherical Earth, the orbital plane of a satellite remains in a fixed orientation in space with the Earth rotating below it. For a satellite with a period of 90 minutes, the Earth rotates though an angle of 22.5 degrees and thus the start and the end of one prograde (west to east) orbit covers a longitude range of 337.5° and not 360°.

The Earth is in fact not a sphere but an oblate spheroid and this causes the satellite orbital plane to rotate up to 8° per day or half a degree per 90-minute orbit. This plus the many other corrections required add to the complexity of the orbital determination program, especially for an angles-only solution. Fortunately, there are a number of orbital determination software packages that are available that include all the required corrections:

1. The Satellite Tool Kit (STK) has orbit determination procedures.
2. The Goddard Space Flight Center (GSFC) of NASA has available an Open Source software for orbit determination and analysis at https://opensource.gsfc.nasa.gov/projects/ODTBX/. It is composed of both MATLAB® and Java code.

In practice, orbit determination is done from a large number of observations that come from different sources and have varying accuracies. Methods are then needed to best combine the results from these diverse measurements to produce one output with the highest possible accuracy. This requires that each measurement be assigned an appropriate weighting and then combined using a modified least squares method, such as Kalman filtering [6].

5.5.1 Simplified Orbit Determination of a Low Circular Satellite

To give a very simplified idea of what is involved in orbit determination we will consider a satellite in a low Earth circular orbit, orbiting around the center of a spherical Earth. In a circular orbit, there is no perigee and we will specify the time parameter as the time of ascending node passage. In this case we have not six but four orbital elements. These are the orbital radius and three angles to give the orientation of the orbit, the inclination i, the longitude of the ascending node Ω, and the position of the satellite along the orbit ν at the time of the observation. Because there are only four orbital elements rather than six, we should only need to make two observations of the satellite at two separate times rather than three. In this analysis we will further restrict the observations to be close enough together in time that the rotation of the Earth during that time is insignificant.

The geometry is indicated in Figure 5.12. For each observation we have a vector equation:

$$\mathbf{r} = \mathbf{R} + \boldsymbol{\rho}$$

where **r** is the vector from the center of the Earth to the satellite;

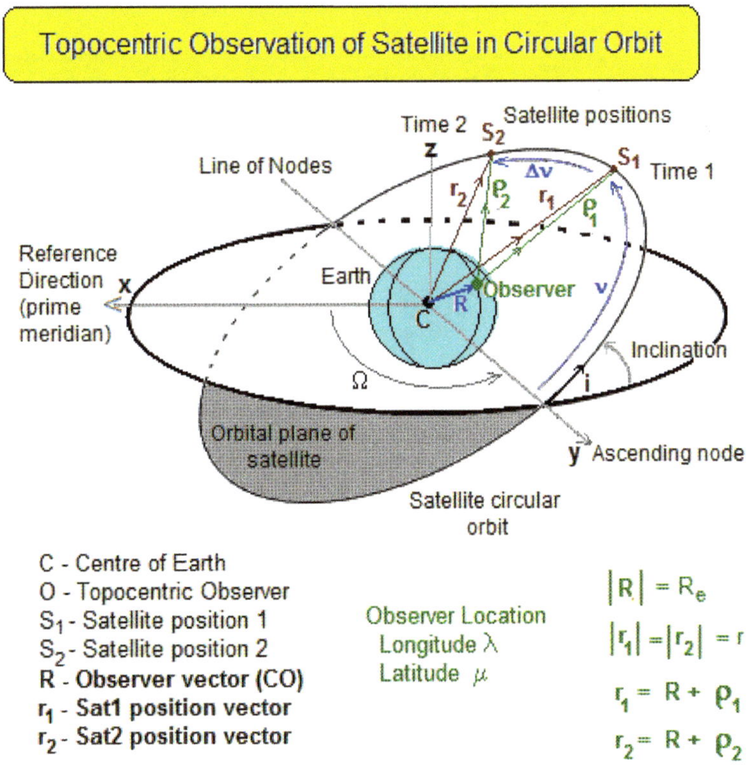

Figure 5.12 Geometry of a circular satellite orbit and observer for orbit determination.

R is the vector from the center of the Earth to the observer;

ρ is the vector from the observer to the satellite.

This is the fundamental equation to all orbit determination.

The problem from here is how to solve this equation at several observation times and determine from these the required orbital elements, and thus a preliminary orbit. As we know the location of the observer (latitude, longitude, and distance from the Earth's center) we can specify the **R** vector. If we have access to a precision radar we can also determine the elevation, azimuth, and distance to the satellite, and thus specify the vector **ρ**. We thus have very simply found the solution to the equation; that is, the determination of vector **r**. And we can do this a second time and thus determine the orbit.

Unfortunately, most of us do not have access to a precision radar, and we must make do with the determination of look angles only. That is, we can measure the elevation and azimuth of the satellite at two times, but not its range. Thus we do not know the magnitude of the vector **ρ**, but we do know its direction.

There are several ways in which we may proceed, but we will choose to determine the Cartesian components of each vector in the geocentric reference frame. The observer vector **Re** in this frame is given in Figure 5.13.

Figure 5.14 gives the satellite position vector in geocentric Cartesian coordinates in terms of the same unit vectors **i**, **j**, and **k**, used to specify the observer position.

Now the components of the topocentric vector from the observer to the satellite are:

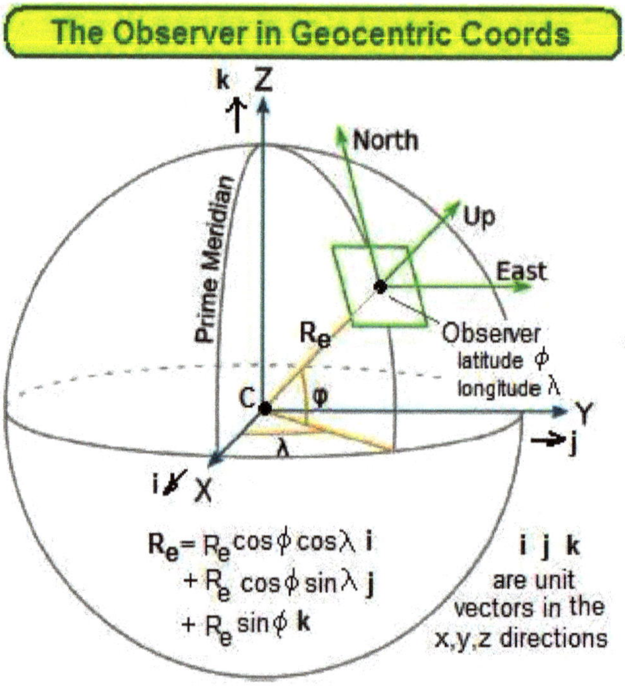

Figure 5.13 The observer vector in geocentric Cartesian components.

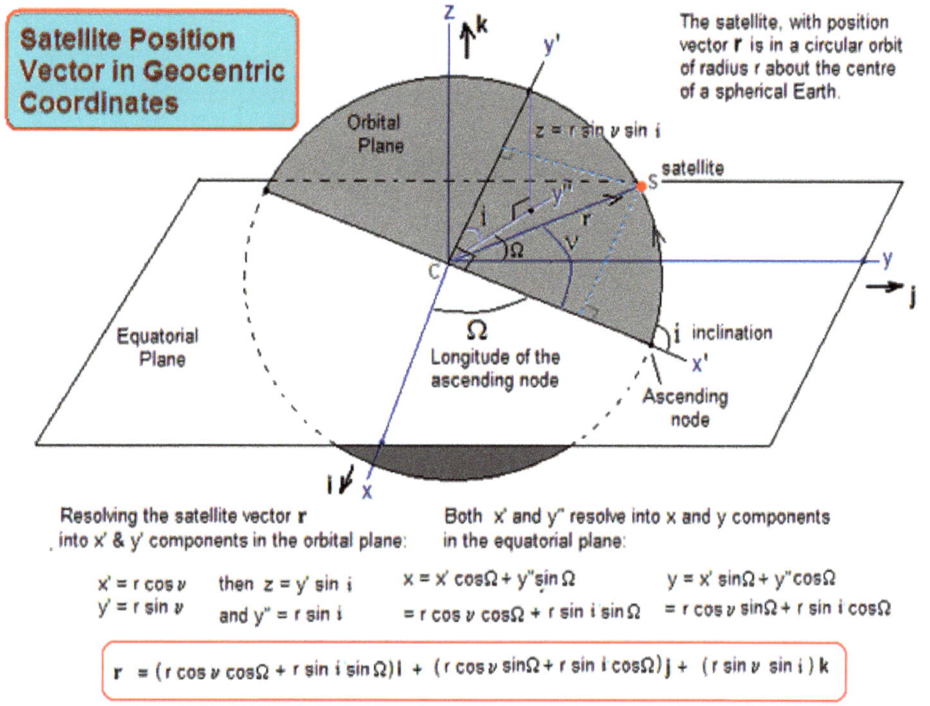

Figure 5.14 The satellite position vector in geocentric Cartesian coordinates.

$$\rho_S = -\rho \cos \text{El} \cos \text{Az}$$
$$\rho_E = \rho \cos \text{El} \sin \text{Az}$$
$$\rho_Z = \rho \sin \text{El}$$

where ρ is the (unknown) length of the vector, El is the elevation of the satellite, and Az is the azimuth of the satellite. We can thus write the vector $\boldsymbol{\rho}$ as:

$$\boldsymbol{\rho} = \rho_s \mathbf{i}' + \rho_E \mathbf{j}' + \rho_Z \mathbf{k}'$$

where $\mathbf{i}', \mathbf{j}', \mathbf{k}'$ are the unit vectors in the topocentric SEZ coordinate system. Note that SEZ forms a right-handed system that NEZ does not.

Unfortunately we cannot use these components in our orbit determination equation because $\mathbf{r}$ and $\mathbf{R}_E$ are in the geocentric coordinate system whereas $\boldsymbol{\rho}$ is in the topocentric coordinate system with unit vectors $\mathbf{i}'$, $\mathbf{j}'$, and $\mathbf{k}'$ in totally different directions to $\mathbf{i}$, $\mathbf{j}$, and $\mathbf{k}$. We must thus rotate the $\boldsymbol{\rho}$ vector from topocentric to geocentric coordinates. The appropriate rotation matrix is:

$$\text{ROT} = \begin{bmatrix} \sin\phi\cos\lambda & -\sin\lambda & \cos\phi\cos\lambda \\ \sin\phi\sin\lambda & \cos\lambda & \cos\phi\sin\lambda \\ -\cos\lambda & 0 & \sin\phi \end{bmatrix}$$

This is used to change the $\boldsymbol{\rho}$ vector components into the XYZ system:

$$\rho_{XYZ} = [\text{ROT}]\,\rho_{SEZ}$$
$$= \begin{bmatrix} \sin\phi\cos\lambda & -\sin\lambda & \cos\phi\cos\lambda \\ \sin\phi\sin\lambda & \cos\lambda & \cos\phi\sin\lambda \\ -\cos\lambda & 0 & \sin\phi \end{bmatrix} \begin{pmatrix} -\rho\cos\text{El}\,\cos\text{Az} \\ \rho\cos\text{El}\,\sin\text{Az} \\ \rho\sin\text{El} \end{pmatrix}$$
$$= \begin{pmatrix} -\rho\cos\text{El}\cos\text{Az}\sin\phi\cos\lambda - \rho\cos\text{El}\,\sin\text{Az}\sin\lambda + \rho\sin\text{El}\cos\phi\cos\lambda \\ -\rho\cos\text{El}\cos\text{Az}\sin\phi\sin\lambda + \rho\cos\text{El}\,\sin\text{Az}\cos\lambda + \rho\sin\text{El}\cos\phi\sin\lambda \\ \rho\cos\text{El}\cos\text{Az}\cos\lambda + \rho\sin\text{El}\sin\phi \end{pmatrix}$$

We now have all three components for all three vectors for one point in time, say t_1. A second observation is needed a few minutes later at time t_2. At this time the position angle of the satellite v will have increased to $v + \Delta v$ in the time interval $\Delta t = t_2 - t_1$. Thus we will end up with two satellite vectors $\mathbf{r}_1$ and $\mathbf{r}_2$ at the times t_1 and t_2. The angular distance moved by the satellite is then given through dot product of the two vectors:

$$\cos\Delta v = \frac{\mathbf{r}_1 \cdot \mathbf{r}_2}{(r_1 r_2)}$$

The angular velocity of the satellite is then given by:

$$\omega = \frac{\Delta v}{\Delta t}$$

from which can be computed the satellite radius and period.

The cross product of the two vectors gives a vector normal to the orbital plane. If this is divided by its magnitude we have a unit vector $\mathbf{n}$ defining the satellite orbital plane:

$$\mathbf{n} = \frac{\mathbf{r}_1 \times \mathbf{r}_2}{|\mathbf{r}_1 \times \mathbf{r}_2|}$$

This can then be used to determine the angles i, Ω, and v, and together with ω, the time of ascending node passage.

Note that it has not been the purpose of this subsection to present a workable orbit determination method for even a simplified case, but to give an idea of the complexity associated with orbit determination, something that historically has occupied the minds of the world's foremost mathematicians, such as Johann Carl Friedrich Gauss in the early 1800s.

5.5.2 Visual Estimation of Low Earth Satellite Orbits

To get a feel for the appearance of satellite motion from the ground, and to what orbits they relate, we can observe the sky during the twilight hours, when the sky darkens sufficiently to allow satellites (or large debris objects) illuminated by the Sun to be seen. Furthermore, we can make crude measurements to estimate the orbital parameters of these objects. An astrolabe (Figure 5.15) was traditionally the instrument used to measure positions and motions of sky objects. The most important ground measurement of a satellite is its angular velocity (degrees per second) across the sky. This can be rapidly estimated by an astrolabe with a small fixed angle of say 10 degrees. The design for such an instrument is given in Figure 5.15.

In operation the eye is placed close to the eyelet screw through which the two light emitting diodes (LEDs) are sighted. The use of low intensity LEDs makes it much easier to use than trying to see unilluminated markers. The instrument needs to be firmly held against a post or other mount, or maybe used with a tripod. The idea is to try and align the LEDs along the trail of the satellite it is measuring, and with a stopwatch measure the time taken for the object to move through the astrolabe angle (10 degrees).

The use of an astrolabe for determining the orbits of artificial satellites was first discussed by C. L. Stong in the Amateur Scientist column of *Scientific American* in 1958 [3]. The formula for orbital height determination was:

$$h = 455\left(\frac{t}{a}\right)\sqrt{\left[1 - 0.0705\left(\frac{t}{a}\right)\right]}$$

where h is the height of the satellite in kilometers;

t is the time of the transit in seconds through angle a;
a the angular distance in degrees.

The article comments that "the equation assumes that the satellite is observed within 15 degrees of the zenith. If the satellite is more than 15 degrees from the zenith, multiply the distance given by the equation by the cosine of the angle between the orbit and the zenith. The equation also assumes that the orbit is a circle."

Many bright satellites in low Earth orbit have orbits that are closely circular. However, some of the large debris pieces, such as rocket bodies, visible to the naked

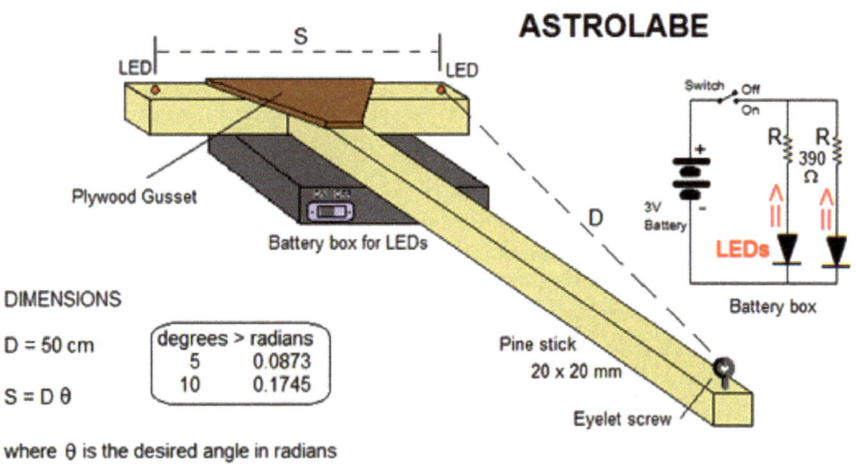

Figure 5.15 Construction of a simple astrolabe for measuring satellite angular velocity.

eye are highly elliptical. These will be travelling very rapidly at perigee where they will be brightest, and need to be identified as noncircular orbits through experience.

Figure 5.16 presents an alternative equation for use with the astrolabe and shows the approximations made in its derivation. It can be seen that the astrolabe angle is actually larger than the orbital angle through which the satellite moves, this difference decreasing with height. This equation is solved iteratively. An initial guess is made for h, say 200 km, and the equation gives an updated value for h that can then be resubstituted on the left-hand side until the difference in h values at each iteration becomes negligible.

Once the satellite height is known, Table 5.1 can be used to determine the satellite period, mean motion, and linear velocity. What remains unknown is the inclination and orientation of the orbit.

These angular parameters can be estimated from the satellite ground track. These can be estimated by observing the sky track and mentally projecting this to the ground, or more accurately by the use of a wide angle or all-sky camera to record the orbital track.

Before we look at the ground track we need to have an idea about how far away we can observe a satellite from a fixed point of the ground. The geometry for this is illustrated in Figure 5.17. The left-hand diagram illustrates the situation for a true zero-degree elevation horizon. This can be solved rapidly using the right angle triangle COS.

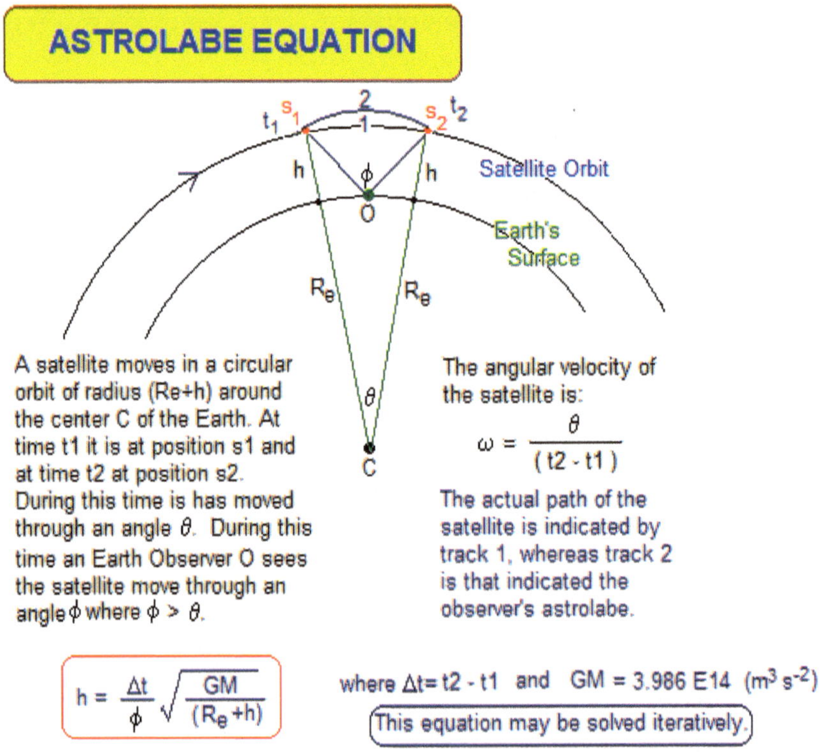

Figure 5.16 One version of the astrolabe equation. The diagram is for near-overhead observations (zenith angle less than 15 degrees or so).

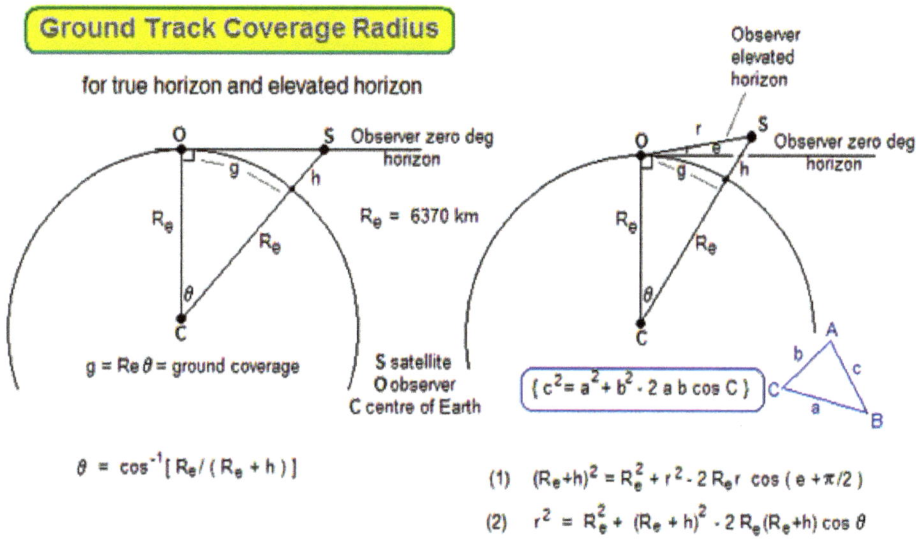

Figure 5.17 The geometry to compute a satellite horizon distance.

Very few sites have clear horizons down to zero degrees, and the right diagram in Figure 5.18 shows the geometry in this case. Unfortunately, when the triangle loses its right angle, the calculations become a more difficult. We must first solve a quadratic equation to find r and then we must use this to solve for the angle theta.

To get an idea of visible satellite sky tracks we can overlay these horizon distances on a satellite ground track map, which is done in Figure 5.18.

Figure 5.19 shows the ground tracks for 24 hours of the International Space Station orbital motion. Overlaid on these are ground observer visibility contours for a Southern Hemisphere observer at latitude −30° and longitude 90° (east), and a Northern Hemisphere observer at latitude +30° and longitude 270° east (90° west). The latitude is the important parameter here.

The visibility contours are a crude approximation to reality because of distortions inherent in the mapping. In particular, great circle distances are accurate in latitude but vary with the inverse cosine of the latitude in the longitudinal direction. As long as we stay with observing latitude less than 30 degrees, the distortions are probably acceptable for visual angle estimations.

Each observer is shown with three visibility horizons, corresponding to satellite heights of 200, 500, and 1,000 km corresponding with the values of Table 5.3.

We need now to consider the satellite tracks within these circles, for it is these that roughly correspond to the ground tracks, and thus sky tracks, in our observing field of view. Extracts of these are shown in Figure 5.19.

Table 5.3 Satellite Visible Horizon Parameters (Zero Degree Horizon)

h (km)	*Re + h*	*Re/(Re + h)*	*θ (degrees)*	*g (km)*
200	6570	0.9595	14.2	1580
500	6870	0.9272	22.0	2446
1000	7370	0.8643	30.2	3356

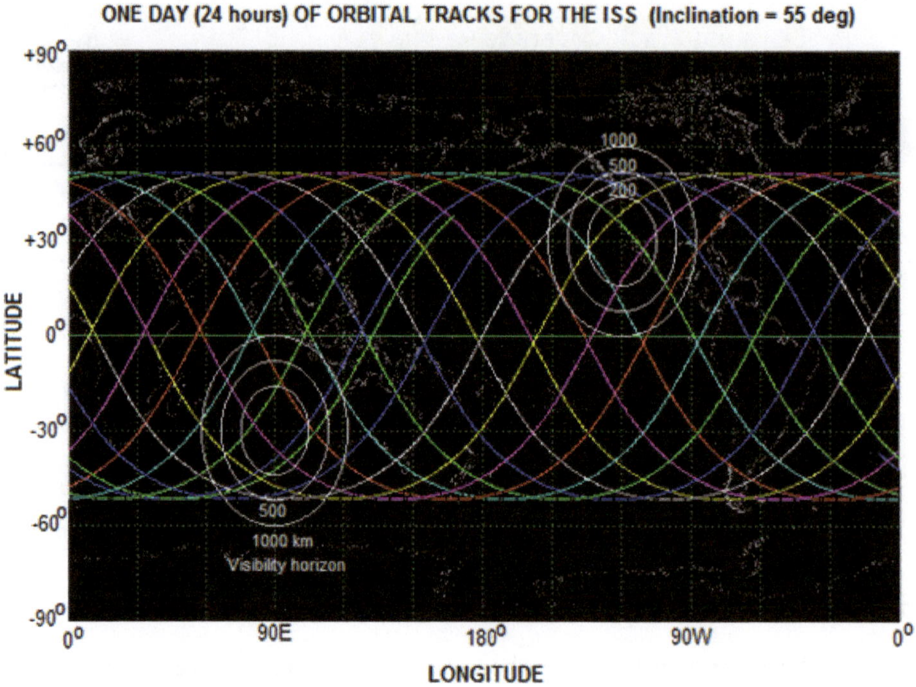

Figure 5.18 Satellite ground tracks for the ISS with southern and northern observing sites shown.

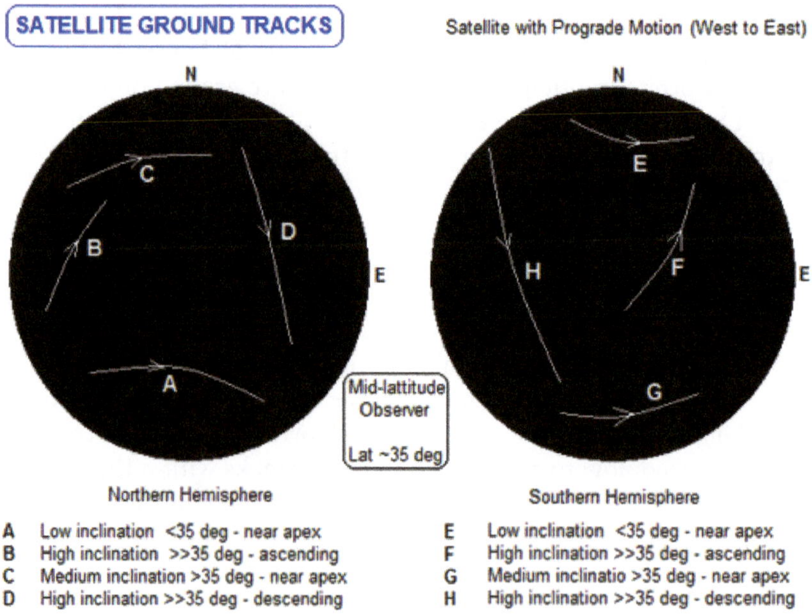

Figure 5.19 Satellite ground tracks in an observer's field of view, for midlatitude Southern and Northern Hemisphere observers.

From these we can make an estimate of the satellite inclination. For an inclination sufficiently greater than the observer's latitude, we can make a more accurate estimation of inclination by evaluating the angle the satellite track makes with the observer's north-south line. This estimate increases in accuracy as the observer moves closer to the horizon.

Knowing the inclination, observer location, and angular velocity we can estimate the longitude of the ascending node, and the time the satellite will pass through it.

To determine subsequent passes of the satellite we must compute the longitude of the next ascending node, which will be 360° minus the angle the Earth rotates in on orbit. If the orbital period is 90 degrees, this angle will be roughly 360 × 90/1,440 = 22.5°.

5.6 Solar Locational Algorithms

The Sun is a broadband transmitter and can mask the transmissions from geosynchronous satellites when it is in the beam of a satellite receiver. This problem is discussed in Sections 11.7 and 12.6. In this section we present algorithms that allow the calculation of solar coordinates.

Simple formulae for the solar declination and hour angle are specified in the *Astronomical Almanac* [1]. The argument of these relations is a modified Julian day number, including a fraction representing the time of day. This particular day number is the number of days from the start of day zero in the year 2000. This can be computed quite readily using a modification of Sinnot (2), to a single line Fortan algorithm by Fliegel and van Flandern. The result is expressed most simply by four lines of BASIC code:

```
N=INT(Y+FIX((M-9)/7))
N=-INT(3*(INT(N/100)+1)/4)
N=N-INT(7*INT(M+9)/12)+Y)/4
N=N-730516.5+D+367*Y+INT(275*M/9)
```

where Y is the four-digit year (e.g., 1989), M is the month number (1–12), and D is the day and part day (i.e., time within the day as a day fraction) of that month. N may then be used as input to the following relations [1], which are all expressed using degrees (not radians!):

- Mean solar longitude L
- L = 280.460 + 0.9856474*N
- Mean anomaly G
- G = 357.528 + 0.9856003*N
- L and G must be reduced modulo 360 before substitution in:
- Ecliptic longitude Le
- $Le = L$ + 1.915*sin(G) + 0.020*sin(2*G)
- Obliquity of the ecliptic e

- e = 23.439-0.0000004*N
- Right ascension (same quadrant as Le) RA
- RA = $a$tan(cos$e$ tanLe)
- Declination dec
- dec = $a$sin(sin$e$ sinLe)
- The Equation of Time EoT
- EoT(minutes) = 4 (L – RA)

The equation of time is the time that the Sun is running fast or slow. That is, local noon will occur at 12.00 – E/60 hours local time. From this the hour angle of the Sun when observed at longitude G_0 (degrees east) is given by:

$$\text{HA (deg)} = 15 * \left(UT - 12 + \frac{\text{EoT}}{60} + \frac{G_0}{15} \right)$$

The term within the parentheses is in hours, which is the more usual unit for hour angle. However, the use of an angular measure is more appropriate in this instance. UT is Universal Time.

The apparent angular diameter of the Sun in degrees is given by:

$$\Pi_s = \frac{0.534}{(1.00014 - 0.01671 * \cos(G) - 0.00014 * \cos(2 * G))}$$

where the term in parentheses is the normalized Sun-Earth distance.

The accuracy of the declination is 0.01 degree and the hour angle 0.03 degree between the years 1950 and 2050. The hour angle accuracy corresponds to a timing accuracy of around 10 seconds.

The azimuth (A) and elevation (e) of the Sun may be computed from its declination and hour angle together with the observer latitude (T_o) using the relations:

$$\sin(e) = \sin(T_o)\sin(\text{dec}) + \cos(T_o)\cos(\text{dec})\cos(\text{HA})$$

$$\sin(A) = \frac{-\cos(\text{dec})\sin(\text{HA})}{\cos(\text{el})}$$

$$\cos(A) = \frac{[\sin(\text{dec})\cos(T_o) - \cos(\text{dec})\cos(\text{HA})\sin(T_o)]}{\cos(\text{el})}$$

The two formulas for the azimuth are necessary to resolve the quadrant ambiguity.

References

[1] *The Astronomical Almanac*, aa.usno.navy.mil/publications/asa.

[2] Bate, Roger, Donald Mueller, and Jerry White, *Fundamentals of Astrodynamics*, New York: Dover, 1970.

[3] Stong, C. L., *How to Track Artificial Satellites*, Amateur Scientist/Scientific American, January 1958, p 98.

[4] Sinnot, R. W., "Taming Our Chaotic Calendar," *Sky and Telescope*, Vol. 70, No. 5, 1984, pp. 454–455.

[5] Vallado, D. A., *Fundamentals of Astrodynamics and Applications*, Third Edition, Microcosm Press and Springer, 2007, pp. 103, 124.

[6] *Application of Kalman Filters in Orbit Determination: A Literature Survey*, 2014, https://indico.esa.int.

Selected Bibliography

Bate, R. R., D. D. Mueller, and J. E. White, *Fundamentals of Astrodynamics*, New York: Dover, 1970.

Boulet, D., *Methods of Orbit Determination for the Microcomputer*, Willmann-Bell, 1991.

Gruntman, M., *Fundamentals of Space Missions*, Rolling Hills Estates, CA: Interstellar Trail Press. 2022.

Heafner, P. J., *Fundamental Ephemeris Computations*, Willmann-Bell, 1999.

King-Hele, D., *Observing Earth Satellites*, MacMillan, 1966.

King-Hele, D., *Observing Earth Satellites*, Van Nostrand Reinhold, 1983.

McCuskey, S. W., *Introduction to Celestial Mechanics*, Addison-Wesley, 1963.

Tattersfield, D., *Orbits for Amateurs with a Microcomputer*, Halsted Press/Wiley, 1984.

Vallado, D. A., *Fundamentals of Astrodynamics and Applications*, Microcosm Press and Springer, 2007.

CHAPTER 6

Collisions

Space debris ranges in size from microns (1 μm = 10^{-3} mm = 10^{-6} m) to tens of meters. There are daily collisions that occur between micron-sized debris pieces and active satellites. The threshold for significant damage lies at around the 1-mm sized objects, with 1-cm debris definitely posing a possible mission termination hazard.

6.1 Collisional Damage

6.1.1 Minor Collisions

Solar UV radiation causes degradation of the outer surface of satellites. If this surface is painted (often for thermal control), small flakes of paint can be detached from the surface. These can pose a collision hazard to active satellites and spacecraft. Of the 135 missions flown by the Space Shuttle, about 100 had to have a windscreen replacement due to collision with a paint flake, at a cost of ~\$100,000 per replacement.

Aerogel is a man-made material that has a density between one tenth and one hundredth that of water. It is the only known substance that can stop a hypervelocity object without destroying it in the process, and it has allowed us to examine and determine the nature of various small particles of space debris. This has shown that the majority of these minor collisions are from degradation fragments of satellites and rocket bodies.

6.1.2 Micrometeoroid Window Impact

Although the majority of minor impacts are due to artificial space debris, it has been found that that natural (not man-made) space debris has a nonnegligible impact hazard, as several Space Shuttle windscreen impacts were found to be due to micrometeorites. That it was natural and not artificial space debris was shown by chemical analysis of the cratering residue, which contained not titanium but minerals found in meteorites.

6.1.3 More Serious Impacts

The impacts shown below show a penetration impact hole discovered in a Hubble Space Telescope solar panel (Figure 6.1) during servicing operations by the Space Shuttle, and on the Hubble downlink antenna (Figure 6.2).

Figure 6.1 An impact crater discovered on a Hubble Space Telescope solar panel. (*Source:* NASA ODPO.)

6.2 Space Debris Penetration

Any impactor over 10 mm in size is bad news for any spacecraft. The graph in Figure 6.3 gives a rough guide to penetration depth as a function of impactor size and

Figure 6.2 A hole found in the Hubble downlink antenna, the result of a space debris impact. (*Source:* NASA ODPO.)

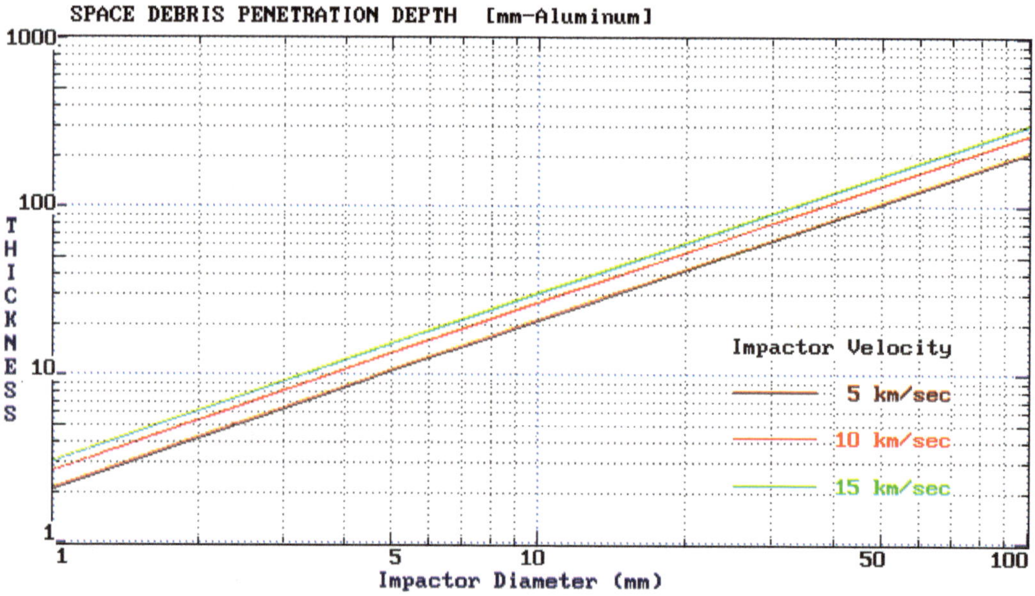

Figure 6.3 Space debris penetration graph.

speed. The ordinate is an aluminum-equivalent thickness in millimeters. Figure 6.4 shows the result of a ground-based test collision.

6.2.1 Spacecraft Shielding

In the early 1950s Fred Whipple suggested the *bumper shield* as a mass effective way to protect satellites from meteoroid damage. This concept has been refined over the years, and applied to the largest and most expensive space assets such as the International Space Station, to provide protection mainly against orbital space debris (Figure 6.5).

6.2.2 ISS Debris Damage

Figure 6.6 shows the location of a small debris impact crater on the handrail of an airlock on the International Space Station.

At first thought one might tend to disregard such a minor event. But imagine if you were an astronaut running your hand around this handrail. And it just happened to tear your glove, allowing your spacesuit to slowly depressurize!

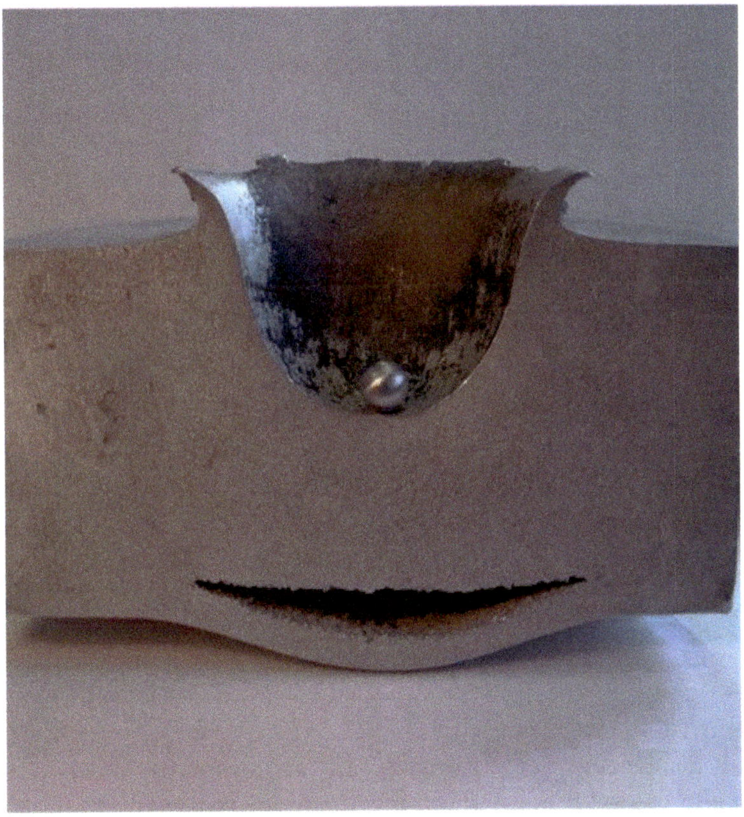

Figure 6.4 A ground-based hypervelocity impact of a 2-mm Al sphere into an Al target 18 mm in thickness. Note crater and incipient spall opposite impact. Also note that the sphere shown in the bottom of the crater is indicative only, the actual impactor, identical in size, was totally destroyed in the impact. (*Source:* NASA OSPO.)

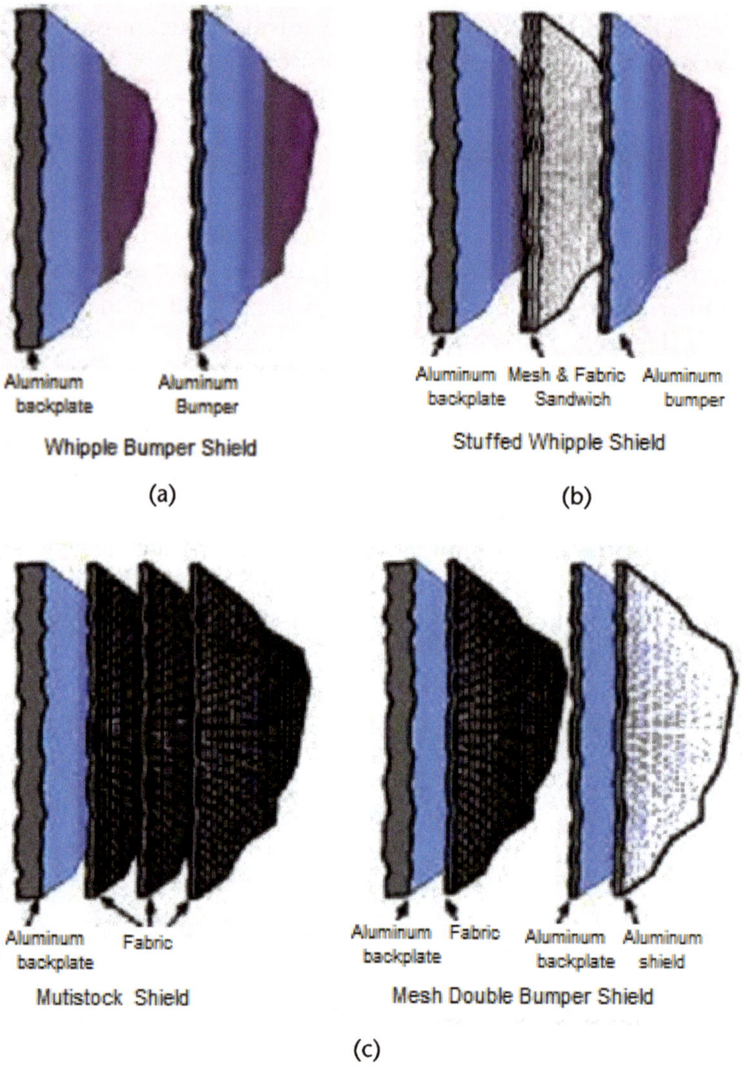

Figure 6.5 Spacecraft debris collision shielding. (*Source:* NASA ODPO space debris report.)

Figure 6.7, from the NASA Orbital Debris Program Office, shows the damage that impactors of various sizes might cause to a manned space mission (e.g., a Space Shuttle).

6.3 Collisional Velocities

In articles on space debris and debris collisions, it is often stated that the average collision velocity between two objects in low Earth orbit is around 10 km/s. This should be contrasted with LEO orbital velocity, which is around 7.5 km/s.

At first sight 10 km/s seems rather high. For a head-on collision, the maximum collision velocity is twice the orbital velocity, or 15 km/s. For a tail collision, the velocity is zero. One might assume that the average collision velocity is halfway

Figure 6.6 ISS crater on handrail. (*Source:* NASA.)

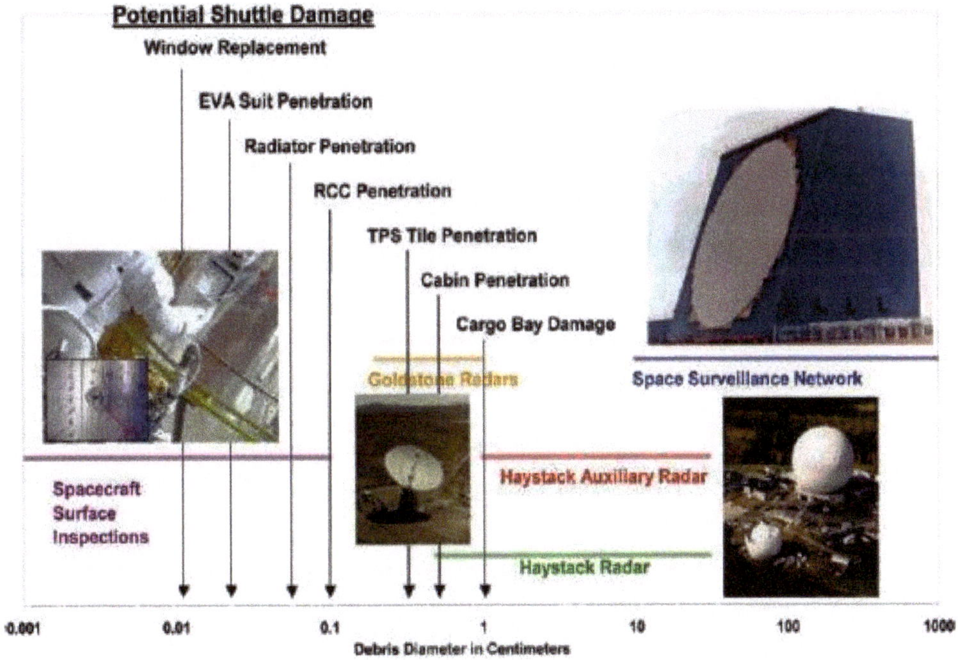

Figure 6.7 Space Shuttle: potential debris damage for various-sized debris. (*Source:* NASA ODPO.)

between these two, or 7.5 km/s. However, there are many more nearly head-on collisions than there are nearly tail-on collisions.

The situation is analogous to driving along a freeway. If you keep to the correct side of the road you and all the other traffic are going at about the same speed, and there are very few collisions. However, if you are unfortunate enough to be traveling on the wrong side of the road, all the traffic is coming directly toward you, and a collision is very likely.

This section will discuss the value of the mean collisional velocity in low Earth orbit.

6.3.1 Population Altitude Distribution

Figure 6.8 from the NASA Orbital Debris Program Office shows the altitude distribution of space objects in low Earth orbit before, and seven years after, the Chinese ASAT test in January 2007.

It can be seen that there is a large distribution of objects at around an 800-km altitude, with a secondary peak around 1,400 km. Collision probability is greatest where the population density is highest. At an 800-km altitude, the orbital velocity is very close to 7.5 km/s, and we will use this velocity in subsequent calculations.

6.3.2 Collisional Velocities

Figure 6.9 shows two orbital objects with velocity vectors $\mathbf{v}_1$ and $\mathbf{v}_2$. Near the collision point these can be represented as straight lines making an angle θ between them. For a head-on collision $\theta = 180$ degrees (π radians), and θ is zero for a *tail* collision.

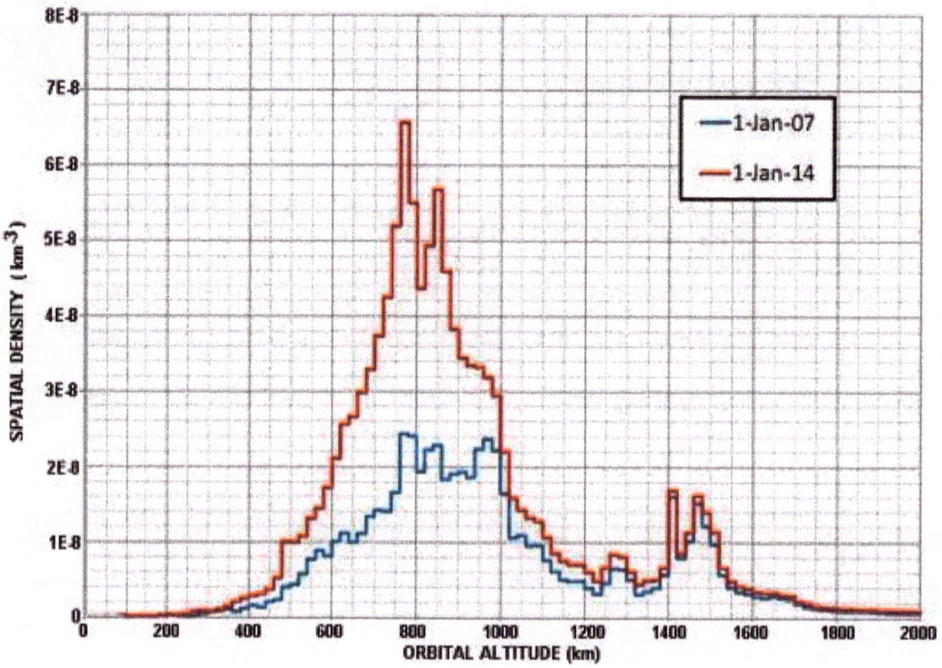

Figure 6.8 Altitude distribution of orbital space objects in LEO. (*From:* NASA ODPO.)

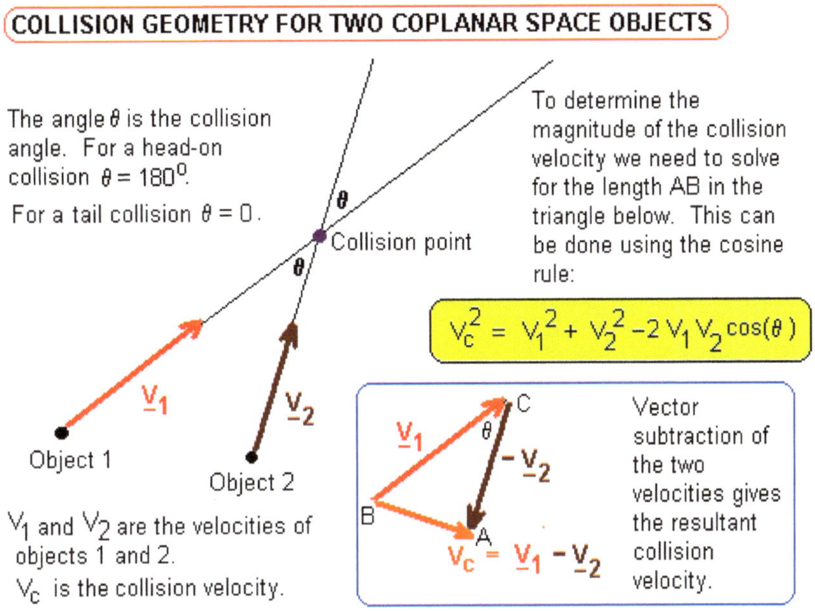

Figure 6.9 Space object collision geometry.

The collision velocity, vector $\mathbf{v}_c$, is found by subtracting vector $\mathbf{v}_2$ from vector $\mathbf{v}_1$. This is accomplished geometrically by reversing the direction of $\mathbf{v}_2$ and laying it next to $\mathbf{v}_1$ as shown in the diagram. The resultant velocity $\mathbf{v}_c$ is drawn from the tail of the first vector to the head of the second.

To calculate the magnitude of the collision velocity v_c we use the cosine rule on triangle ABC:

$$v_c^2 = v_1^2 + v_2^2 - 2v_1v_2\cos(\theta)$$

If we consider two circular orbits at the same altitude (but with different inclinations) then

$$v_1 = v_2 = v_o \quad \text{(the orbital velocity)}$$

and we can then write the collision velocity as

$$v_c = \sqrt{2v_o(1-\cos(\theta))}^{1/2}, \text{ or}$$

$$v_c = 1.41v_o\sqrt{[1-\cos(\theta)]}$$

Figure 6.10 is a plot of the above formula for v_c as a function of the collision angle θ.

Note that the collision velocity rises steeply as the collision angle increases from zero, and then it flattens out as θ approaches a head-on collision.

If there is a uniform distribution of possible collision angles from 0 to 180 degrees, we might expect that the mean collision velocity occurs at an angle $\theta =$

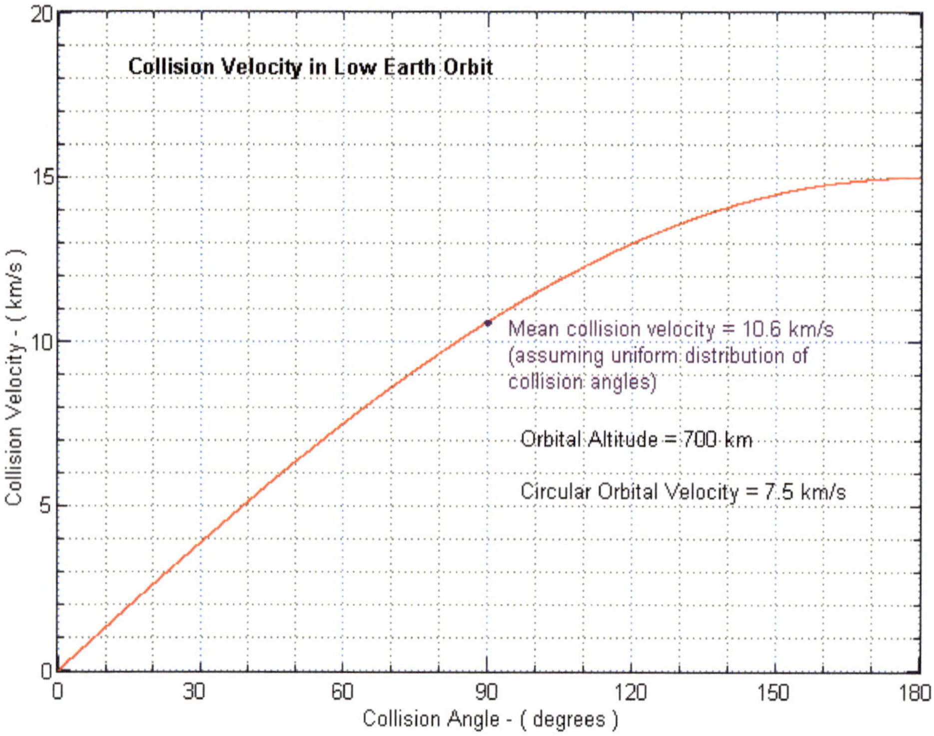

Figure 6.10 Collision velocity in LEO as a function of collision angle.

90 degrees, and indeed it does, as shown in the following section. For a circular orbit at 700 km with an orbital velocity of 7.5 km/s this mean collisional velocity is 10.6 km/s.

6.3.3 Mean Collision Velocity

The mean collision speed v_{cm} can be approximated by the formula:

$$v_{cm} = \frac{\sum_{i=0}^{N} v_c\left(\frac{i\pi}{N}\right)}{N+1}$$

In other words, we are computing the value of the collision speed v_c at $N + 1$ different points between 0 and 180 degrees (e.g., 0, 10, 20, 30, ..., 180, here $N = 18$), and dividing by the number of points (19 in this case) to obtain the average value. As N is made larger, this approximation becomes more accurate. Using a computer to evaluate, we find that

$$\frac{v_{cm}}{v_o} = 1.27$$

If $v_o = 7.5$ km/s then $v_{cm} = 10.6$ km/s.

The above solution was found by *numerical integration*. We can also find an exact analytical solution by evaluating the integral

$$
\begin{aligned}
v_{cm} &= \frac{\int_0^\pi v_c(\theta)\,d\theta}{\int_0^\pi d\theta} \\
&= \left(\sqrt{2}\frac{v_o}{\pi}\right)\int_0^\pi (1-\cos\theta)^{1/2}\,d\theta \\
&= \left(2\frac{v_o}{\pi}\right)\int_0^\pi \sin\frac{\theta}{2}\,d\theta \\
&= \left(4\frac{v_o}{\pi}\right)\left[-\cos\frac{\theta}{2}\right]_0^\pi \\
&= 4\frac{v_o}{\pi} \\
&(= 1.27v_o)
\end{aligned}
$$

And as we noted previously, this is the value of v_c for a collision angle θ = 90 degrees.

6.3.4 Nonuniform Populations

The above results are for an ideal space object population where there is a uniform distribution of possible collision angles, which implies that each collision angle (or equal ranges of collision angle), have an equally likely probability of occurring. This is not the situation in real life, where there are few objects with an inclination close to zero (at least in LEO). The histogram in Figure 6.11 shows the distribution of inclinations for orbital space objects in low Earth orbit in 1994.

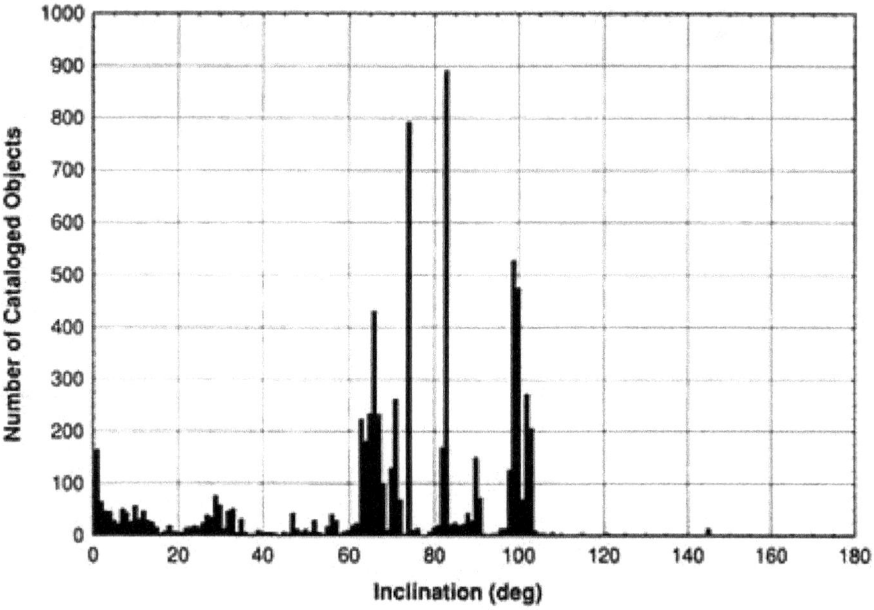

Figure 6.11 Inclination distribution of bright OSO in LEO (1994) [4].

To calculate the mean collision speed in the case of nonuniform populations, we must compute a weighted average. Such a computation by Rossi and Farinella [1] came up with a mean v_c of 9.7 km/s, so our simple approach using the random approximation did not do so badly.

6.4 Hypervelocity Collision Analysis

Collisions between objects in space are invariably hypervelocity collisions. Such a collision is defined as one where the velocity (v) of the impactor (relative to the target) is so great that its kinetic energy ($K = ½mv^2$) is greater than the energy released in the detonation of the same mass (m) of high explosives.

Given that the energy density of a typical high explosive (TNT) is 4.2×10^6 J/kg, we can compute the transition to hypervelocity from:

$$\frac{1}{2}mv^2 = 4.2 \times 10^6 \mathrm{m}$$

which gives us $v \sim 3$ km/s. It is thus usual to regard a hypervelocity collision as one in which the relative velocity between target and impactor is 4 km/s or greater.

In low Earth orbit, the velocity of a satellite is between 7 and 8 km/s, and the typical collision velocity between two space objects in LEO is around 10 km/s. Thus most accidental LEO collisions are hypervelocity collisions. The deliberate destruction of a satellite with a terrestrially launched missile will also be a hypervelocity collision. The missile does not need to carry an explosive warhead, as the destruction from the kinetic impact is greater than what could be achieved by a high explosive.

A hypervelocity collision may be categorized as catastrophic or noncatastrophic. A catastrophic impact is one in which both the target and the impactor are totally destroyed, producing a range of fragments. In a noncatastrophic impact the impactor is destroyed and the target is damaged (or cratered), but not totally destroyed. Hypervelocity experiments have shown that target destruction usually results when the impactor kinetic energy is greater than 40,000 M_t Joule, where M_t is the mass of the target. If we designate the mass and relative velocity of the impactor as M_i and v, respectively, we can calculate the ratio of target to impactor mass (the factor $\Gamma = M_t/M_i$) which divides the two classes of collision behavior:

$$\Gamma = \frac{v^2}{80{,}000}$$

For a relative collision velocity of 10 km/s, $\Gamma \sim 1{,}250$. Other data indicates that Γ may be as high as 2,500 for a catastrophic collision.

6.4.1 Noncatastrophic Collision

For this case:

$$M_i < \frac{M_t}{\Gamma}$$

In this case the impactor is totally destroyed and the ejected mass M_e is given by:

$$M_e = \gamma M_i$$

where experiment has shown that $\gamma \sim 120$ for collision velocities around 10 km/s.

6.4.2 Catastrophic Collision

For a catastrophic collision:

$$M_i \geq \frac{M_t}{\Gamma}$$

Both the impactor and the target are totally destroyed and the ejected mass is the sum of the two masses:

$$M_e = M_i + M_t$$

In many cases the impactor mass is such a small fraction of the target mass that the ejected mass is essentially the target mass.

6.5 Fragment Mass Distribution

A simplified model developed at the NASA Orbital Debris Program Office allows the approximate calculation of the mass distribution of the fragments ejected in a hypervelocity collision:

$$N(>m) = 0.8\left(\frac{m}{M_e}\right)^{-0.8}$$

where

$N(>m)$ is the number of fragments ejected with mass m or greater

Thus a lot more smaller fragments are produced than are large ones. This expression produces an infinite number of very small fragments that is physically unrealistic. There is a limit at which the production of small fragments drops off. A modified form of the above expression, which takes this into account, could be written as:

$$N(>m) = 0.8\left(\frac{m + \delta m}{M_e}\right)^{-0.8}$$

where δm is related to the smallest sized particles produced. There is not much data on this size, but we have used $\delta m = 5 \times 10^{-6}$ in simulations below. This is below the size of particles that can be detected individually by current surveillance sensors.

The size (s) of a fragment of mass (m) depends of course on the fragment shape and the density (ρ) of the material of which it is made. As a very simple approximation, fragments are taken to be spherical particles. While this may be close to reality for the small particles, it is not true for the larger mass fragments. It does, however, make analytical calculations possible that would otherwise not be so, and which would require a detailed knowledge of fragment shapes.

For spherical particles, where s is taken to be the diameter of the particle:

$$s = \left(\frac{6m}{\pi\rho}\right)^{1/3}$$

In a further simplification, we take $\rho \sim 2{,}000$ kg/m^3 (the density of aluminum, the most common material used in spacecraft construction, is 2,700 kg/m^3), giving:

$$s \sim \left(\frac{m}{1{,}000}\right)^{1/3}$$

A fragment of mass 1 kg thus has a rough size of 0.1m or 100 mm, whereas a 10-mm fragment will have a mass of 0.001 kg or 1 gram. Because of the cube law dependence of mass on size, an order of magnitude change (i.e., a power of ten) in size results in a three-order magnitude change in mass.

6.6 Fragment Velocity Distribution

Small mass fragments will on average be ejected at greater velocities than large fragments. For sizes greater than 100 mm (i.e., masses greater than 1 kg), 80% of the fragments will be ejected with velocities under 250 meters/s. Reynolds has developed an empirical formula relating modal ejection velocity (ΔV) to fragment size:

$$\log(\Delta V) = 1.486 - 0.804\log(s) - 0.0676\log^2(s)$$

where ΔV is in meters/s and s is in meters (i.e., strict SI units).

Note that ΔV is the most probable velocity in a distribution that extends sensibly from about 0.1 ΔV to about 1.3 ΔV.

Table 6.1 presents results computed using the previous formulae for a catastrophic satellite collision of a hypothetical target of mass one ton.

The column of fragment numbers $N(> m)$ is computed using the modified mass fragment formula that limits the production of very small particles. The column marked N(unbounded) lists the results of the unmodified formula for comparison. The size is computed on the spherical particle approximation with a density of 2,000 kg/m^3 and dVel is the modal ejection velocity.

It is interesting to compare these results with the fragments observed and estimated from the Chinese ASAT (antisatellite) test of January 12, 2007 where the target (an old weather satellite FY-1C) had a mass close to the Section 6.5 model mass (i.e., 1,000 kg), see Table 6.2.

Table 6.1 Satellite Breakup Model (Catastrophic)

Target Mass = 1000 kg				
Size (mm)	Mass (kg)	N(>m)	dVel (m/s)	N(unbounded)
1	0.000001	3023912	1947.1	12679149
2	0.000010	1452838	1409.5	2009510
5	0.000100	306294	985.7	318486
10	0.001000	50276	665.9	50477
22	0.010000	7997	434.6	8000
46	0.100000	1268	274.0	1268
100	1.000000	201	166.8	201
215	10.000000	32	98.1	32
464	100.000000	5	55.8	5

Note that only the >10 mm are regularly trackable by the US Space Surveillance Network. Other sensors employed at irregular times can provide information down toward the 1-mm size, and the last row is a best estimate.

While the $N(>0.1\text{ mm})$ value agrees well with the model, the >10 mm numbers are an order of magnitude higher than the model, and the >1 mm numbers are three times the model prediction. One possible explanation is that the model is based on fragment mass. The fragment size is derived from the mass on the basis of a relatively compact spherical particle. If the Chinese test produced relatively planar fragments (or bent plates), the mass of such fragments will be a lot less than the mass of a compact sphere of the same size (the size of the plates being measured in the plane of the plate, not its thickness). In this case, a given real fragment size refers to a smaller mass than in the model, and the smaller mass will be present in larger numbers. The discrepancy becomes less as the size of the plate approaches its thickness, and the fragment becomes a better approximation to a compact object.

Various formulas have been used to relate the size and mass of a debris fragment. The above table used a spherical particle model with a diameter s and a density $\rho = 2{,}000$ kg/m^3. In this model the size and mass are related by the previously quoted formula:

$$s = \left(\frac{6m}{\pi\rho}\right)^{0.333} \qquad [s1]$$

Another approach is to note that the average space object has a mass to cross-sectional area ratio of $m/A = 100$. This produces a mass-size relationship of:

Table 6.2 Fragment-Size Distribution in Chinese ASAT Test

Size s (mm)	N(>s)
10	2,600
1	150,000
0.1	3,000,000

$$s = 0.11m^{0.5} \qquad [s2]$$

or alternatively, a slightly more accurate relation $m = 62A^{1.13}$, which gives:

$$s = 0.18m^{0.44} \qquad [s3]$$

Table 6.3 lists size values for $s1$ using densities of 2,000 and 1,000 kg/m^3 (the latter is the density of water) plus the two size values computed using $s2$ and $s3$:

The use of the lower density relationship ([$s1$] with $\rho = 1000$) is the only one that helps reduce the discrepancy between the model and the Chinese ASAT data, but not sufficiently to warrant its use. Even lower densities help bring the >10-mm results into line, but not the >100-mm values. Only a variable density with size (as suggested) appears adequate, with very low apparent densities for large fragments, increasing to more typical values for the small sizes. This can be explained by realizing that the larger sizes of debris contain significant interior space—they are not solid objects and hence have lower densities.

6.7 Orbital Parameter Changes

As each fragment is ejected from the collision it assumes a new orbit dependent upon the magnitude of the ejection velocity, and the direction of this velocity vector with respect to the velocity vector of the satellite at the time of collision. The momentum of the impactor is generally so much smaller than the momentum of the target that it can be ignored.

The component of the velocity in the direction of the original target orbit will affect the shape and size of the orbit. An initially circular orbit will be turned into an elliptical orbit with a different period than the circular orbit. Whether the ellipse is larger or smaller than the original circle is determined by whether the velocity component is in the same or opposite direction as the original target motion. This is illustrated in Figure 6.12.

Table 6.3 Space Debris Fragment Size

Mass (kg)	Size (mm) rho=2000	Size (mm) rho=1000	Size (mm) m/A=100	Size (mm) m=62A^1.13
0.000001	1.0	1.3	0.1	0.4
0.000010	2.2	2.7	0.2	1.1
0.000100	4.6	5.8	0.8	3.1
0.001000	10.0	12.6	2.6	8.6
0.010000	21.5	27.1	9.1	23.7
0.100000	46.4	58.5	31.7	65.4
1.000000	100.0	126.0	110.0	180.0
10.000000	215.4	271.5	381.4	495.8
100.000000	464.2	584.8	1322.5	1365.4

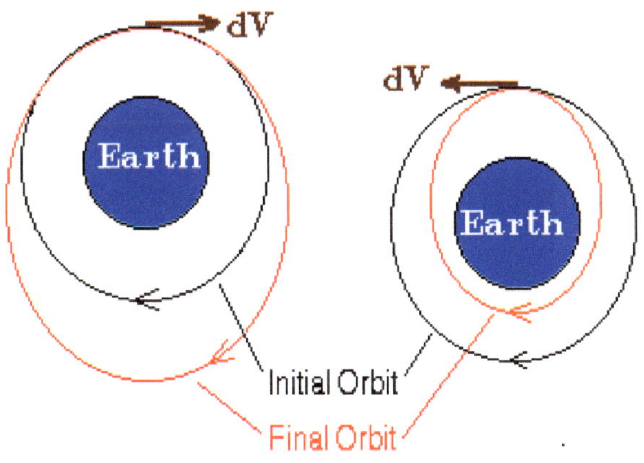

Figure 6.12 (a) The orbital change that results from a velocity increment component along the direction of the satellite track. The originally circular orbit becomes an ellipse with a perigee equal to the circular radius, and a larger apogee. (b) The velocity component is in the direction opposite to the satellite motion. The orbit becomes elliptical, with the apogee now equal to the circular radius, and a smaller perigee.

The perigee (q) is the closest distance of the orbit to the Earth and the apogee (p) is the furthest distance of the orbit from the Earth. The sum of the two is twice the semimajor axis (a) of the ellipse. The perigee and apogee are also related to the semimajor axis by the eccentricity (e) of the ellipse:

$$p = a(1+e)$$
$$q = a(1-e)$$
$$p + q = 2a$$

Given the velocity (v_c) and radius (r_c) of an initial assumed circular orbit, as well as a tangential velocity component δv, we can compute the size parameters and the period of the new orbit. We need to consider the positive (along track) and negative (opposing track) velocity increments separately.

Positive velocity increment:

$$a = \frac{GM}{v_c^2 - 2v_c\delta v}$$
$$q = r_c$$
$$e = 1 - \frac{r_c}{a}$$
$$p = a(1+e)$$
$$T = 2\pi\sqrt{\frac{a^3}{GM}}$$

Negative velocity increment:

$$a = \frac{GM}{v_c^2 + 2v_c\delta v}$$

$$p = r_c$$

$$e = \left(\frac{r_c}{a}\right) - 1$$

$$q = a(1 - e)$$

$$T = 2\pi\sqrt{\frac{a^3}{GM}}$$

(Note that δv is the magnitude of the velocity increment.)

The period of the original circular orbit is:

$$T = 2\pi\sqrt{\frac{r_c^3}{GM}}$$

Tables 6.4 and 6.5 give an idea of the magnitude of orbital changes that may occur as a function of incremental velocity. The first table is for positive velocity increments, and the second is for negative velocity increments, both along the original orbital path. The first two columns give the height (km) and period (minutes) of an initial circular orbit. The next five sets of two columns give the same quantities (height and period) for different incremental velocity changes. In the positive increment case, the height is the apogee of the ellipse, the perigee being equal to the circular height. In the negative case, the height is the perigee of the ellipse, the apogee being equal to the circular height.

A negative value for the perigee means that it is below the surface of the Earth. A fragment traversing such an ellipse will obviously not complete one orbit, but will either burn up on reentry or hit the Earth within less than one-half of the orbital

Table 6.4 Orbit Change: Delta V Along Track Positive

Initial		V=50m/s		V=100m/s		V=150m/s		V=200m/s		V=250m/s	
Ht	Per	Ht	Per	Ht	Per	Ht	Per	Ht	Per	Ht	Per
200	88.4	371	90.1	546	91.9	726	93.7	911	95.6	1102	97.6
300	90.4	475	92.2	654	94.0	839	95.9	1028	97.9	1223	99.9
400	92.4	579	94.3	762	96.1	951	98.1	1145	100.1	1344	102.2
500	94.5	683	96.4	871	98.3	1063	100.3	1262	102.4	1465	104.6
600	96.5	787	98.5	979	100.5	1176	102.6	1379	104.7	1587	107.0
700	98.6	891	100.6	1087	102.7	1289	104.8	1496	107.1	1709	109.4
800	100.7	995	102.8	1195	104.9	1401	107.1	1613	109.4	1831	111.8
900	102.8	1099	105.0	1304	107.1	1514	109.4	1730	111.8	1953	114.2
1000	105.0	1203	107.1	1412	109.4	1627	111.7	1848	114.2	2075	116.7
1100	107.1	1307	109.3	1521	111.7	1740	114.1	1966	116.6	2198	119.1
1200	109.3	1412	111.6	1629	113.9	1853	116.4	2083	119.0	2321	121.6

Table 6.5 Orbit Change: Delta V Along Track Negative

Initial		V=50m/s		V=100m/s		V=150m/s		V=200m/s		V=250m/s	
Ht	Per	Ht	Per	Ht	Per	Ht	Per	Ht	Per	Ht	Per
200	88.4	33	86.7	-129	85.1	-287	83.5	-442	82.0	-593	80.5
300	90.4	130	88.6	-36	87.0	-198	85.4	-356	83.8	-511	82.3
400	92.4	226	90.6	56	88.9	-110	87.2	-271	85.6	-429	84.1
500	94.5	322	92.6	148	90.9	-21	89.2	-186	87.5	-347	85.9
600	96.5	418	94.7	241	92.8	68	91.1	-100	89.4	-265	87.7
700	98.6	514	96.7	333	94.8	157	93.0	-15	91.2	-183	89.5
800	100.7	610	98.7	425	96.8	245	94.9	70	93.1	-101	91.4
900	102.8	706	100.8	518	98.8	334	96.9	155	95.0	-20	93.2
1000	105.0	802	102.9	610	100.8	422	98.9	239	96.9	61	95.1
1100	107.1	898	104.9	702	102.9	511	100.8	324	98.9	143	97.0
1200	109.3	994	107.0	794	104.9	599	102.8	409	100.8	224	98.9

period. (Even were the fragment to hit the Earth, only a fraction of its initial mass would survive ablation traveling through the atmosphere.)

Velocity increments out of the original orbital plane tend not to change the shape of the orbit significantly but do change its orientation in space. In particular, the inclination of the orbit may be changed. The change depends on where the velocity increment occurs. The greatest change to inclination is when the velocity increment occurs around the equator, and is least when the sublatitude of the target is near its inclination value (positive or negative).

6.8 Gabbard Diagrams

John Gabbard, a NORAD employee in the 1960s and 1970s, developed a diagram named after him [2] that is very useful for illustrating the orbital changes we have just been discussing. A Gabbard diagram is a scatter plot of height versus period. The apogee and perigee of each ejected fragment is shown as a point on the diagram. The resultant plot looks like two asymmetrical boomerangs joined at their apices. An example is shown in Figure 6.13.

The data used to make this plot is the data in the aforementioned example of a 1,000 kg target. Only mass fragments of 1 kg and larger have been used to compile the diagram. The catastrophic collision is assumed to have taken place at an altitude of 1,000 km. The point where the two lines intersect is the original circular orbit at an altitude of 1,000 km and with a period of 105 minutes. Points to the right of this location represent fragments that have been scattered in the forward direction of motion, while points to the left represent fragments scattered opposite the initial target motion. Note how the perigee heights (red points) in the first case become the apogee heights (blue points) in the second.

Figure 6.14 illustrates how the smaller fragments are scattered into larger and smaller orbits than are their more massive counterparts.

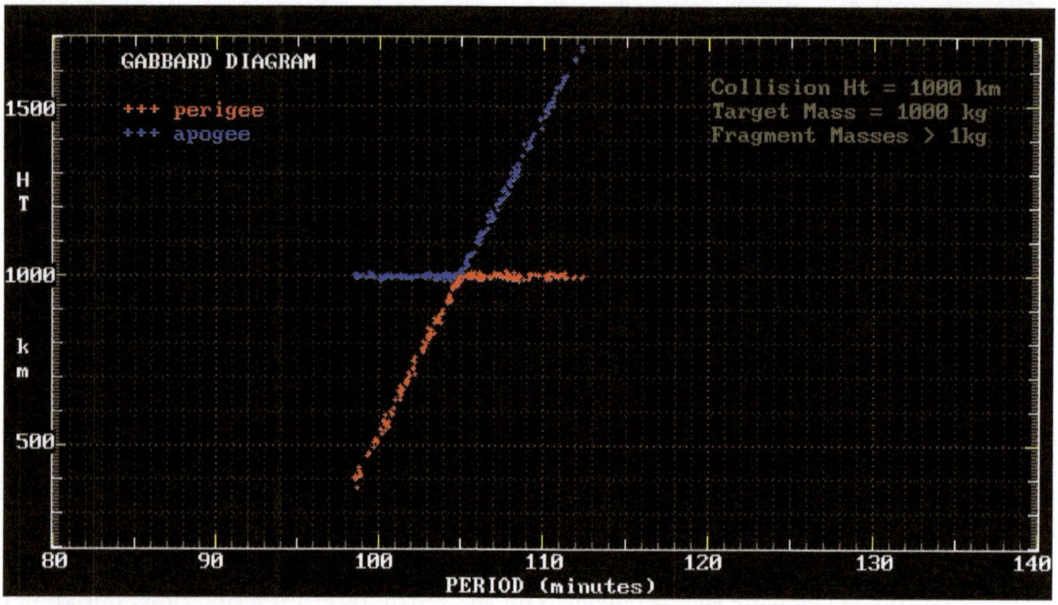

Figure 6.13 An example of a Gabbard diagram that plots the height versus period of the fragments tracked after an orbital collision. This is a simulated collision, not an actual event.

In this case apogee (+) and perigee (–) points are distinguished by different symbols, and the different colors are used to indicate different masses. It can be seen that the largest masses (≥100 kg) are relatively close to the original orbit, whereas the smallest masses (≥1 kg) are the most dispersed.

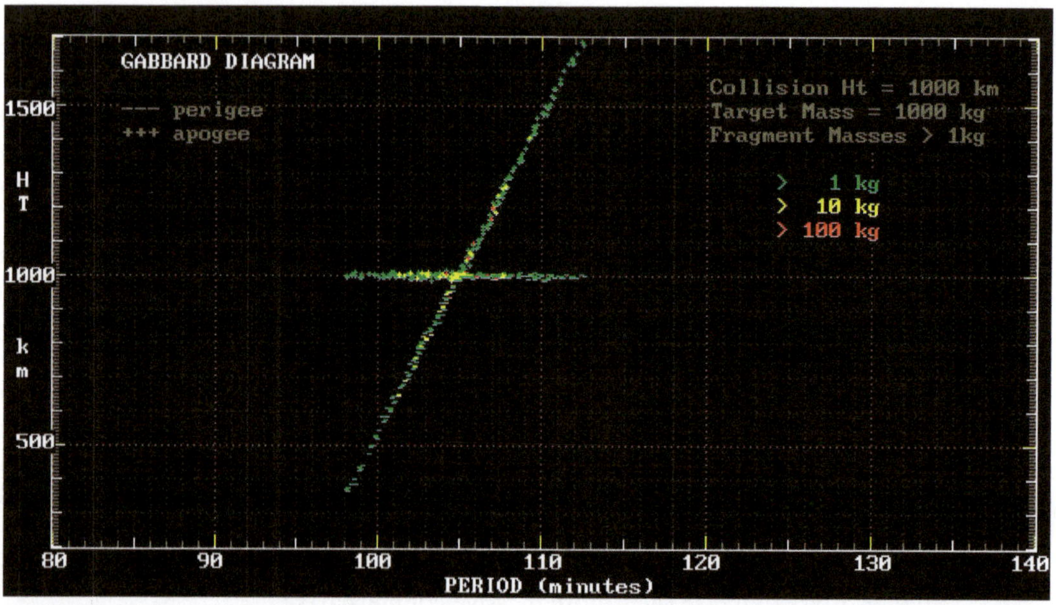

Figure 6.14 A Gabbard diagram for a simulated collision showing that the smaller mass fragments are scattered to larger distances.

6.8.1 Debris Orbital Decay

Gabbard diagrams are also useful for indicating the orbital decay of debris fragments. We have already seen in one of the tables above that when a collision occurs at a sufficiently low altitude some of the fragments that are ejected in the opposite direction to the original motion will be forced into orbits with a perigee below the Earth's surface. These will undergo atmospheric entry in less than half the period of their orbit. Because most of these particles will be those of smaller mass, they will in general burn up during reentry. Heavier fragments that may have a chance of depositing at least some of their mass on the Earth's surface will generally be the last fragments to decay due to atmospheric drag.

It is not necessary for a fragment to have a negative perigee to survive less than one orbit. Any fragment with a perigee of less than about 100 km will encounter so much air resistance (drag) that it will never make it back to apogee, and will deorbit in less than one period.

Fragments above this height will lose energy, mostly at perigee, and this will reduce their apogee height. Interestingly enough, their perigee height changes a lot less, until the apogee is brought down close to the perigee. The lower the perigee height of a fragment the larger the reduction in apogee height. This is seen on a Gabbard diagram as a drooping of the apogee plot at the left of the diagram.

Figures 6.15 and 6.16 illustrate this decay. Figure 6.15 is a plot within one orbit of the collision, which this time has taken place at a height of 300 km.

The yellow points represent those fragments with perigees below 100 km, and which are thus not present after one orbit.

Note the progressive droop in apogee as the period gets smaller (Figure 6.16). This plot simulates the fragment population some days after the collision, in conditions

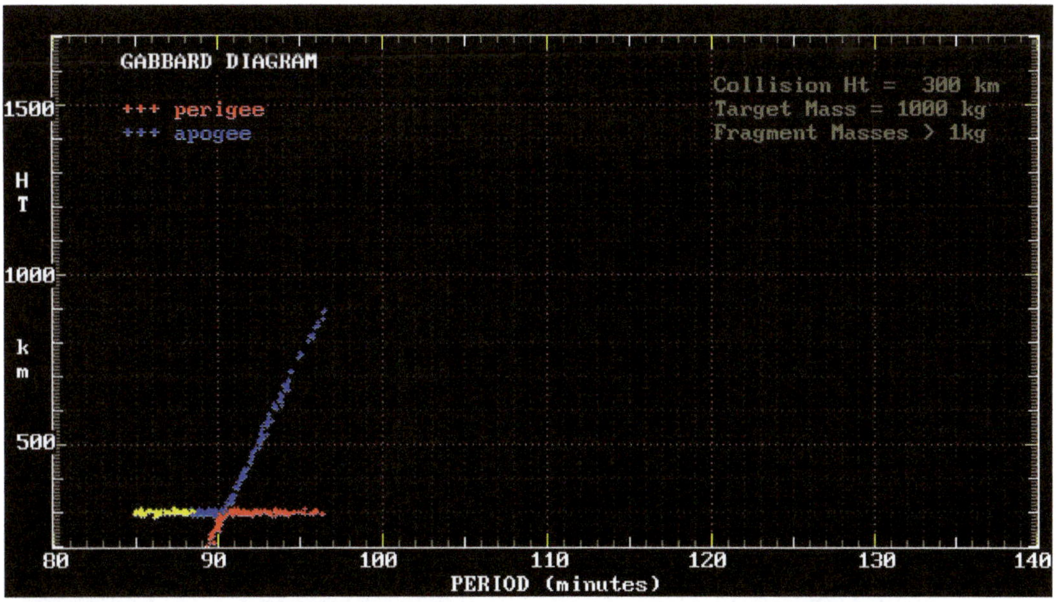

Figure 6.15 Gabbard diagram for a collision at very low altitude. This diagram has been drawn for fragments at a time less than one orbit after the collision (in practice it would not be possible to obtain the relevant data in this time frame).

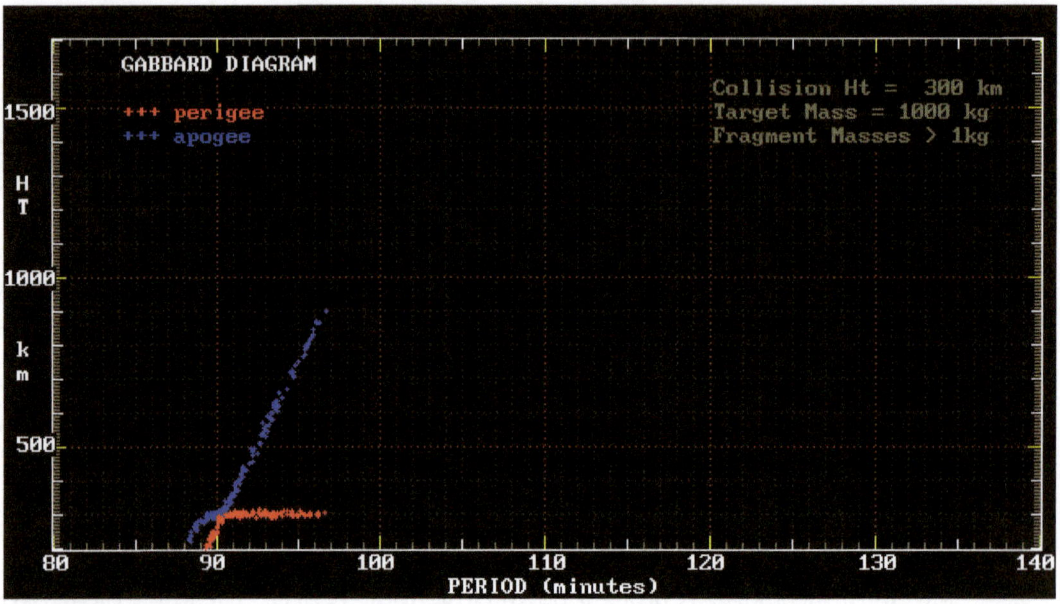

Figure 6.16 The Gabbard diagram several orbits after the low altitude collision. The fragments with perigees below 100 km have all gone (atmospheric reentry), and the low apogee particles show substantial reduction (upper-left arm droop).

of sunspot minimum (i.e., minimal solar X-ray and shortwave UV output, which implies low atmospheric densities in the upper atmosphere).

6.9 CONASS and COLA: Protecting Current Assets

Conjunction assessment leading to possible collision avoidance action are affirmative results that not only look toward protecting current space assets, but also look toward the possible prevention of further accidental collisions that generate even more space debris (Figure 6.17).

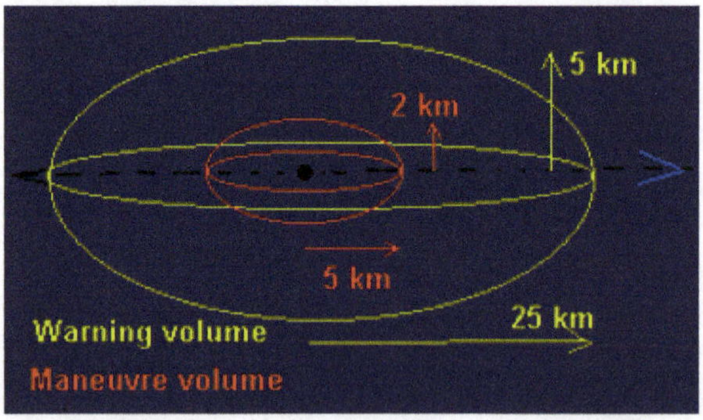

Figure 6.17 A typical CONASS warning area might be 5-km radial and 25 km in-orbit. An associated COLA maneuver area might be 2-km radial and 5-km in-orbit. As SSA tracking accuracies improve these volumes are reduced.

The Iridium-Cosmos collision should not have happened. It did so only because of insufficient computer resources devoted to CONASS. Prior to this collision CONASS was applied only to high-value resources (e.g., ISS). To remedy this deficiency the US Air Force set up and staffed a new conjunction assessment office. This office was given a mandate to liaise with space industry to try and prevent the repetition of the Iridium collision, and to improve space situational awareness in general.

The Space Shuttle had to perform a COLA maneuver about every 10 flights. In today's more congested orbital environment, the ISS is generally performing COLA maneuvers every month [3]. A short notice to the ISS has resulted in astronauts sheltering in a Soyuz return capsule. Several commercial firms now offer a CONASS service for the space industry. These include Exoanalytics, COMSPOC, ElectroOptic Systems, ArianeSpace, Slingshot, and LEOlabs. Foreign governments also operate their own SSA and STM services. This includes the Japanese Space Agency (JAXA), Polish Space Agency (POLSA), European Space Agency (ESA), and many others.

CONASS and COLA are now the most important elements of a burgeoning industry that goes by the name of space traffic management. This is covered in Chapter 19. Figure 6.18 from a NASA report gives the TNT explosive equivalent of LEO hypervelocity collisions by particles in the size range from 0.2 to 10 cm.

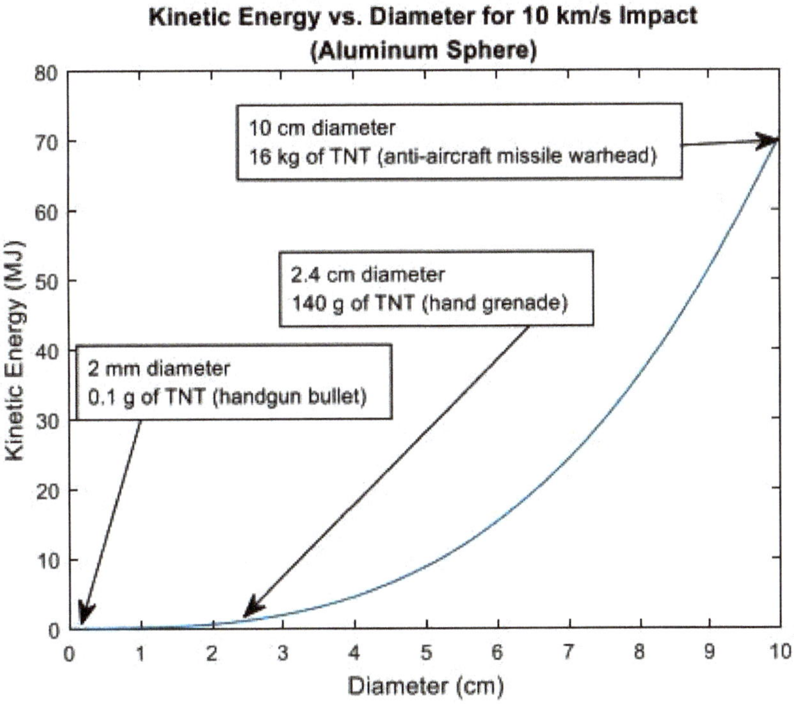

Figure 6.18 This graph provides some realism to space debris hypervelocity collisions by equating them to their explosive equivalents in TNT mass. Anything larger than 2 cm is likely to disable a spacecraft. (*From:* NASA report.)

References

[1] Rossi, A., and P. Farinella, *Collision Rates and Impact Velocities for Bodies in Low Earth Orbit*, *ESA Journal*, Vol. 16, January 1992, pp. 339–348.

[2] Matney, Mark, *Gabbard Plot Discussion*, ntrs.nasa.gov/api/citations/20150009502/downloads/20150009502.pdf.

[3] *NASA Orbital Debris Quarterly News*, Vol. 27, No. 4, October 2023.

[4] National Research Council, *Orbital Debris: A Technical Assessment*, National Academy Press, 1995, Figure 3.6.

Selected Bibliography

Johnson, Nicholas L, and McKnight, Darren S, *Artificial Space Debris*, Orbit (1987).

National Research Council, *Orbital Debris—A Technical Assessment*, National Academy Press (1999).

CHAPTER 7

Orbital Decay in Low Earth Orbit

7.1 Introduction

Orbital decay due to atmospheric drag is the major natural sink for objects in low Earth orbit [1].

Atmospheric density is thus a critical component for studies of meteor flight, artificial space debris reentry, and space object (satellite and debris) lifetime.

Atmospheric density varies with altitude, geographical latitude and longitude, time, season, and atmospheric dynamics. In the upper atmosphere (>100 km) it also depends strongly on the soft x-ray and extreme ultraviolet (EUV) output of the Sun (which varies with the solar cycle/activity). It also varies according to the geomagnetic activity. In the lower atmosphere a mean density as a function of altitude is normally sufficient for models of meteors and space debris reentry. In the upper atmosphere for lifetime studies, which consider atmospheric drag, the density must be specified as a function of altitude, solar 10-cm flux F_{10}, and the geomagnetic activity index A_p.

7.2 An Isothermal Atmosphere

An isothermal atmosphere is one that has a constant temperature. In such an atmosphere the density (ρ) can be written as

$$\rho = \rho_o \exp\left(-\frac{h}{H}\right)$$

where

ρ_o is the density at the surface of the planet
h is the height above the surface
H is the scale height

The scale height H is the height at which the density falls to a value of $1/e$ of the surface density. It is furthermore the height at which the entire atmosphere would fit if it had a constant density ρ_o up to this height H, and is also a function of the atmospheric temperature T, and the mean molecular mass m:

$$H = \frac{kT}{mg}$$

where

$k = 1.38 \times 10^{-23}$ J/K is Boltzmann's constant

g is the acceleration due to gravity (~9.8 at Earth's surface)

Although the temperature of the Earth's atmosphere does vary with height, it is still possible to use the isothermal approximation over limited height ranges. Below 100 km the composition of the atmosphere is relatively constant (20% molecular oxygen and 80% molecular hydrogen) so the mean molecular mass is constant. This region of the atmosphere is referred to as the homogenosphere. Above 100 km the constituents of the atmosphere change and this is referred to as the heterogenosphere.

Below 100 km the variation of gravity with altitude is insignificant. Thus, the only factor that varies significantly is the temperature.

For studies of natural (meteoroid) and artificial debris reentry in the lower atmosphere, it is generally sufficient to use the isothermal approximation as given Section 7.3.

7.3 The Real Atmosphere

Figure 7.1 shows an idealized mean temperature variation with height for mid-latitudes. There are only small variations below an altitude of 100 km where the atmospheric composition is homogeneous. Above this altitude atmospheric composition changes and solar EUV and X radiation become predominant in determining temperature variation.

From the surface the temperature initially decreases at what is called the standard adiabatic lapse rate of about 7 degrees per kilometer. It then plateaus at −55°C for almost 10 km, before rising again to another plateau just below 0°C. It decreases again in the mesosphere before finally rising when it reaches the thermosphere.

The changing temperature causes the scale height to change. From the surface to a 100-km altitude H varies between about 5.5 and 8.0 km, rising and falling along with the temperature.

Despite the changing scale height, a simple isothermal model with a fixed scale height can represent the atmosphere up to 100 km with an error less than 50%, and up to 130 km with an error no larger than a factor of 2. These errors are quite comparable to the variation that the real atmosphere shows from day to day from the mean standard model. This model is quite suitable for demonstrating the physics of meteor flight and space debris reentries. High velocity meteoroids (~60 km/s) start ablating around a 130-km altitude, but space debris reentries, with the much lower velocity of ~8 km/s, only start to show significant atmospheric interaction below 90 km.

The simple isothermal model is specified by the formula:

$$\rho = 1.3\exp\left(\frac{-h}{7{,}000}\right) \text{ kg m}^{-3}$$

where h is the altitude in meters. The comparison of the density from this model with the US Standard Atmosphere model is given in Table 7.1.

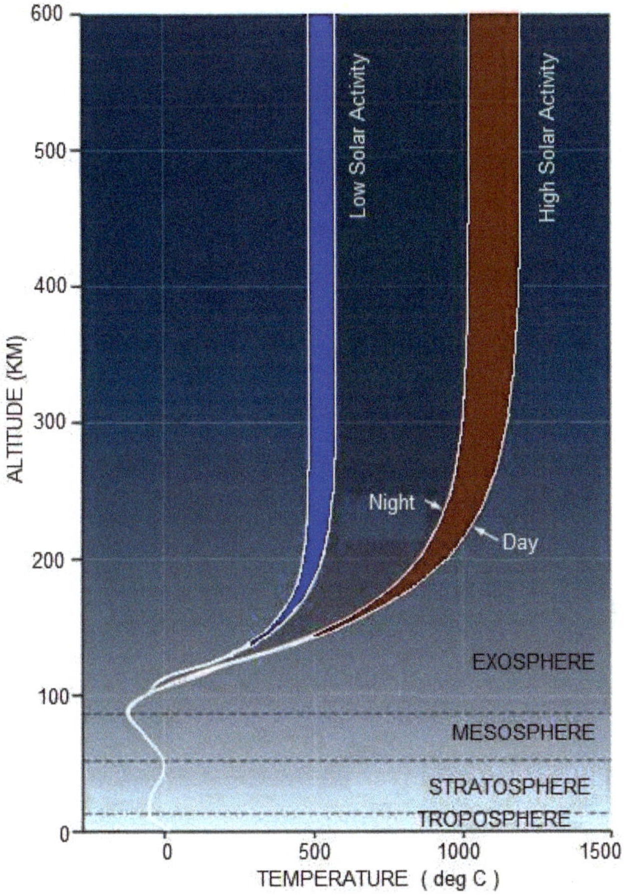

Figure 7.1 Idealized variation of atmospheric temperature with altitude. (*Source:* ESA.)

Table 7.1 Isothermal Versus US Standard Model Densities

Height	*Isothermal*	*Standard*
0	1.30E+00	1.17E+00
10	3.12E-01	4.62E-01
20	7.47E-02	8.67E-02
30	1.79E-02	1.66E-02
40	4.29E-03	4.05E-03
50	1.03E-03	1.17E-03
60	2.46E-04	3.38E-04
70	5.90E-05	8.58E-05
80	1.41E-05	1.79E-05
90	3.39E-06	3.20E-06
100	8.12E-07	5.47E-07
110	1.95E-07	1.03E-07
120	4.67E-08	2.49E-08
130	1.12E-08	8.23E-09

7.4 The CIRA Model

The CIRA model is the COSPAR International Reference Atmosphere. COSPAR is the United Nations Committee for Space Research. Graphs of CIRA atmospheric temperature and density to 500-km altitude are shown in Figure 7.2.

The changing temperature is the predominant parameter that changes the scale height and hence the density. However, this is reinforced by the decrease in gravity (g), with increasing altitude and also the reduction in mean molecular mass above

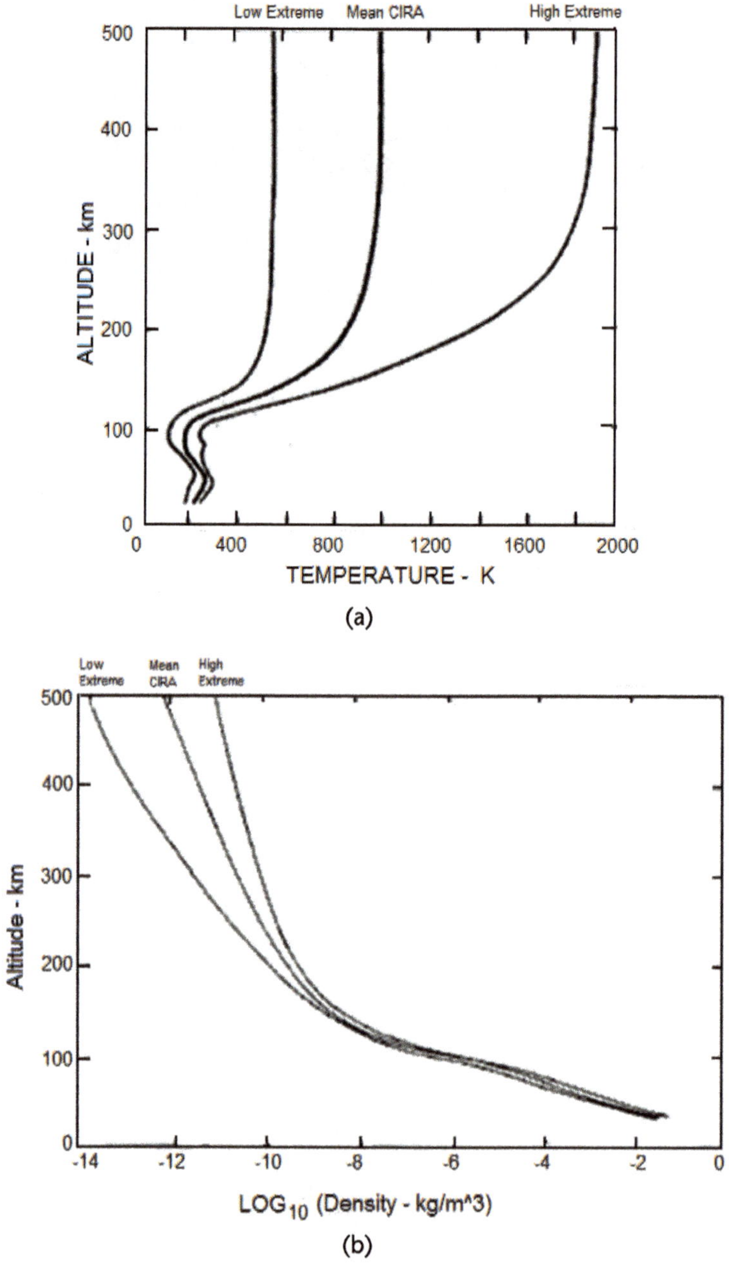

Figure 7.2 COSPAR International Reference Atmosphere. (*Source:* Jursa.)

100-km altitude. Below 100 km the Sun has only a minor influence on the density. Above 200 km it becomes quite significant. This is because the solar EUV and soft X-rays, which affect the upper atmospheric density, are absorbed around the 100-km altitude level.

7.5 Models to 180 Km

We can use the mean CIRA model to 180 km by specifying the density values every 5 or 10 km. Because the density varies over many orders of magnitude over this range, it is more useful to use the logarithm (base 10) of these values. Linear interpolation can be used to calculate the values intermediate between the specified points at the 10-km intervals. The actual density can then be calculated by raising the interpolated log values as a power of 10. That is, $\rho = 10^{\text{logval}}$.

An alternative is to fit a polynomial to the log points such that:

$$\log_{10}(\rho) = \text{polyfn}(h)$$

which gives:

$$\rho = 10^{\text{polyfn}(h)}$$

A seventh-order polynomial fit with the polynomial coefficients is shown in Figure 7.3. The function is given by:

$$\text{polyfn}(h) = a0 + a1 * h + a2 * h^2 + a3 * h^3 + a4 * h^4 + a5 * h^5 + a6 * h^6 + a7 * h^7$$

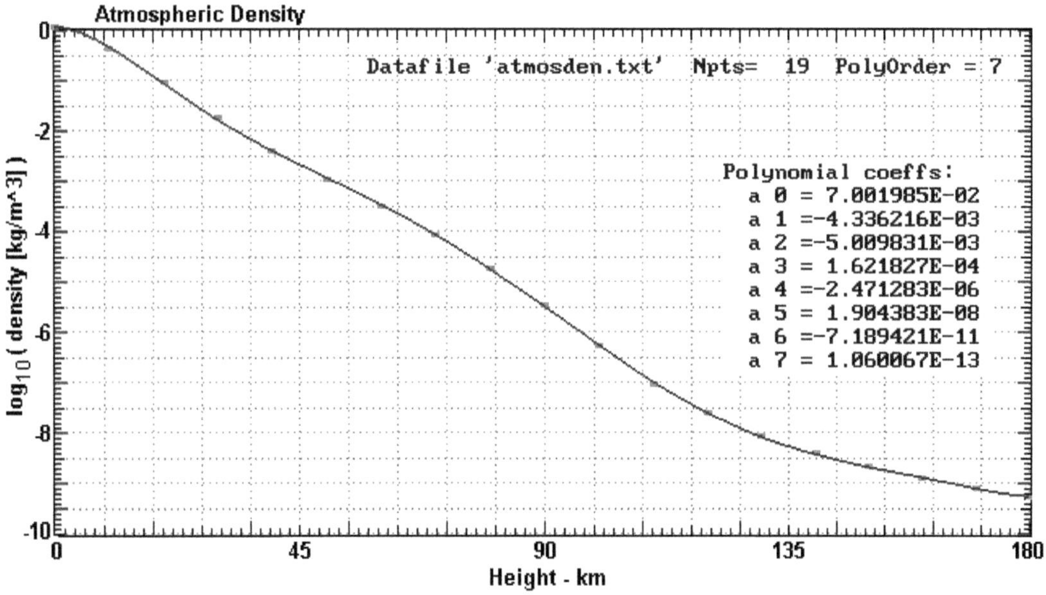

Figure 7.3 Polynomial fit for CIRA atmospheric density profile.

7.6 A Model From 180 to 500 Km

Above 180 km we can no longer use a mean reference model. We need a model that varies according to the space weather conditions at the time, specifically the solar EUV flux and the geomagnetic activity. We use the solar radio 10-cm flux (F_{10}) as a proxy for the solar EUV output, and the geomagnetic A_p index as a proxy for the geomagnetic activity. The model we have developed is suitable for investigating orbital decay and satellite and debris lifetimes in the altitude range from 180 to 500 km.

The algorithm used to calculate the atmospheric density in this range is given by the four formulae:

$$T = 900 + 2.5\,(F_{10} - 70) + 1.5\,A_p \quad \text{kelvin}$$
$$\mu = 27 - 0.012\,(h - 200) \qquad 180 < h(\text{km}) < 500$$
$$H = T/\mu \qquad (\text{km})$$
$$\rho = 6 \times 10^{-10}\,\exp(-(h - 175)/H) \qquad (\text{kg m}^{-3})$$

Note that T, μ, and H should be regarded as intermediate values in the computation of density. They are pseudo parameters and do not necessarily correspond to temperature, molecular weight, and scale height at the specified altitude. Only the final parameter ρ should be regarded as a model output.

Note: It has been found that an average F_{10} value is the more appropriate parameter to use rather than the daily value. Typically the last 90 days are used to form the average.

Above a 500-km altitude the lifetime of an orbital body is larger than the solar period of ~11 years, and so once again an average or mean atmospheric model may be appropriate.

7.7 A LEO Orbital Decay Model

The atmospheric density model described in the last chapter for the height range 200–500 km can be used, together with the appropriate forces acting on a satellite, to produce a simple model to predict reentry times for satellites in the lower altitude range of low Earth orbit.

The model has been confined to satellites with orbits totally below about 500 km. Such orbits are essentially circular, with the use of the semimajor axis in place of the orbital radius. The atmospheric density ρ is specified by a simple exponential with variable scale height H. For a fixed exospheric temperature T, H is made to vary with altitude h through the use of an effective atmospheric molecular mass μ. This μ includes both the actual variation in molecular mass with height, and a compensation term for the variation in temperature over the considered altitude range. The variation in density due to the space environment is introduced through T, which is specified as a function of the solar radio flux $F_{10.7}$ and the geomagnetic index A_p. For time intervals of a year or more, the average A_p can generally be taken as 0. That is, the 10-cm solar radio flux is the dominant factor, and is generally used in an averaged form, and the average preferred is that of the last 90 days prior to the specified date. Sometimes a small correction is made to weight the current flux more strongly. The data in Figure 7.4 was used to develop this model.

7.8 Model Equations

Figure 7.5 shows the forces acting on a satellite traveling through a small but non-negligible atmosphere in low Earth orbit.

For a circular orbit, period P and semimajor axis a are related by:

$$P^2 G M_e = 4\pi^2 a^3$$

where G is the universal gravitational constant and M_e is the Earth's mass (SI units).

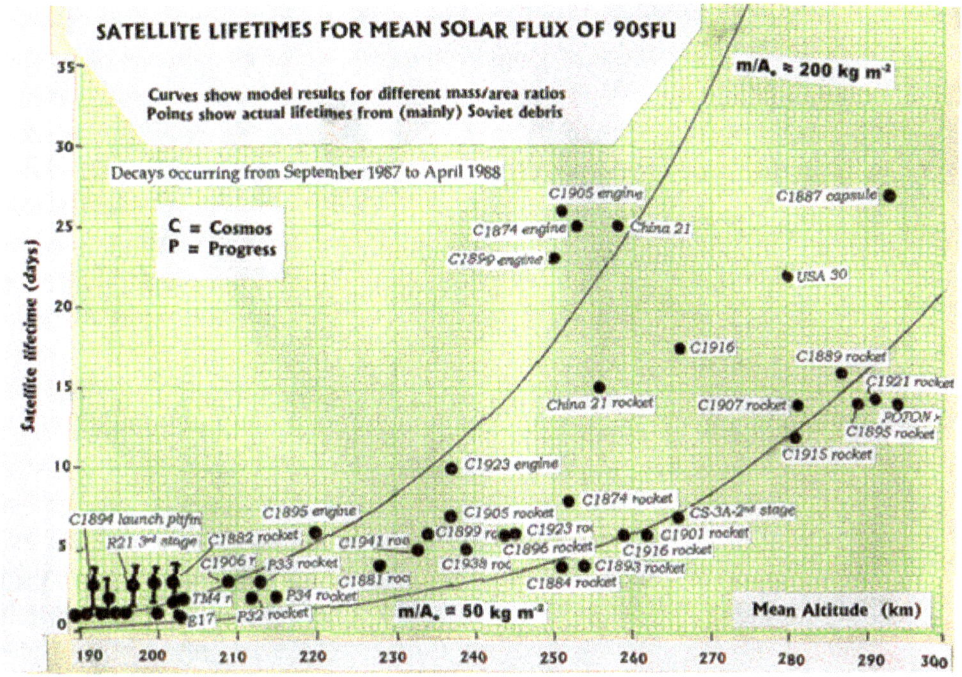

Figure 7.4 Data used to derive the satellite decay model.

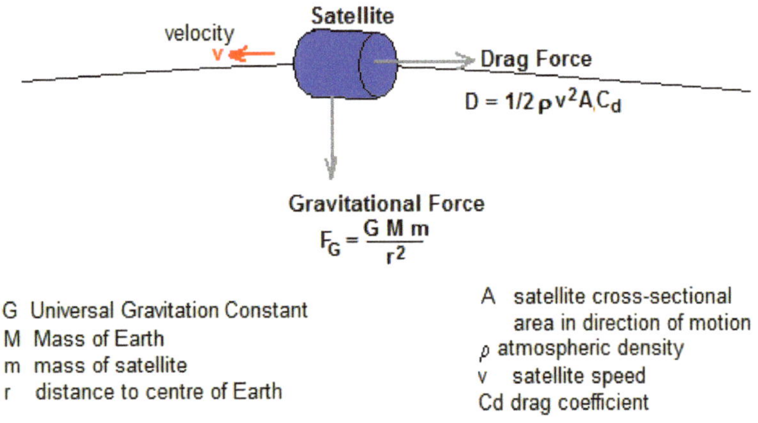

Figure 7.5 Forces on an orbiting satellite in low Earth orbit.

The reduction in the period of an object in a circular orbit is given by:

$$\frac{dP}{dt} = -3\pi a\rho\left(\frac{A_e}{m}\right)$$

where A_e (= AC_d) is the effective drag cross-sectional area of the object and m is its mass.

Typical values of m/A_e vary from 50 to 200 kg m^{-2}, and if unknown should be taken as 100.

The last two equations may be iterated, with a new density calculated at each step, from a starting altitude and time. This can be used to estimate the object lifetime until reentry, or to calculate the step time from one altitude regime to the one below. The first might be used in a simple debris evolution model, whereas the second will be needed in a more advanced model that considers altitude-dependent densities.

7.9 Model Output

Using the following parameters:

Mass = 100.0 kg

Area = 1.0 m^2

Initial height = 400.0 km

$F10.7 = 85 \quad A_p = 2$

The model produces a timeline of orbital decay given by Table 7.2.

The columns on the printout should be relatively evident. The first column (time) is the elapsed time from when the satellite was at the initially specified height. The height is the current height at that elapsed time. The period is the orbital period of the satellite in minutes. The last two columns are the mean motion in revolutions per day and the decay rate in the mean motion (units are revolutions per day per day). These last two parameters are the standard parameters that appear in the two line element sets that NORAD and others issue for various satellites, and that may be found on several websites. Mean motion is simply calculated by dividing the period (in minutes) into 1,440, the number of minutes in a day. Decay is the change in mean motion per day.

7.10 Lifetime Predictions

The basic scheme to the use with the above model is to find appropriate values for the space environmental parameters (solar 10-cm radio flux, and geomagnetic activity index), and estimate or calculate a value for the satellite mass-to-area ratio. In the absence of any information on the latter quantity it is usual to employ a value of 100 kg per square meter. This is an average value for many satellites. Extreme values do not vary a great deal from this mean, with a low of 50 and a high of 200 covering 90% of relevant spacecraft. Peculiar orbiting objects may, however,

Table 7.2 Satellite Decay Model Output

Time (days)	*Height (km)*	*Period (mins)*	*Mean Motion (rev/day)*	*Decay (rev/day^2)*
0.0	400.0	92.5	15.5630	2.58E-04
105.1	390.0	92.3	15.5939	3.27E-04
199.3	380.0	92.1	15.6285	4.16E-04
273.6	370.0	91.9	15.6633	5.30E-04
332.0	360.0	91.7	15.6981	6.76E-04
377.9	350.0	91.5	15.7332	8.66E-04
413.8	340.0	91.3	15.7683	1.11E-03
441.9	330.0	91.1	15.8036	1.43E-03
463.8	320.0	90.9	15.8391	1.85E-03
480.7	310.0	90.7	15.8745	2.39E-03
493.9	300.0	90.5	15.9103	3.10E-03
504.1	289.9	90.3	15.9462	4.04E-03
511.9	280.0	90.1	15.9820	5.26E-03
518.0	269.8	89.9	16.0185	6.90E-03
522.6	259.9	89.7	16.0546	9.03E-03
526.2	249.7	89.5	16.0917	1.19E-02
528.9	239.6	89.3	16.1284	1.57E-02
530.9	229.9	89.1	16.1640	2.06E-02
532.5	219.7	88.9	16.2017	2.75E-02
533.7	209.5	88.7	16.2392	3.66E-02
534.6	199.4	88.5	16.2765	4.89E-02
535.3	189.0	88.3	16.3153	6.60E-02
535.8	179.1	88.1	16.3522	8.81E-02

Reentry after 536 days (1.47 years).

have values outside this range. Having substituted the above values in the model, the next step is to compare the output with the actual behavior of the satellite in which you are interested. This will involve obtaining a list of actual satellite periods, altitudes (semimajor axes) or mean motions for a short period (maybe a few weeks if the satellite is near a 500-km attitude, or a few days if the altitude is 300 km). If the model underestimates the actual decay, then you must decrease your mass-to-area ratio. If the program overestimates the actual decay you must increase the ratio. Eventually you should arrive at an output that closely approximates the actual behavior of the satellite. The future predictions of the model, together with the lifetime estimate, will then be (assuming the parameters remain constant) the best estimate possible with this model.

Note: When comparing the model output with mean motion values obtained from TLEs, do not worry too much about comparing the decay values from the TLE, particularly if the satellite is a new one. There can initially be quite a discrepancy in the decay values (rev/day/day) as listed in the TLEs.

7.11 Practicalities

There are practicalities that need to be considered in undertaking the steps in the previous section. The most obvious is obtaining appropriate space environmental parameters. Current values for F10 and Ap may be obtained from the US Space Weather Prediction Center [2], and the Australian Space Weather Service [3]. Older data may be obtained from the US National Geophysical Data Center [4].

The other problem with the simple model presented here is that it does not allow for the space environment to vary. One solution is to write your own program using the simple model as a guide, and allow for the environmental values to vary with time.

Another difficulty sometimes encountered in matching model outputs to the real world is that predictions may suddenly cease to be valid (even after working well for some time). One possibility is that the altitude of the satellite has undergone a substantial change. This can cause a change to the effective cross-sectional area. One reason this might occur is due to a ground control command. It might also be possible that the satellite has run out of energy to keep it stabilized (through use of momentum reaction wheels and/or thrusters) in a desired altitude. This can result in a change of altitude, and even a tumbling. As long as the tumbling period is short compared to its orbital period, it is still possible that the average cross-sectional area it presents to the atmosphere will settle to a new constant value. The biggest changes in cross-sectional area due to altitude changes are likely to occur in the last phases of the satellite's life, just before it is due to reenter and burn up in the lower atmosphere. Unfortunately, this is normally the time when one would like to make the best prediction!

So far we have only treated orbits that are circular, or very close thereto. Satellites with very elliptical orbits travel to altitudes right outside the sphere of the current atmospheric model, and must be treated quite differently (Figure 7.6). They are also subject to other perturbations (variations of their orbit) due to the combined gravitational attractions of the Sun and the Moon. However, orbits with only a slight eccentricity can be accommodated by the use of an effective height in the model. This effective height is given by the formula:

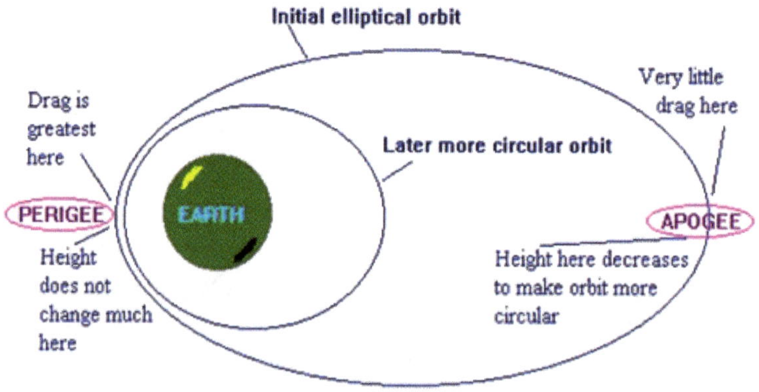

Figure 7.6 Atmospheric drag for an elliptical orbit.

$$h_e = q + 900e^{0.6}$$

where q is the perigee (lowest) height of the orbit (in km) and e is the orbital eccentricity.

For example, the lifetime of a satellite in an elliptical orbit with a perigee height of 400 km and an eccentricity of 0.01 is the same as the lifetime of a satellite in a circular orbit of height:

$$400 + 900(0.01)^{0.6}$$

Note that this formula is only a rough approximation with reasonably stringent applicability constraints. It should not be used for orbits with eccentricities in excess of 0.1, and it is really only applicable for conditions of a constant space environment. Thus it might be used for orbits with lifetimes in excess of 10 years (for which we can use an average solar cycle activity), or for orbits that are expected to decay within a few months, over which solar activity is expected to be relatively constant.

It is probably fortunate that most satellites nearing the end of their lives will have orbits with very low eccentricities (i.e., nearly circular). The reason for this is that atmospheric drag acts to circularize orbits. The apogee height is decreased whilst the perigee height is little affected until the orbit becomes close to circular (Figure 7.7).

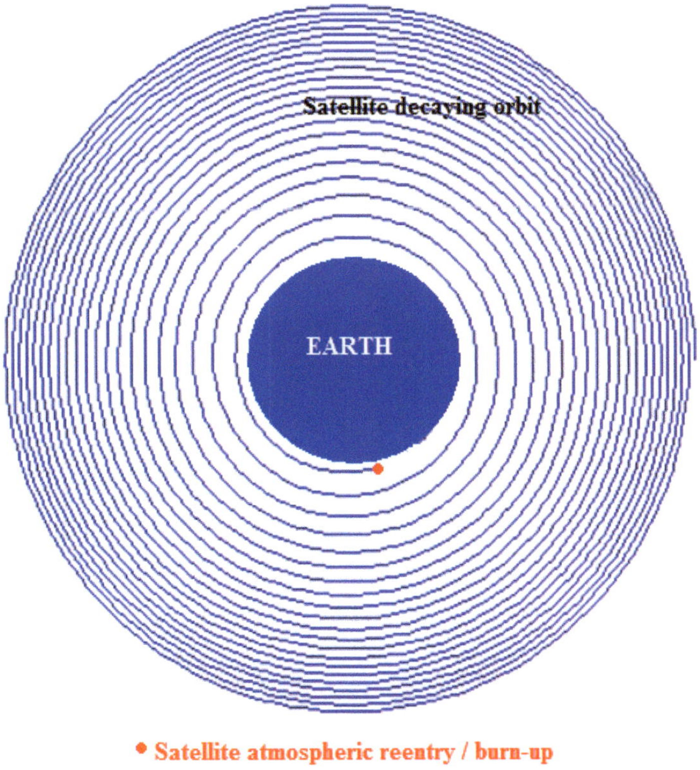

Figure 7.7 Time-compressed representation of orbital decay due to atmospheric drag.

References

[1] Tobiska, W. K., R. D. Culp, and C. A. Barth, "Predicted Solar Cycle Twenty-Two 10.7-cm Flux and Satellite Orbit Decay," *Journal of the Astronautical Sciences*, Vol. 35, No. 4, October–December 1987, pp. 419–433.

[2] US Space Weather Prediction Center, www.swpc.noaa.gov

[3] Australian Space Weather Service, sws.bom.gov.au.

[4] US National Geophysical Data Center, ngdc.noaa.gov.

Selected Bibliography

Books

Hargreaves, J. K., *The Upper Atmosphere and Solar-Terrestrial Relations*, New York: Van Nostrand Reinhold, 1979.

Jursa, A. S. (ed.), *Handbook of Geophysics and the Space Environment*, US Air Force Geophysics Laboratory, 1985. (This text can be obtained from the National Technical Information Service, 5285 Port Royal Road, Springfield, VA, 22161, Document Accession Number: ADA 167000.)

King-Hele, D, *Observing Earth Satellites*, Van-Nostrand Reinhold (New York, 1983)

King-Hele, D, *Satellite Orbits in an Atmosphere: Theory and Application*, Glasgow: Blackie, 1987. (This very detailed mathematical text is the bible for anyone seriously interested in the effect of the atmosphere on satellite orbits. Analytical expressions are derived with approximations where appropriate for the changes in the orbital elements.)

Newton, I., *Principia*. (This classic text is not only the basis of our everyday fundamental physical dynamics but it also appears to be the first text to discuss the effect of atmospheric drag on Earth satellite orbits.)

Tascione, T. F, *Introduction to the Space Environment*, Malabar, FL: Krieger, 1994.

Journals

Planetary and Space Science, journal published by Pergamon Press.

CHAPTER 8

Space Debris Reentry Hazards

8.1 Introduction

There are tens of thousands of pieces of space debris over 10 cm in size that are in low Earth orbit. These are causing, and will continue to pose, significant hazards to space operations. The vast majority, however, pose no hazard to the Earth's surface, as they will totally ablate (burn up) during atmospheric reentry, when this occurs.

8.2 Satellite Lifetimes

A satellite in low Earth orbit (below an altitude of 1,500 km) will lose energy due to air resistance, and drop slowly back toward the Earth. The upper atmosphere is very tenuous and its density decreases dramatically with altitude. It thus can take a long time to decay back into the Earth's lower atmosphere, and this time increases semiexponentially with height above the Earth's surface. Table 8.1 indicates the approximate lifetime of a satellite versus its initial altitude.

These lifetimes depend upon atmospheric density, which is driven by the solar EUV flux into the atmosphere at about the 100-km level, and also the mass and effective cross-sectional area of the satellite in the direction of motion. Figure 8.1 shows the average expected lifetimes for satellites (or debris objects) below an initial altitude of 500 km. Multiples curves indicate how these vary with the solar 10-cm flux (which is a good proxy for solar EUV flux) and for some typical values m/A, the mass to cross-section ratio.

Table 8.1 Typical Satellite Lifetimes for Circular Orbits

Satellite Altitude	*Satellite Lifetime*
200 km	1 day
300 km	1 month
400 km	1 year
500 km	10 years
700 km	100 years
900 km	1000 years

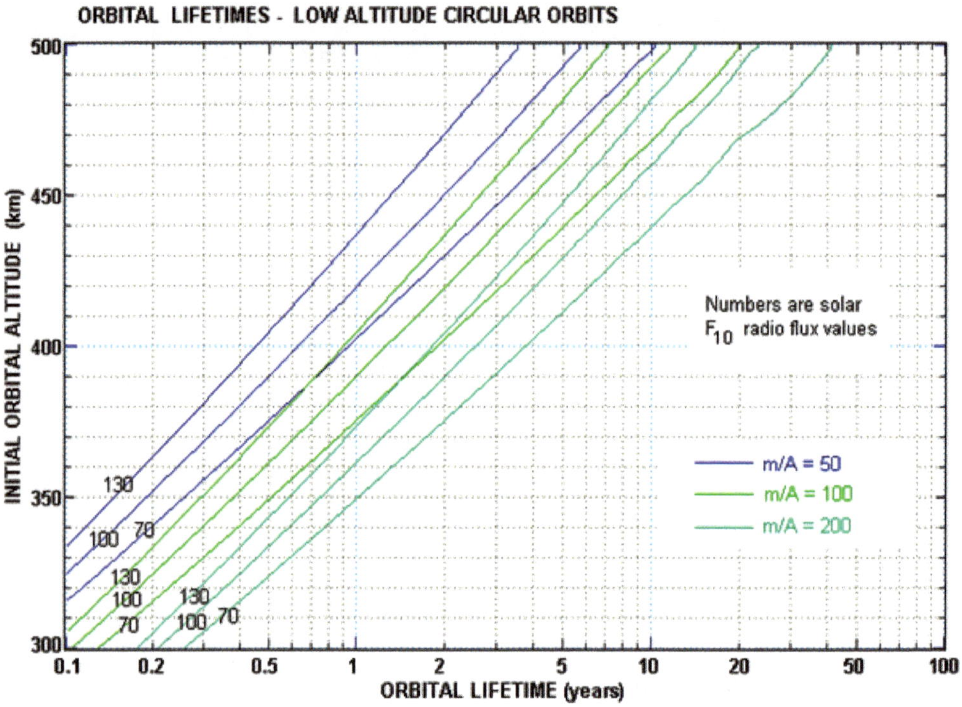

Figure 8.1 Estimated orbital lifetime of space debris as a function of initial altitude, solar flux, and debris mass to cross-sectional area.

8.3 Only Large Bodies Pose a Reentry Hazard

In general, a space object has to be over 1,000 kg (1 ton) in mass to be likely to partially survive the reentry process. This includes derelict satellites, space stations, and rocket bodies. Only small pieces of these will hit the ground.

8.4 Reentry Statistics

The histogram in Figure 8.2, compiled and drawn by The Aerospace Corporation's Center for Orbital Reentry Debris Studies (CORDS) in California, shows that on the average one or two pieces of debris (greater than about 100 mm in size) reenter the Earth's atmosphere each day.

More recent reentry statistics can be found at aerospace.org/reentries, including a CSV file of reentries later than 2000.

8.5 Survival Statistics

It has been estimated that of the 200 to 600 orbital reentries each year, about 20% may be large enough to have partially survived reentry and dropped at least some fragments on the Earth's surface.

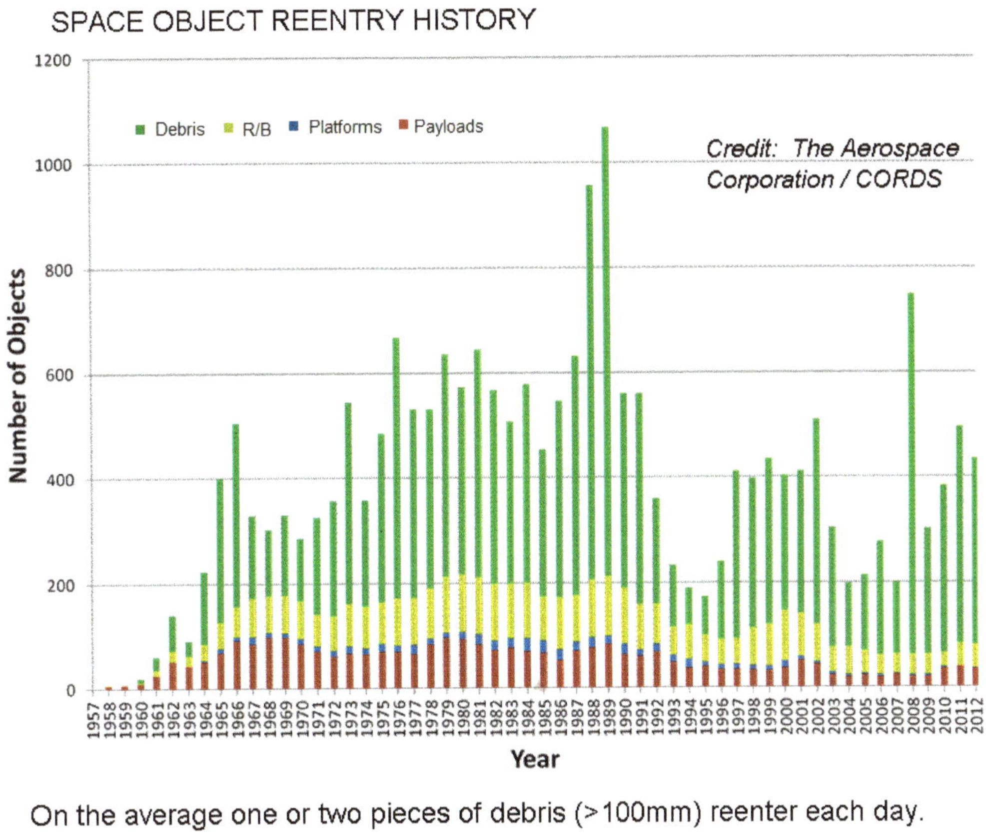

Figure 8.2 Annual reentry statistics from the start of the space age to 2012. (*Source:* Aerospace Corporation/CORDS.)

This equates to an average of about 80 per year. Most of these will be small pieces, and most will drop into thc ocean.

On the average, Australia might be hit once in about 2 years by some type of artificial space debris (Figure 8.3). This is less than the meteorite hit rate.

Figure 8.3 An Australian considers whether this collection of metal pieces may have survived reentry through the Earth's atmosphere. (*Source:* Reuters.)

8.6 Reality

As far as we are aware, no one, as of 2025, has been killed by any piece of reentering space debris, although there may have been a few minor injuries (Figure 8.4).

8.7 Estimated Relative Risks

The Aerospace Corporation has compiled a table (Table 8.2) showing the estimated relative risks of common terrestrial events in comparison to deaths or injuries from meteorite falls and reentry events.

This means that once in 100 years someone, somewhere in the world, may be injured or killed by a piece of reentering space debris. So we are dealing with a very low risk event at current population levels of space objects in low Earth orbit. And we should also note that the risk of being injured or killed by a piece of natural

Figure 8.4 One woman in the southern United States was hit by a piece of insulation from a US military satellite. (*Source:* Tulsa World.)

Table 8.2 Daily Casualty Estimates for Various Events

Events	*Daily Casualty Expectation*
Natural disasters (worldwide), 2009	29 deaths
Natural disasters (worldwide), 2010	813 deaths
Work accidents (United States), 2009	12 deaths
Motor vehicle accidents (United States), 2009	99 deaths
Unintentional poisonings (United States), 2009	84 deaths
Homicide (United States), 2009	45 deaths
Falls (United States), 2009	68 deaths
Influenza (United States), 2009	8 deaths
Meteorite falls (worldwide), 2009	1.1×10^{-4} deaths or injuries
Reentry events (worldwide), 2009	2.7×10^{-5} deaths or injuries

(*From:* [4].)

space debris (i.e., a meteorite) is greater than the risk of being injured or killed by a piece of man-made debris.

However, with the increasing launch rate, particularly after 2020, the expected casualty rate has been increasing and a new study [5] has estimated at least one casualty every decade.

8.8 Debris Reentry Survival

Figures 8.5 and 8.6, from the NASA Orbital Debris Program Office, show some large and small objects that have made it to ground. The small objects only survived because they were part of larger objects and did not become separated until sometime into the reentry process when their kinetic energy had already been dissipated.

Two basic types of debris have been found to survive the reentry process. These are refractory materials with very high melting temperature, and interior or protected components that are not exposed to the heat of reentry until a fairly late stage in the reentry process. These are listed below.

REFRACTORY MATERIALS

- Titanium;
- Steel;
- Ceramics and some glasses.

INTERIOR and PROTECTED COMPONENTS

- Insulation;
- Compact components.

Figure 8.5 Some large reentered space objects. (*From:* NASA ODPO.)

Figure 8.6 Small objects that have survived reentry. (*From:* NASA ODPO.)

8.8.1 Two Western Australian Reentries

8.8.1.1 *SKYLAB*, 1979

In July 1979, the US Space Station *Skylab* reentered over southeastern Western Australia with a large number of objects impacting the ground (Figure 8.7). The Esperance Museum contains the greatest collection of these items, but several museums in Australia have some *Skylab* debris items on display.

8.8.1.2 MARBLE BAR TITANIUM SPHERE, 1988

The object shown in Figure 8.8 was found by an employee of the Western Australia Water Corporation near the town of Marble Bar. We analyzed it using an X-ray microprobe. The object had about a 380-mm diameter, was almost pure titanium, and showed melt flow around the attachment point. It had a CO2 laser weld around the equator and Cyrillic writing was visible on the surface. It was sold in 2005 in New York for $100K to a collector!

Nicholas Johnson of the NASA Orbital Debris Program Office identified this for us as a pressure vessel from a Russian Foton rocket. The composition of the vessel was probably responsible in large part for its survival, and it is not infrequently that space debris of this type is found around the world. (There are several other instances of this type we are aware of in Australia.) The hole was probably first formed when the vessel was violently detached from the rocket, possibly due to an explosion of unspent fuel.

Figure 8.7 A large piece of *Skylab* where it landed.

Figure 8.8 A pressure vessel from a Russian Foton rocket. (*Source:* John A. Kennewell.)

8.8.2 *Skylab*

In 1973, a Saturn 1B rocket lofted *Skylab*, the US space station, into a low Earth orbit.

Because of the low orbit, atmospheric drag was working to make the orbit even lower. NASA was aware of this, but was hoping to reboost the station into a higher orbit with a rocket assist package carried into orbit by the Space Shuttle. Unfortunately, delays in the Shuttle program combined with greater than predicted drag conspired to prevent this option from being carried out (Figure 8.9).

At 12:40 a.m. on the 12th of July 1979, the residents of the coastal town of Esperance in Western Australia were awoken by loud booming noises. Outside a spectacular display of multiple moving lights lit up the night sky for several minutes. *Skylab* had reentered the Earth's atmosphere, broken up and deposited multitudinous pieces of itself along about 1,000 km of its final orbit. The day after the reentry, and for some months (and even years) after the event, people recovered thousands of pieces that originally constituted the space station.

Many of these pieces ended up in the Esperance Museum where they can be seen today (Figure 8.10). A must visit for anyone interested in looking at the diverse types of objects that can survive the reentry process.

8.9 The Reentry Process

Uncontrolled reentries are always at very low grazing angles (<1 deg). The reentering object is traveling at about 8 km/s down to below 80-km altitude (Figure 8.11).

Breakup starts at around 78-km altitude. This is termed the reentry interface. Time from breakup to impact is from 6 to 30 minutes. The length of the reentry footprint mostly depends on the reentry process above 50-km altitude. The width is also affected by tropospheric winds.

External items that have high drag (large area and low mass), such as solar panels, will be shed first. After this catastrophic breakup begins, where the spacecraft begins to come apart, the individual items will then begin to experience their own

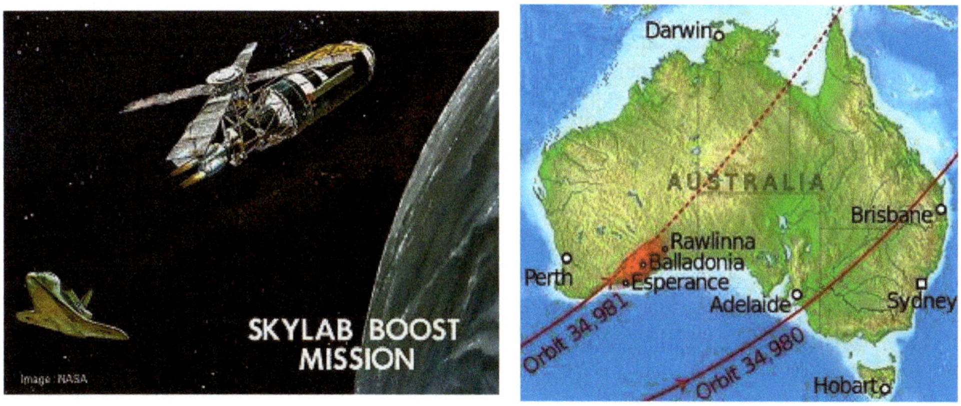

(a) (b)

Figure 8.9 (a) The planned *Skylab* boost mission that never occurred, and (b) the last two orbits of *Skylab* before reentry over the Esperance area in Western Australia. (*Source:* NASA.)

Figure 8.10 Some of the recovered pieces of *Skylab* debris displayed at the Esperance Museum in Western Australia. (*Source:* John A. Kennewell.)

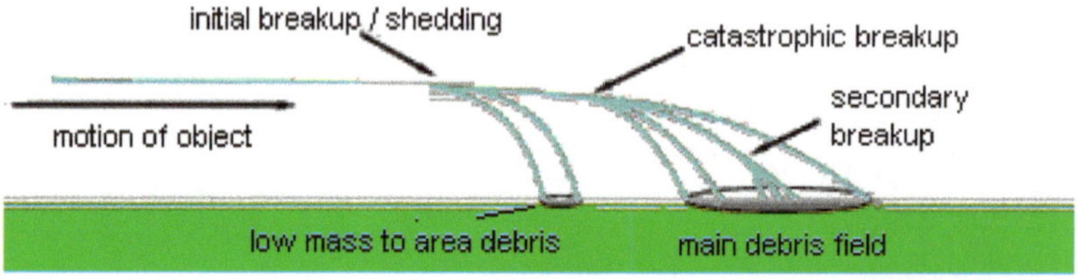

Figure 8.11 General overview of the uncontrolled reentry of a large defunct spacecraft.

drag. Drag is always greatest for items with low mass to area (m/A) ratios. These will lose their horizontal velocity more quickly than pieces with a high m/A ratio (i.e., massive compact items). The latter will retain their horizontal velocity for longer and thus will travel further. We expect these to end up at the furthest end of the ground footprint. This course of events may be changed if a low m/A item is shielded within a high mass compact item.

Items can be spread out over almost 2,000 km in the direction of travel. Orthogonal to this direction the footprint may be only 20 km, although with high and variable lower winds it might extend up to 70 km, particularly for low m/A objects (Figure 8.12). By the time a piece of debris makes it to the surface of the Earth (and most don't) it will have lost all of its space hypervelocity, and will be traveling at a terminal velocity determined by the lower atmosphere. This is around 50 m/sec.

8.9.1 The Reentry Footprint: *Columbia*

The US Space Shuttle *Columbia* tragically made an uncontrolled reentry back to Earth in February 2003 due to damage it sustained at launch.

Before reentry, the mass of the *Columbia* was around 60 tons. A massive search recovered about 85,000 pieces, which comprised 38% of the initial mass of the spacecraft. These were spread over a 1,000-km × 20-km footprint area. The total search cost was $300 million (US dollars).

Despite coming down over a well-populated area of the Earth there were no ground injuries. This in itself is testimony to the very low probability of being hit by a piece of space debris, even for the largest space objects in orbit.

8.10 Reentry Predictions From Agencies

The US Strategic Command usually issues reentry predictions at 4, 3, 2, and 1 days, and 12, 6, and 2 hours before expected reentry, and within a few hours after reentry.

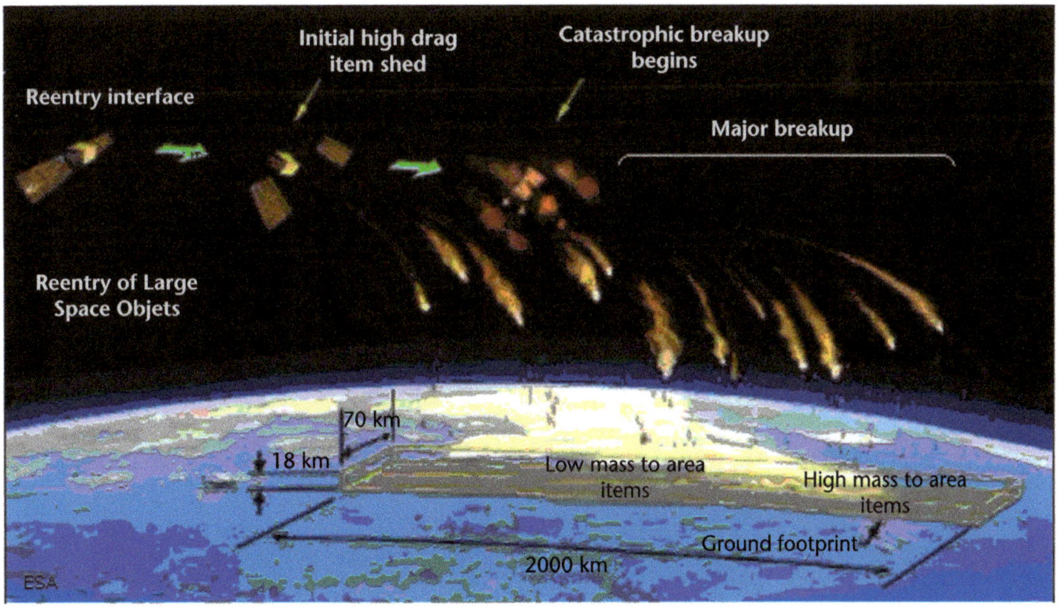

Figure 8.12 Expected reentry details of a large spacecraft in an uncontrolled reentry. This shows the typical range and cross-range distances over which reentering objects may land. This process starts at the reentry interface, which is at a height just below 80 km [4].

The Aerospace Corporation CORDS, located in Los Angeles, issues predictions for large object reentries at least 5 days in advance [1].

Some private groups or individuals publish reentry predictions. One group specializes in military satellite predictions (which are generally not provided by US government agencies).

None of the above prediction sources provide 100% coverage of reentries.

8.11 Reentry Predictions From Models

The NASA ODPO now routinely conducts a survivability analysis on all large satellites. NASA uses two models in this process: the simplified Debris Assessment Software (DAS), which is freely available, and the higher fidelity Object Reentry Survival Analysis Tool (ORSAT). Both of these are highly theoretical with very limited calibration data. Fortunately, most objects reenter over the ocean and tests of the predictions are few and far between. ORSAT graphical output is shown in Figure 8.13.

NASA performed a detailed analysis using the NASA ORSAT model for the Upper Atmosphere Research Satellite (UARS) with an initial mass of 5,668 kg that reentered in 2002.

From this analysis, we note that:

1. Only about 10% of the initial mass is expected to survive;
2. Approximately 26 objects are expected to survive reentry;
3. Individual impact masses are generally less than 250 kg;
4. Impact velocities generally range from 5 to 200 m/s with 50 m/s being the average (180 kph).

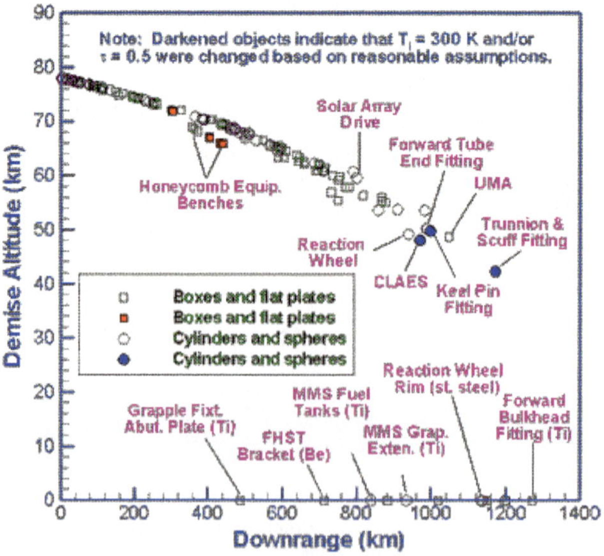

Figure 8.13 Output from the NASA ODPO reentry model. (*From:* NASA ODPO.)

8.12 Reentry Prediction Accuracy

Timing predictions for reentries have a minimal error of 10% (and sometimes up to 25%). This means that a week before the event we can only know the day of reentry. As a satellite nearing reentry makes over 16 orbits of the Earth during this time, a knowledge of where the reentry will occur is impossible.

A day before the event, we can confine the reentry path to a little over a single orbit. This can be used to eliminate large areas of the Earth, but anywhere under that single orbital path is a possible target.

Even two hours before the event the error is 12 minutes. For the satellite traveling at 8 km/s this equates to a positional error in the reentry interface point of 5,760 km! And this does not include the 1,000 km or so interface footprint over which the debris may be scattered.

8.12.1 Long-Term Predictions

Long-term (one month to one year) predictions can provide a statistical likelihood of the impact area (as shown in Figure 8.14) based purely on the inclination of the satellite orbit (which does not change over these time periods).

A computer survival analysis for a particular satellite can also estimate the number of objects expected to survive, and their total mass. A model to calculate this prediction is given in Section 9.9.

8.12.2 Medium-Term Predictions

Predictions of a week or so can provide the day of reentry and a plot of the 16 or so orbits that the space object will travel over during that time. This can be used to exclude some areas from concern (Figure 8.15).

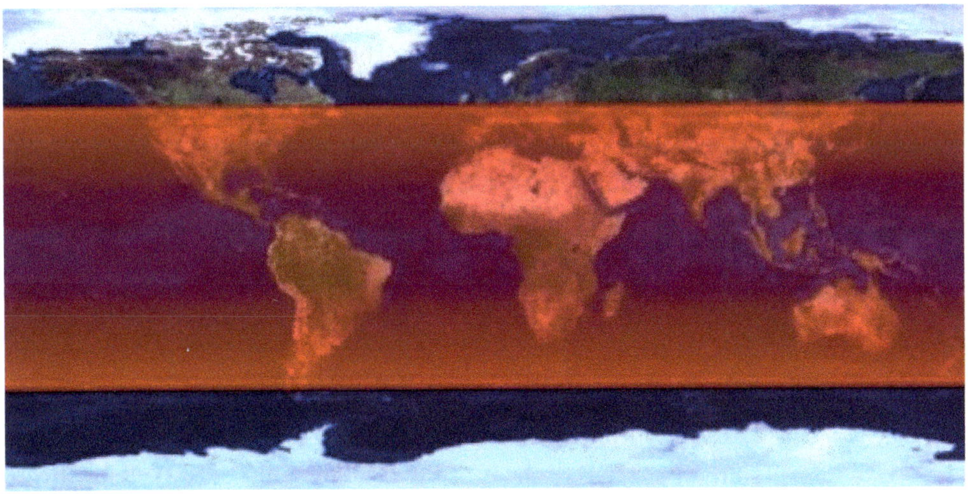

Figure 8.14 The intensity in the red region gives a probabilistic estimate of the likely of reentry at that point, for an object with an inclination of 55 degrees. (*Image credit:* Ailor/The Aerospace Corporation.)

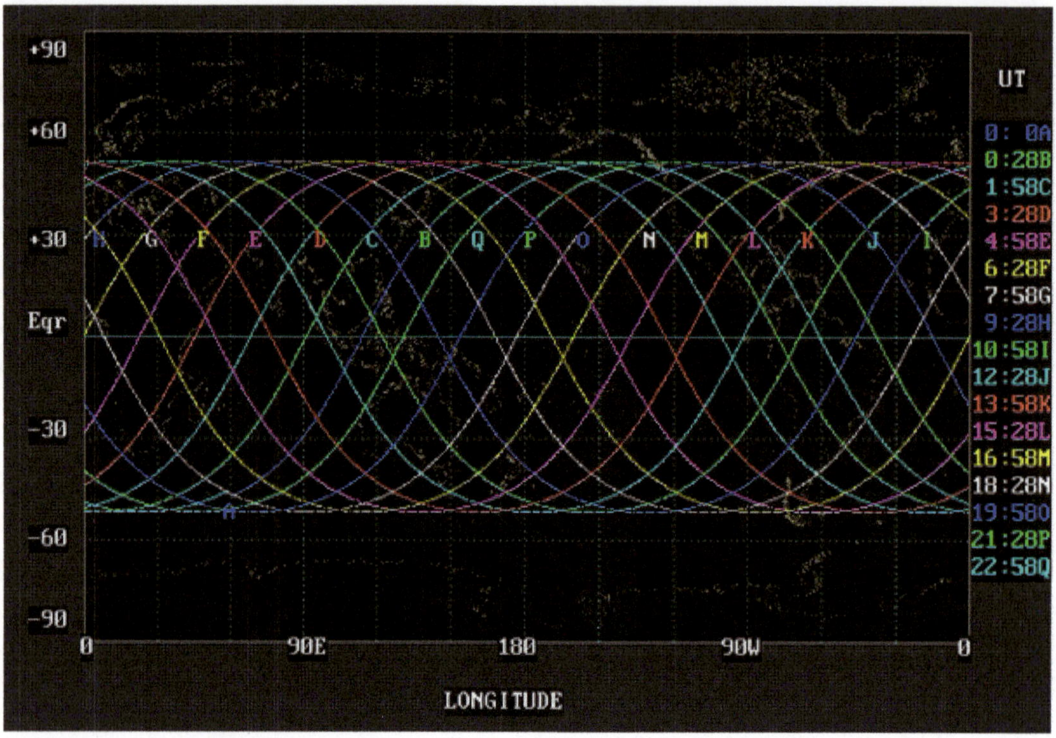

Figure 8.15 Predictions made a week before reentry can indicate the day on which reentry is expected to occur and the paths of the last orbits.

8.12.3 Short-Term Predictions

A short-term reentry prediction is generally one made when the object is expected to reentry within the next day. As predictions are usually accurate to within 10%, in this case the prediction will generally indicate the orbit over which reentry will occur (Figure 8.16). This can be used to exclude large areas of the world, and a possible warning for the locations under the final orbit.

8.13 Debris Reentry Hazard Criteria

There are several instances in which a debris reentry is considered to be hazardous. These are:

- Injury or death will typically result if the impacting energy of a piece of debris exceeds 15J. This is equivalent to a one-kilogram mass (e.g., one-liter milk carton) hitting someone at 6 m/s (20 kph), or a tennis ball hitting a person at 22 m/s (80 kph).
- A reentry hazard is declared if the estimated probability of one death exceeds 1 in 10,000. This essentially means that there would have to be 10,000 reentries of this exact type to cause one death anywhere in the world.

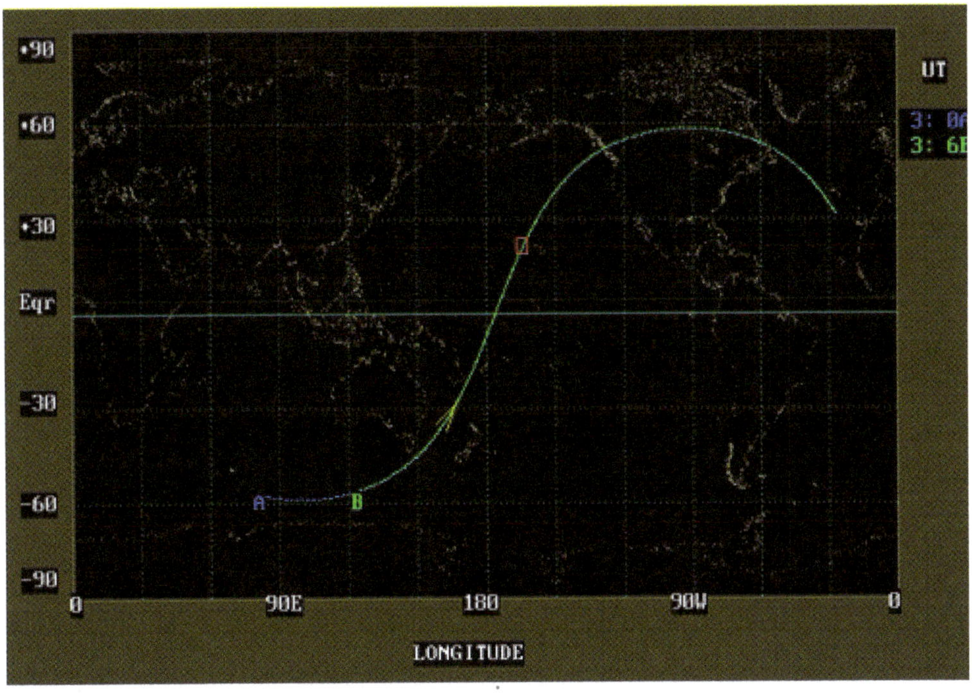

Figure 8.16 This map indicates the expected location on a single orbit in which the satellite is expected to reenter (red rectangle). The prediction error is such, however, that the reentry may occur anywhere along the green path. This error is due to an uncertain knowledge of the atmospheric conditions at the predicted time.

- An aircraft will experience catastrophic damage if it is hit by an object with a mass over 300 gm. The object velocity is generally not important as it will be substantially less than the velocity of a commercial airliner.
- An airspace hazard zone is declared if more than 27 objects are expected to impact in the zone. This is transmitted to pilots by the existing NOTAM (Notices to Airmen) system.

8.14 Special Reentry Events

The probability of injury or death from a space object reentry and the number of injuries or deaths that can occur are both so low that in most cases local emergency responses are the only necessary response to cope with a reentry event, and then only after it has occurred. There is only one exception to this, and that is if the reentering object is known, suspected, or found to contain radioactive (or very possibly) other hazardous materials, such that it presents no increased impact hazard, but a hazard if an unsuspecting person were to discover the reentered objects.

There are two types of radioactive components that have been carried by satellites or spacecraft in the past:

1. *Radioisotope thermoelectric generators (RTGs).* These are used on deep-space craft that venture to parts of the solar system where solar power is inadequate to supply the operating power needed.
2. *Nuclear reactors.* These have been used mainly by the Soviets to supply the large power requirements of radar reconnaissance satellites in low Earth orbit.

8.14.1 Radioisotope Thermoelectric Generators

An RTG (Figure 8.20) converts the heat from radioactive decay into electrical energy. RTGs are frequently used on deep-space missions to the outer planets where solar energy does not provide adequate power (too far from the Sun).

The radioactive isotope of choice is plutonium-238 (Pu-238). RTGs are normally only a reentry hazard if the launching rocket fails to reach its nominal orbit. Pu-238 is nonfissionable and US RTGs are very robust.

The maximum amount of Pu-238 is about 10 kg (for 200W of power). Plutonium is biologically toxic and is mainly an alpha particle emitter with some gamma.

8.14.2 Nuclear Reactors

Nuclear reactors have been used in space applications where high power is required (Figure 8.17). The United States has placed one reactor in space and the Soviets approximately 33. The Soviet reactors were used to power radar ocean reconnaissance satellites used to track movements of US Naval vessels. They had very low orbits and solar panels could not be used because the drag would have caused reentry in a few days. All remaining Soviet space nuclear reactors have been boosted into high orbits where they will not reenter for around a thousand years. The last RORsat was launched in 1988. Russia has announced it is building a new 500-kW space reactor for a mission to the planets in 2030.

8.14.3 Case Study: *COSMOS 954*

The *COSMOS 954* was a Soviet military radar ocean reconnaissance satellite that carried a nuclear reactor (Figure 8.18). The reentry footprint occurred over Canada.

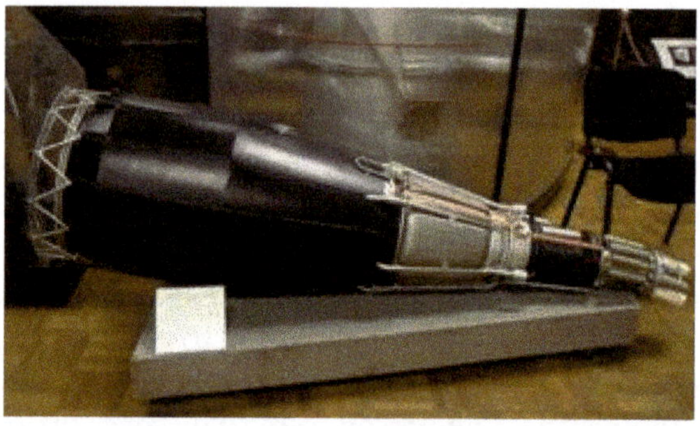

Figure 8.17 This is a scale model of a Soviet space nuclear reactor (Topaz) in a Moscow museum. The actual reactor was 4.6m in length, weighed 980 kg, and had an electrical output of 5 kW. All past reactors have used highly enriched uranium-235 oxide (90%), with a maximum fuel mass of <50 kg. (*From:* Moscow Museum.)

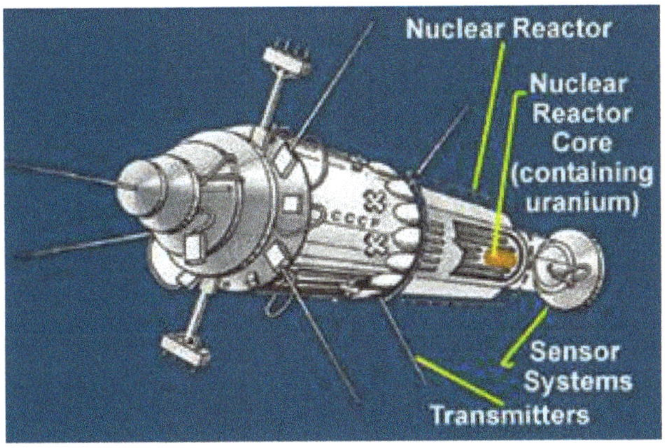

Figure 8.18 Representation of a Soviet radar ocean reconnaissance satellite with nuclear reactor. (*Source:* Wikipedia.)

The main part of the reactor was deposited on a frozen lake, but the containment vessel of the reactor was breached and other radioactive contamination occurred across a large area. The cleanup cost to the Canadian government was $6 million of which the Soviets paid $3 million. No one was injured. Reentry occurred January 24, 1978, over the Canadian Northwest Territories in a footprint covering 100,000 square kilometers.

The highest radiation levels encountered (~ 200 R/hr) could cause severe radiation sickness and death from only a few hours exposure. The cleanup operation, which the Canadians designated Operation Morning Light (Figure 8.19), lasted about 3 months.

Figure 8.19 Canadian cleanup after Soviet radioactive RORSAT reentry.

8.14.4 Case Study: Russian MARS 96

The Russian MARS 96 was a Russian mission to Mars that failed to reach the correct Earth orbit. It carried an RTG. US Air Force Space Command determined that the reentry footprint could occur over Australia, but did not issue a warning to the Australian government, although a warning was issued to some individuals (mostly US government employees) in Australia via the US Embassy in Canberra.

The reentry actually placed debris not over Australia but over Bolivia and Chile. All evidence points to the fact that the RTG probably landed in the Andean mountains, but no one has yet recovered it. And no directed search has ever been made!

This incident was of concern to the then-Australian Prime Minister, and resulted in the negotiation of a warning agreement between the United States and Australia, as well as the generation of an Emergency Management Plan mainly involving ARPANSA (the Australian Radiation Protection and Nuclear Safety Agency).

8.15 Planned Reentries

Where possible, it is obviously appropriate to direct a reentry so that it occurs over both a very low population area, and also an area where aircraft flights are very infrequent. Such an area exists in the South Pacific Ocean and there is now a coordinated effort to direct rocket bodies and large spacecraft at their end of life to this region. Such maneuvers are of course only possible for intact rocket bodies with planned additional fuel for this, and for intact satellites that have sufficient maneuvering fuel left at their end of life.

8.16 Suggested Debris Reentry Hazard Response

Australia now has several emergency management plans to cope with reentering space debris. However, because reentries are almost impossible to predict, and because damage probabilities are small and confined to very small areas, the necessary response in most circumstances can be coordinated after the fact, and will only involve local resources. Only in the instance of a known or suspected reentry of radioactive material will a wider response be necessary. Figure 8.20 indicates what we believe is an appropriate response under the various indicated circumstances.

The issuance of any advice and alerts should always be tempered by the large inaccuracies of any prediction (particularly with regard to possible area of impact), the very low probability of damage, injury, or fatalities and should have regard to the anxiety and panic they may engender.

8.17 Postimpact Object Location

Most reentries will go unnoticed. Some may be stumbled upon years after the event. Only very bright reentries near populated areas generally have sufficient reports to give the necessary triangulation accuracy such that recovery may be attempted.

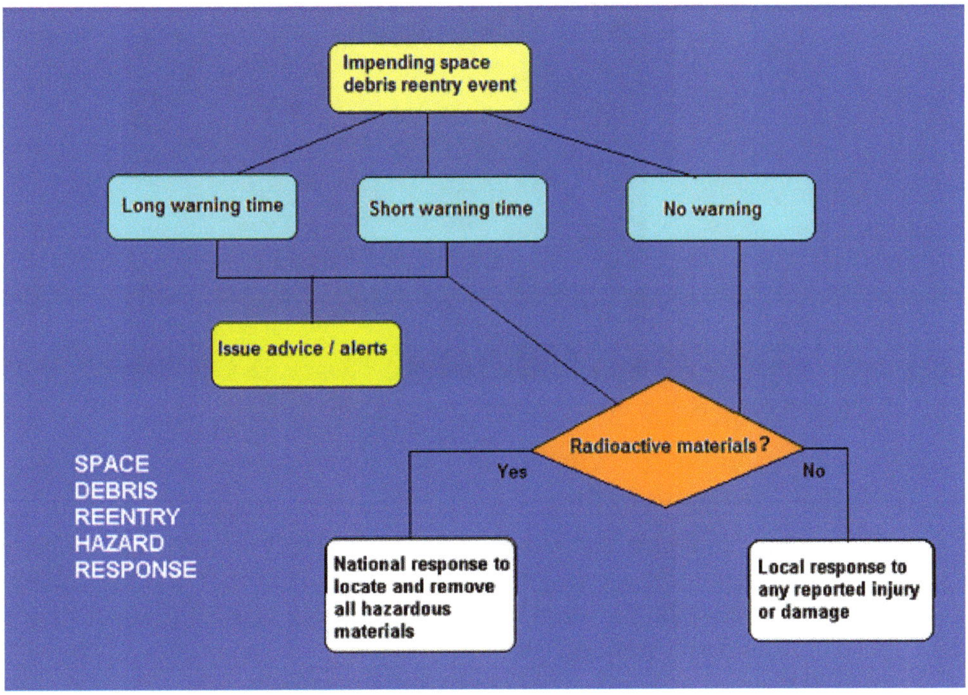

Figure 8.20 Suggested response to space debris reentry impact.

The Desert Fireball Network is a research project based at Curtin University under the direction of Dr. Phil Bland. They have deployed over 50 sky sensors across Western and Central Australia. This network has the potential of locating reentered space objects to within a hundred meters. It also encourages public reports of fireballs to augment their sensors [2].

8.18 Satellite, Meteor, or Reentry?

There are a few simple points you can use to determine whether an object you have seen in the sky is an orbiting satellite or large piece of orbiting space debris, a meteor or fireball, or a piece of reentering space debris.

8.18.1 Orbiting Satellite or Large Orbiting Debris Object

- A point of light, like a star, that moves slowly across the sky. The image shown here is a trail of light because of the time exposure. A satellite will be visible for a few minutes as it moves across the sky.
- Satellites and large pieces of debris orbiting the Earth shine by reflected light from the Sun. Because of this, and because any visible space objects are in low Earth orbit, they are only visible for one to two hours after sunset, and one to two hours before sunrise. They may fade from view as they enter the shadow of the Earth.

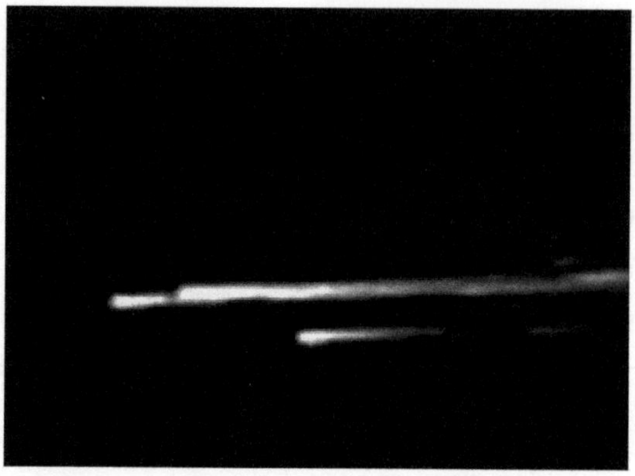

Figure 8.21 A reentry event. (*Source:* NASA.)

8.18.2 Meteor or Fireball

- A point of light that moves very rapidly across the sky. The motion may be so rapid that the eye sees it as a line or trail of light. It will be visible for only a few seconds at most since most meteors are visible for less than one second. A fireball is simply a very bright meteor.
- Meteors and fireballs shine by their own light, which is created due to the intense heat of entry into the Earth's atmosphere. Thus a meteor may be seen at anytime and anywhere across the sky.

8.18.3 Reentering Space Debris

See Figure 8.21.

- A single point or several points of light moving moderately quickly across the sky. Because a debris reentry has a lower velocity than a meteor/fireball it will be visible for longer, normally longer than 10 seconds.
- A reentering space object also is visible by the light (and heat) created in the reentry process, and thus may be visible at any time of night (or even in the daytime if it is very bright).
- Reentry of a large object will be visible as multiples traces moving together. Reentry always occurs at grazing incidence. This does not mean it is parallel to the horizon, as grazing incidence passing overhead will appear to ascend into the sky and then descend.

References

[1] aero.org/capabilities/cords.

[2] Desert Fireball Network, fireballsinthesky.com.au.

[3] FireOPAL Project, Curtin University, Western Australia.

[4] Ailor, W., *tech-39E.pdf*, The Aerospace Corporation, STS UNCOPUOS.

[5] Byers, M., et al., "Unnecessary Risks Created by Uncontrolled Rocket Reentries," *Nature Astronomy*, Vol. 6, 2022, pp. 1093–1097.

Selected Bibliography

Australian Space Debris Reentry Response Plans.

adrc.asia/management/AUS/spred.pdf.

extranet.fesa.wa.gov.au/sites/emwa/Lists/StateEmergencyManagementDocumentLibrary/State%20EM%20Plans/Hazard%20Plans/Westplan%20SPRED.pdf.

CHAPTER 9

A Debris Reentry Model

The model described here can be used to follow the flight of a piece of natural space debris (i.e., a meteoroid) as it enters the Earth's atmosphere, or it can be used to examine the reentry of orbital space debris. The difference is primarily in the initial velocity and angle of entry.

9.1 Space Debris Flight Through the Atmosphere

The observation of a meteor during its transient existence can provide a wealth of information about the meteoroid and the atmosphere through which it travels. In the late sixteenth century, observations of meteors from separated sites gave proof that the Earth's atmosphere extended to a height of at least 100 km. Air densities and temperatures at these altitudes were derived from photographic records in the first half of the twentieth century. Later research provided insights into the nature of the particles that produced the meteor phenomenon.

The model presented here simulates the flight of an object through the upper atmosphere using the equations of classical meteor ablation theory. This theory was developed by many people, including Fred Whipple and Ernst Opik. Although the theory has some definite limitations, it not only provides a good insight into the deceleration, mass loss, and visible radiation of the entry/reentry process, it also forms the basis of more extended theories that deal with the fragmentation of the object as it succumbs to the large deceleration experienced during atmospheric flight.

This model traces the variation of the object velocity, deceleration, mass, and visual radiation with time. By varying the appropriate model parameters such as initial conditions or the atmospheric environment, the effect of each can be readily evaluated. The model is valid not only for the Earth but can also be used for other bodies with an atmosphere such as Mars.

9.2 Classical Debris Ablation Theory

When an object first encounters a significant atmosphere it undergoes collisions with the air molecules. By this process, momentum is transferred from the meteoroid to the air. The resultant deceleration of the object is proportional to the cross-sectional area it presents to the direction of travel and to the drag coefficient. The cross-sectional area is dependent on the shape of the body, its physical orientation, and whether or not it is rotating.

High in the atmosphere, where there are few air molecules, the drag coefficient lies close to one. All molecules within the cross-sectional forward projection undergo collisions. Lower down, where the atmospheric density becomes greater, an air-cap forms around the incandescent body, shielding it in part from some of the air column. This effectively reduces its effective cross section, and the drag coefficient drops to around one-half (0.5).

The kinetic energy of the object is continually reduced as it traverses the atmosphere to lower altitudes. This energy is converted into heat, light, and energy of ionization. Most of the energy is in the form of heat, which causes vaporization of surface material from the meteoroid, a process known as ablation. The mass of the object is reduced at a rate proportional to the loss of kinetic energy. Two parameters are important in this transfer of energy: the heat of ablation of the object material, and the heat transfer coefficient. This latter quantity is a measure of the efficiency of the collision process in converting kinetic energy to heat.

The visual power that is radiated by the meteor is proportional to the loss of kinetic energy of the ablated material. The proportional constant is termed the luminous efficiency, and this has a strong dependence on the nature of the meteoroid. The visual power can be transformed into an absolute magnitude through a simple experimentally derived formula. By definition, the absolute magnitude of a meteor is the observed magnitude at a distance of 100 km. The apparent magnitude then includes the actual object-observer distance ratio as a correction term.

9.3 The Flight Equations

As the object encounters significant atmosphere, its momentum is decreased through collisions with air molecules. This momentum transfer can be quantified with the help of Newton's second law of motion giving the resistance or drag equation:

$$\frac{dv}{dt} = \frac{\Gamma A \rho_a v^2}{m^{1/3} \rho_m^{2/3}}$$

where

v is the speed of object (m s^{-1});

dv/dt is the deceleration of the object (m s^{-2});

Γ is the drag coefficient;

A is the object shape factor;

m is the object mass (kg);

ρ_m is the object density (kg m^{-3});

ρ_a is the atmospheric density (kg^{-3}).

A second equation, the differential mass equation, is derived by assuming that the rate of loss of mass is proportional to the kinetic energy transferred to the mass of air intercepted by the meteoroid:

$$\frac{dm}{dt} = \frac{\Lambda A \rho_a v^3 m^{2/3}}{2\zeta \rho_m^{2/3}}$$

where

Λ is the heat transfer coefficient;
ζ is the heat of ablation of the object.

The visual power radiated by the meteoroid is given by:

$$I = 0.5 T_I \frac{dm}{dt} v^2$$

where T_I is the luminous radiation efficiency. This may be converted into a rough apparent visual magnitude for a ground-based observer through the empirical relation:

$$M_v = 6.8 - 2.5\log_{10}\left(I \frac{h_o^2}{h^2}\right)$$

where h_o is a reference height of 100 km. The squared altitude ratio takes into account the apparent magnitude change due to the distance from the observer, but is only correct for an observer directly below the meteor. Absolute magnitude is obtained by setting $h = h_o$.

Table 9.1 lists typical initial values for the above parameters and a few others used in the computer implementation of the model.

9.4 Atmospheric Density

Atmospheric density varies with altitude, and it also varies to a lesser extent with location (latitude and longitude), time (season and hour of the day), and a few other minor factors. And of course it varies enormously from planet to planet. With the

Table 9.1 Parameters Used in the Debris Reentry Model

Parameter	*Description*	*Typical Initial Values*
A	Object shape factor	0.8–1.2
Γ	Drag coefficient	0.5–1.0
ρ_m	Object density	500–8000 kg m^{-3}
ζ	Object heat of ablation	(0.1–10) × 10^6 J kg^{-1}
Λ	Heat transfer coefficient	0.5–6
T_I	Luminous efficiency	2×10^{-2}–2×10^{-4}
v	Object velocity	11,000–72,000 m s^{-1}
m	Object mass	~10^{-8}–1000 kg
Z	Entry zenith angle	0°–99°
h	Object height	130–180 km

appropriate expression for air density this model can be used to compare the differences in meteors that occur on Mars and the Earth.

The accuracy of the expression to be used for air density really depends on the purpose for which the model is being used. If the model is only used to illustrate the basic nature of meteor ablation, a simple exponential expression of the form $\rho = \rho_0 \exp(-h/H)$ is quite adequate (see Section 9.7 for a more comprehensive discussion of atmospheric models). This assumes that the atmosphere is isothermal and homogeneous (composition does not change with altitude). A suitable expression for the Earth is given by:

$$\rho = 1.25 \exp\left(\frac{-h}{7{,}000}\right)$$

where the height above the surface (h) is in kilometers and the density is in kilogram per cubic meter. For meteors, only altitudes below 150 km need to be considered.

9.5 Model Results

A typical output of this model is shown in Table 9.2.

All of the parameters of the model can be changed to show what effect each has. Other parameters may be output. Kinetic energy might be an appropriate quantity to add to the list of output parameters.

Table 9.2 Model Output for a Meteor Flight

```
NATURAL DEBRIS REENTRY FLIGHT
  Initial Mass (kg)? 0.001
  Density (kg/m^3)? 3500
  Speed at Entry (km/s)? 20
  Zenith Angle (deg)? 45
```

TIME (S)	HEIGHT (KM)	SPEED (KM/S)	DECEL (M/S/S)	MASS (%)	VISUAL MAG
1.50	120.0	20.0	1	100.0	9.7
2.00	110.0	20.0	4	99.9	8.0
2.50	100.0	20.0	17	99.7	6.2
3.00	90.0	20.0	71	98.8	4.5 vis
3.50	80.0	19.9	297	95.0	2.7 vis
4.00	70.1	19.6	1245	80.7	1.1 vis
4.50	60.6	18.2	5197	42.0	0.2 vis
5.00	52.6	13.0	16377	5.5	1.9 vis
5.50	48.1	5.5	9838	1.0	7.0
6.00	46.2	2.6	3170	0.7	10.9
6.50	45.2	1.6	1339	0.7	13.5
7.00	44.6	1.1	699	0.7	15.4
7.50	44.1	0.8	418	0.7	16.9

9.6 Exploring the Model

Examination of the model output can give an idea of the basics of object ablation during flight through the atmosphere:

1. Notice the extremely high deceleration experienced by the meteoroid as its kinetic energy is reduced. The maximum shown in the above example is 1,600g (i.e., 1,600 times the gravitational acceleration).
2. A 1-gram meteoroid typically produces a zeroth magnitude meteor.
3. The meteor is visible for 2 seconds; in fact, the typical visual meteor (which has a magnitude less than 0) will be visible for less than a second.
4. The ablation process consumes mass in preference to speed. Only when two-thirds of the mass is gone does the speed really start to significantly reduce.

We need to plot some graphs to show the effect of varying different meteor parameters.

9.6.1 Meteoroid Velocity

The curves in Figure 9.1 show the effect of varying the meteoroid velocity, keeping mass and density constant.

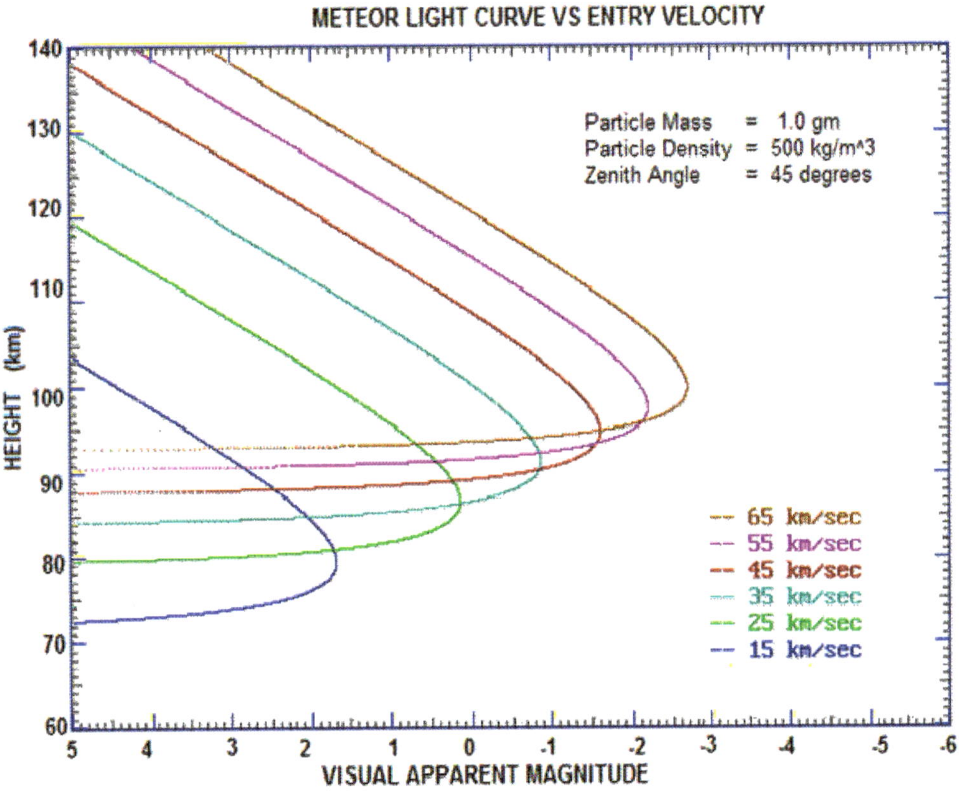

Figure 9.1 Output of a meteor flight model showing the variation of flight profile height with initial (entry) velocity.

Higher velocity meteoroids start to ablate earlier in the entry process. They also produce brighter meteors per unit mass.

9.6.2 Meteoroid Density

The curves in Figure 9.2 show the effect of varying the meteoroid density, keeping mass and velocity constant.

Meteoroids with larger densities will be more compact objects. These travel further down into the atmosphere during ablation and produce a slightly brighter meteor.

9.6.3 Meteoroid Mass

Figure 9.3 shows the effect of varying the meteoroid mass, keeping velocity and density constant.

The predominant effect is on meteor brightness. We can see that a cometary meteor (low density–500 kg/m^3) with a meteoroid mass less than 10 mg will not be visible. There is a minor decrease in height of maximum brightness as the meteoroid mass increases.

There are many more such plots that can be made to learn about the ablation process.

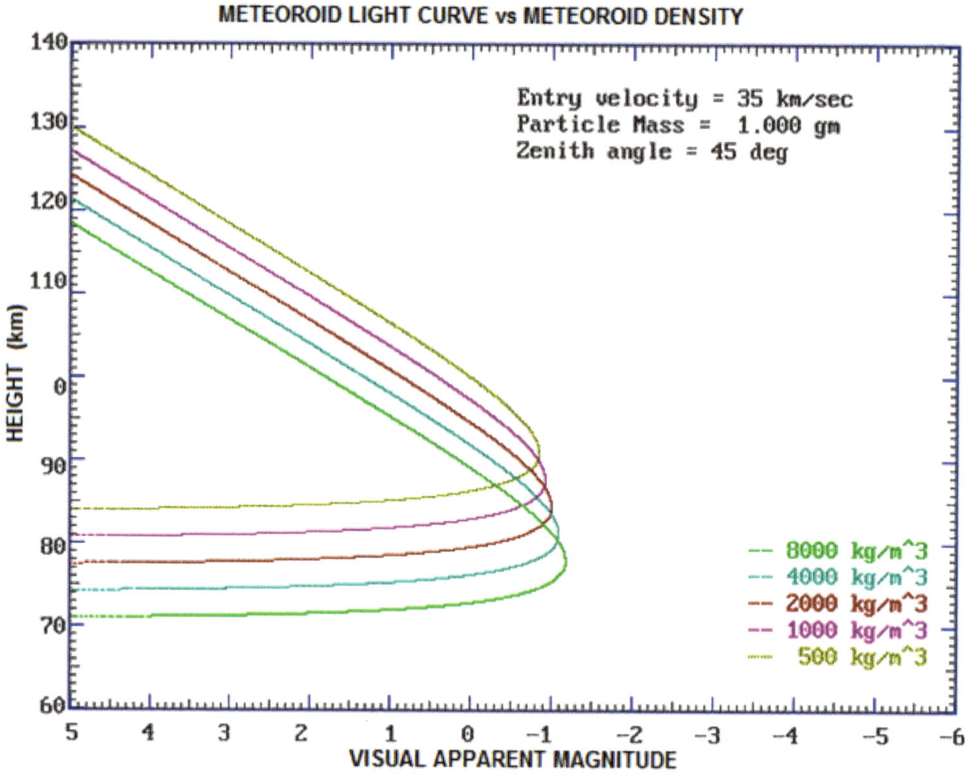

Figure 9.2 Meteor model showing how the height profile varies with object density.

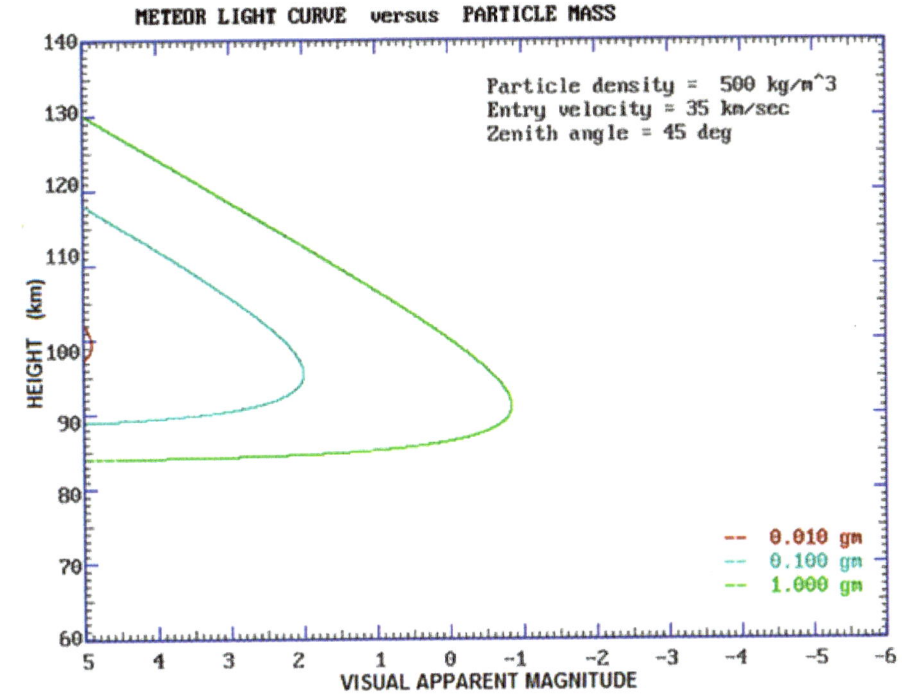

Figure 9.3 Meteor model showing brightness versus meteoroid mass.

9.7 Theory and Observations

How does the above theory correspond to reality? Consider the three light curves in Figure 9.4 from L. Jacchia as reproduced by Hawkins [1].

The solid lines are the observed light curves for three bright meteors, whereas the dashed lines are the classical ablation model that best match the observed curves. The velocities of the meteors are, respectively, (a) 36 km/s, (b) 25 km/s, and (c) 69 km/s.

The model fits the observed curve in (a) quite well, except at the very end of the flight. The situation in (b) is still not too bad but shows substantial brightening near time of maximum light, whereas in (c) the model fails totally before the time of maximum light.

The resolution of the problem is that meteoroids don't always remain as one smoothly ablating mass. The enormous deceleration experienced in meteor flight may cause the meteoroid to implode or fragment. This is particularly the case for a cometary meteor where the meteoroid is of a low density, rather fragile material. This produces many pieces, each of which then undergoes ablation as a separate entity. This results in more light being produced than for a single object. In the case (c), the object has undergone terminal implosion into many pieces, creating a flare which quickly fades away.

In summary, meteoroid fragmentation produces shorter trails, wider trails, and a brighter meteor. Fragmentation is often not a single event but something that occurs intermittently along the reentry path.

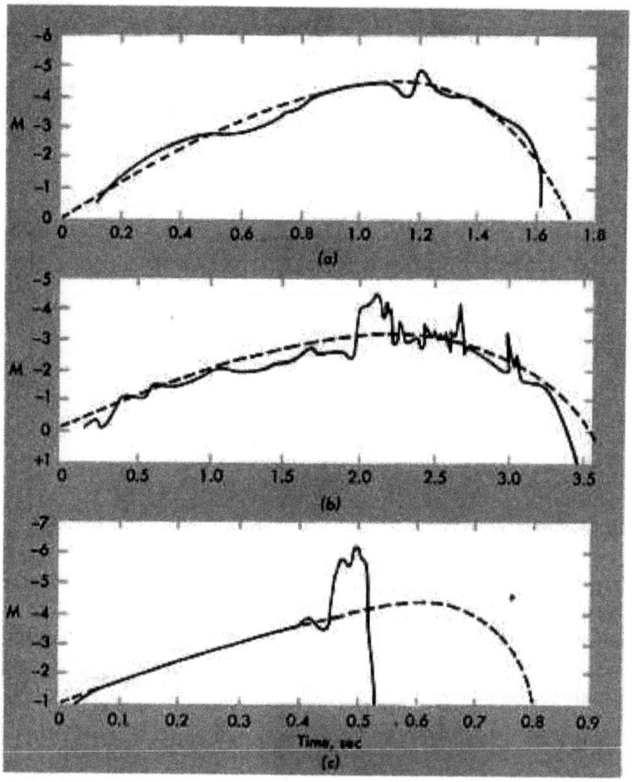

Figure 9.4 Actual meteor magnitudes with time showing deviation from meteor model due to meteoroid fragmentation. (*Source:* Hawkins.)

As long as the meteoroid remains intact, we have every reason to believe that classical ablation theory provides a close estimate to the process that occurs in meteor flight. Extension of the theory to include fragmentation can demonstrate the features of this process.

9.7.1 The Shape Factor

When a large piece of space debris reenters the Earth's atmosphere we often have knowledge of its shape and size. This is not so in the case of a meteoroid. In both cases, it is the effective cross section of the body in the direction of travel that is the important parameter in the deceleration process. Because of our ignorance of this parameter for a meteoroid it is conventional to define a dimensionless shape parameter A where:

$$S = A\left(\frac{m}{\rho_m}\right)^{2/3}$$ is the effective cross-sectional area of the meteoroid.

Figure 9.5 shows the values of A for various shapes.

All these are close to unity, and this value is normally used in meteor ablation theory.

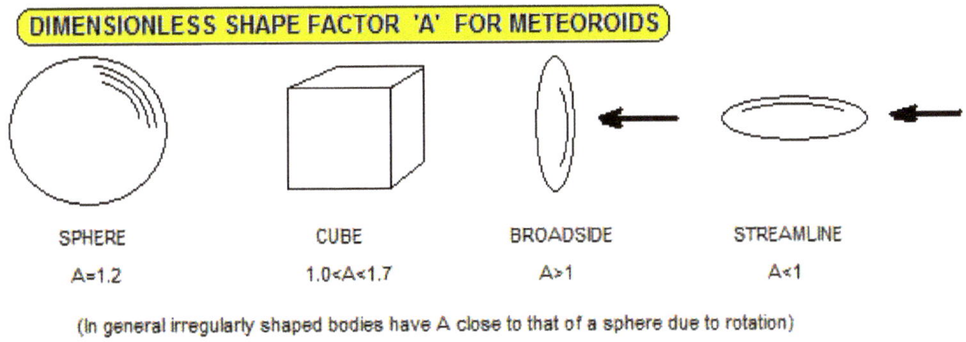

Figure 9.5 Typical A parameters for various shapes.

9.8 Orbital Space Debris Reentries

This theory not only applies to meteoroids entering a planetary atmosphere, but is also applicable to the reentry of man-made space debris.

In the reentry of orbital space debris the zenith angle is around 89 degrees. That is, the debris object enters at nearly grazing incidence to the atmosphere. Such an object has a velocity around 8 km/s, lower than any meteoroid (which has a minimum velocity of 11 km/s, equal to the escape velocity of the Earth).

Gravity is much more important in this case, and the reentering object will travel a substantial distance around the Earth during the ablation process. The model in the previous section therefore needs to incorporate the force of gravity into the equations.

Thus the model as modified for space debris reentries has a fixed entry velocity of 8 km/s which is the orbital velocity of any object close to the Earth's surface. The zenith entry angle is taken to be 89 degrees, and the acceleration of gravity is simply added to the deceleration parameter. This will produce a small anomalous increase in the debris velocity for the first part of the reentry, but as the object curves toward the ground this error becomes less, we find that at peak deceleration, the gravitational component is insignificant.

One of the main differences between a meteoroid reentry is that reentry of space debris occurs over a large horizontal difference. As the model does not include the curvature of the Earth, it is essentially a flat Earth approximation.

A typical output of this model for reentry of a large (200 kg) piece of debris with a bulk density of 2,700 kg per cubic meter is given in Figure 9.6. This density is typical of much of the material (i.e., aluminum) used to produce satellites, and thus is typical of space debris.

A value of 9.9 for the apparent magnitude simply means that the object is not visible.

A graphical presentation of this data is given in Figure 9.7.

Five parameters are displayed: mass, deceleration, velocity, time, and visibility. The first four parameters are quantitatively displayed on a log scale. The visibility curve is related to visual magnitude by the expression value = 3 – visual magnitude

ATMOSPHERIC SPACE DEBRIS RE-ENTRY

Mass = 200 kg Density = 2700 kg/m3

TIME (S)	HEIGHT (KM)	SPEED (KM/S)	DECEL (M/S/S)	MASS (%)	VISUAL MAG
0	80.0	8.001	0.0	100.0	9.9
69	70.0	8.641	0.0	99.2	-2.1 vis
133	60.0	9.101	0.0	95.3	-4.4 vis
196	50.0	8.877	16.0	77.7	-6.1 vis
271	40.0	5.857	57.4	37.0	-5.5 vis
522	30.0	1.051	2.5	20.4	2.0 vis
1424	20.0	0.429	0.3	17.4	9.9
3478	10.0	0.192	0.1	14.9	9.9
8045	0.0	0.087	0.0	12.8	9.9

Figure 9.6 Orbital space debris reentry model text output.

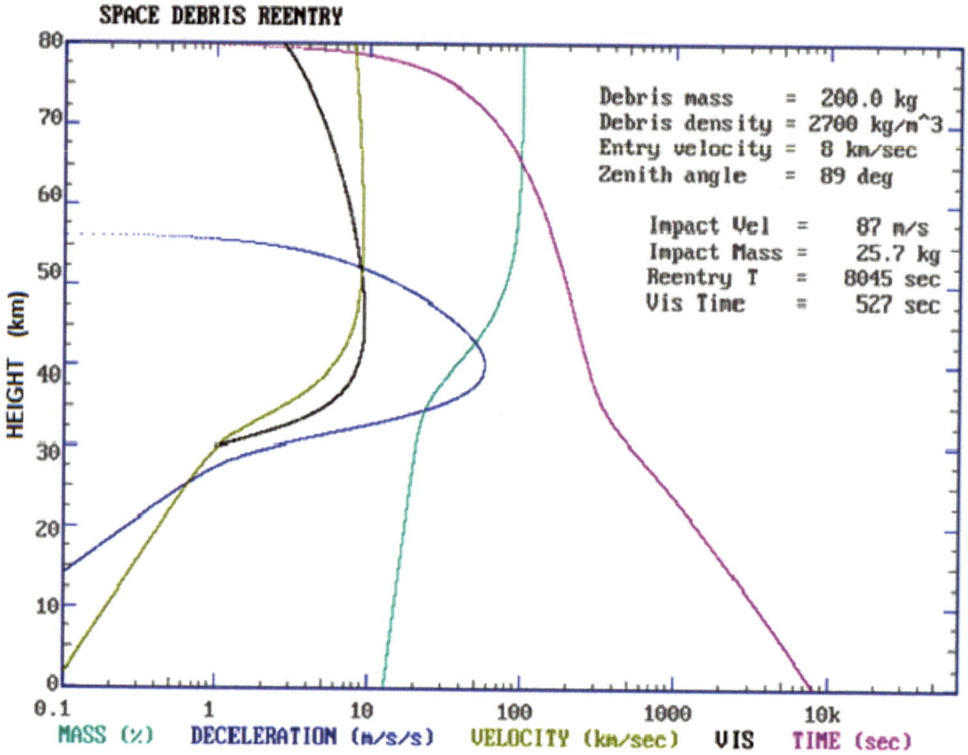

Figure 9.7 Graphical output of orbital space debris reentry mode for a large debris fragment.

(since a log scale cannot display negative values). The spurious large magnitudes that appear at low altitudes are suppressed below 10 km. Negative decelerations, due to the gravity term in the deceleration value are also suppressed below a value of 0.1.

Evidence of the atmospheric interaction is first indicated in the visual spectrum as the object becomes visible due to its rising temperature. This is despite very little mass loss and very little decrease in velocity (the model actually shows a small increase in velocity, but this is not real but an artifact of how the simple model includes the gravitational component of deceleration).

Mass loss occurs preferentially over velocity loss and starts to become apparent just below 70-km altitude. Velocity loss only becomes really apparent around the time of maximum deceleration at 40-km altitude. Maximum brightness also occurs around this time, although we should remember that the apparent brightness is a product of intrinsic brightness (defined as h = 100 km, and inverse altitude squared (assuming a ground observer).

This model is very simple and assumes a bulk material in a compact form (e.g., spherical). A large defunct spacecraft or rocket body will have a much lower density while it is intact and will fragment early in the ablation of reentry. Each fragment will than need to be analyzed separately.

Another graphical presentation of a much smaller (1 kg) fragment is shown in Figure 9.8.

Only 100 grams of this fragment survives, and it impacts the ground with a lower velocity (36 m/s). The reentry is only visible for 170s, but this is much longer than the meteor caused by a similar mass meteoroid. Most reentry debris pieces have impact speeds of between 20 and 200 m/s (70 to 700 kph) and so do not pose a hypervelocity impact hazard. And the majority are in the lower part of this range.

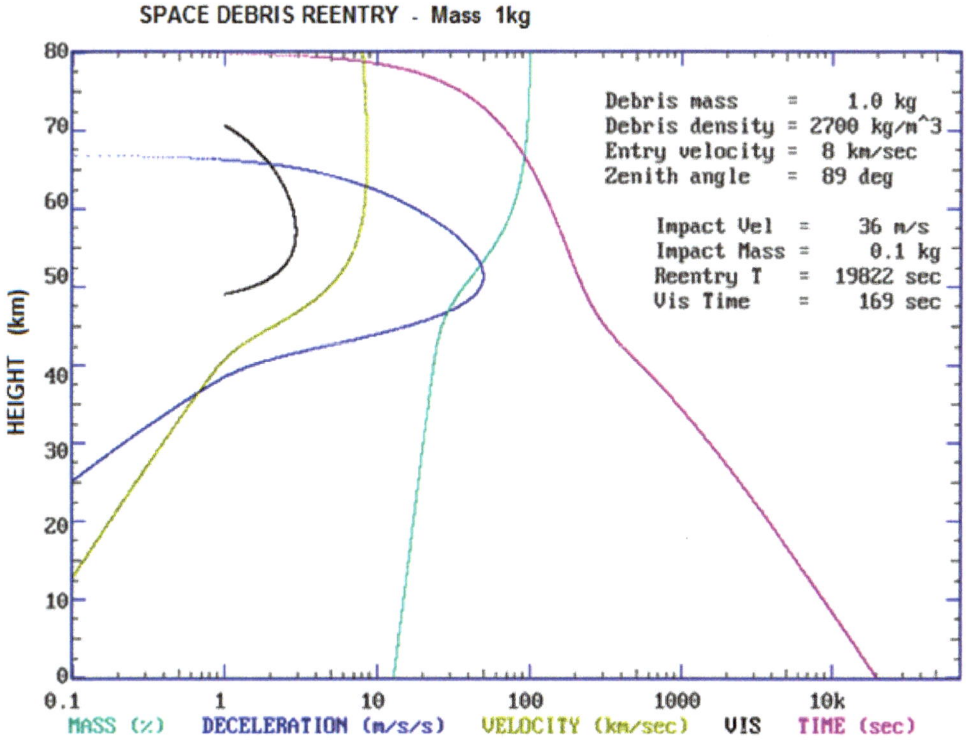

Figure 9.8 Orbital space debris reentry model for a small debris fragment.

9.9 Reentry Latitudinal Probability Prediction

Every object in Earth's orbit below about 1,000-km altitude is subject to atmospheric drag, which will decrease its energy and slowly reduce altitude in a spiral that will eventually cause it to reenter the Earth's atmosphere. Objects with an initial altitude of 1,000 km may take centuries to complete this process. However, objects at 500-km altitude will normally decay within about 10 years, and the average object at 200-km altitude will decay within a very few days.

Most of these objects will burn up in the Earth's atmosphere during reentry and thus pose no hazard to people or property on the ground. However, objects with an initial mass over about 1,000 kg (one ton) will generally not suffer total ablation and will drop fragments on the Earth's surface.

9.9.1 Reentry Predictions

Because of an imperfect knowledge of the Earth's upper atmospheric density in the future (as well as in the present), the most accurate reentry predictions have an error of at least 10% in time. This error is independent of the timescale involved. Thus if a satellite or piece of debris is due to decay within a year, the error in predicting the time of reentry will be greater than one month. On the other hand, if an object is due to decay in the next 24 hours, the best prediction can only specify the actual time to within about two and one-half hours. Since such an object orbits the Earth in less the 90 minutes, the prediction of an actual uncontrolled reentry location is virtually impossible.

Of course we can predict the orbital path of an object very accurately for a few weeks, and knowledge of this path can be used to exclude large areas of the Earth's surface from any reentry danger. It is only locations underneath the orbital path (and a few tens of kilometers each side of the path) that could lie within a hazard zone.

For long-term predictions, there is only one orbital parameter that can guide any locational forecast, and that is the inclination of the orbit. This is the only orbital parameter that changes very little over long periods of time.

9.9.2 Latitudinal Reentry Exclusion

Figure 9.9 shows the 3D path of a satellite in a circular LEO. Note that as a satellite nears reentry its orbit will be close to circular, even though it may have initially been quite elliptical.

This satellite travels around the Earth in a plane that passes through the center of the Earth, and that is inclined at an angle to the Earth's equatorial plane. This angle is called the inclination of the orbit.

From a surface perspective we are more interested in the ground track over which the satellite passes. This is shown in Figure 9.10.

Figure 9.10 shows one complete orbit of the satellite, starting at a point where it crosses the equator traveling north. This point is called the ascending node of the orbit.

Note that the orbit is completed over a longitudinal range of less than 360 degrees. This is because, while the satellite is moving in a plane that is relatively

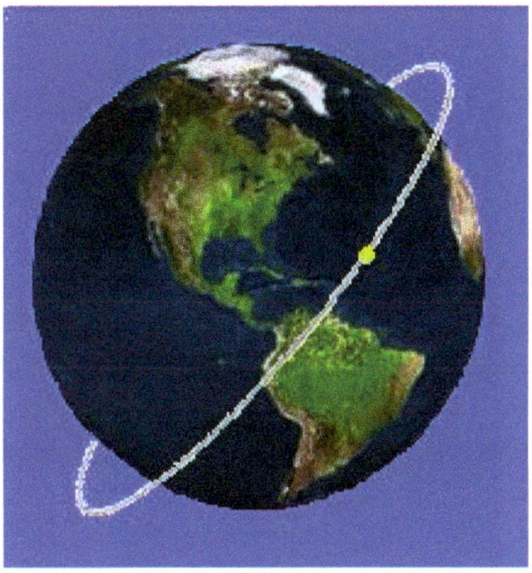

Figure 9.9 Close satellite orbit.

fixed in space (i.e., with respect to the stars), the Earth is rotating. In the roughly 90 minutes that it takes for a LEO satellite to orbit the Earth, it has rotated ~26 degrees about its axis, and so one complete orbit takes place over only degrees of longitude. For this reason, the subsequent ground tracks will not overlie the first, and a plot of many ground tracks into the future will fill most of the longitudinal space around the globe.

What we can note however, is that the maximum latitude reached by the ground track is equal to the inclination of the orbit. Points on the Earth's surface that have a latitude greater than the orbital inclination will never be overflown by the satellite, and thus are excluded from any reentry debris that may make it to the ground during reentry of this object.

Only satellites with an inclination of 90 degrees overfly every point on the Earth's surface. Satellites with inclinations greater than 90 degrees are said to be in

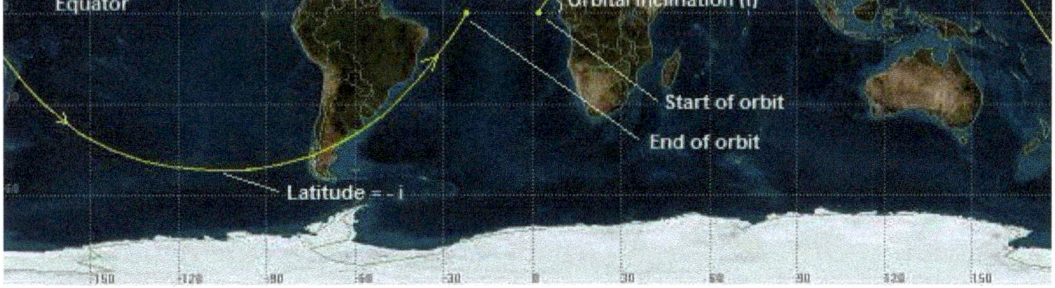

Figure 9.10 One orbit of a LEO satellite.

retrograde orbits (because they orbit in the opposite direction to the Earth's rotation), and can be considered to have an effective inclination of 180 (true inclination) degrees. Thus an object in a retrograde orbit with an inclination of 120 degrees will have a ground track that never exceeds 180 – 120 = 60 degrees in latitude.

9.9.3 Latitudinal Reentry Probability

We can thus see that the probability of reentry over a latitude greater than the object orbital inclination is zero. What about locations at latitudes less than the inclination? Because the satellite ground track generally spends more time at the higher latitudes, the probability of a reentry is dependent on latitude and we can calculate what that probability is for each latitude band.

From the spherical trig diagram in Figure 9.11 we have that:

$$\sin(\mu) = \sin(\iota)\sin(\theta)$$

where

μ is the satellite latitude;

ι is the satellite orbital inclination;

θ is the angle that the satellite has moved around its orbit (from the ascending node).

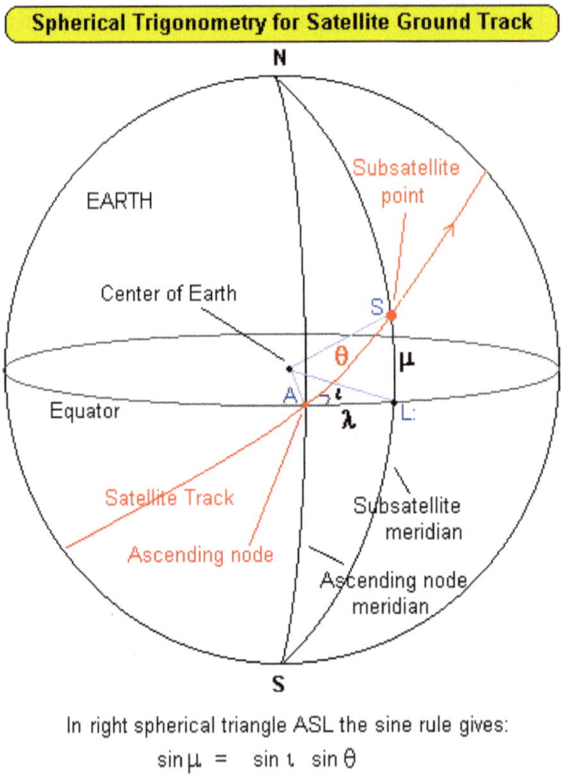

Figure 9.11 Spherical trigonometry of satellite ground track.

If we differentiate this equation we obtain:

$$\cos(\mu)\,\delta\mu = \sin(\iota)\cos(\theta)\,\delta\theta$$

giving

$$\delta\theta = \frac{\cos(\mu)\,\delta\mu}{\sin(\iota)\cos(\theta)}$$

Now the probability that reentry will occur in the latitude range from μ to $\mu + \delta\mu$ is given by

$$p(\mu\,\delta\mu) = \frac{2\delta\theta}{2\pi} = \frac{\delta\theta}{\pi}$$

The two in the numerator of the first expression on the right of the equality sign is because the satellite passes through the same latitudinal band on the descending phase of its orbit. The 2π in the denominator is the total possible orbital range (angle) over which the satellite moves. Because a satellite in a circular orbit moves linearly with time (i.e., equal angles of θ correspond to equal time intervals), the probability is the amount of time during which it is in the specified latitudinal band as a fraction of the total orbital time. Thus

$$p(\mu\,\delta\mu) = \frac{\cos(\mu)\,\delta\mu}{\pi\sin(\iota)\cos(\theta)}$$

To obtain a probability only in terms of latitude and inclination we must eliminate θ from the above formula. We note that

$$\sin(\theta) = \frac{\sin(\mu)}{\sin(\iota)}$$

$$\cos^2(\theta) = 1 - \sin^2(\theta)$$

$$\cos(\theta) = \sqrt{\left[1 - \left(\frac{\sin(\mu)}{\sin(\iota)}\right)^2\right]}$$

giving

$$p(\mu\,\delta\mu) = \frac{\cos(\mu)\,\delta\mu}{\pi\sqrt{\left[\sin^2(\iota) - \sin^2(\mu)\right]}}$$

This gives the probability for reentry in a latitudinal band from μ to $\mu + \delta\mu$ in the Northern Hemisphere. There is an equal probability for reentry in the Southern Hemisphere over a latitudinal range from $-\mu$ to $-\mu - \delta\mu$.

Figure 9.12 plots this probability for various orbital inclinations from 30 to 90 degrees, over a latitudinal band one degree in width.

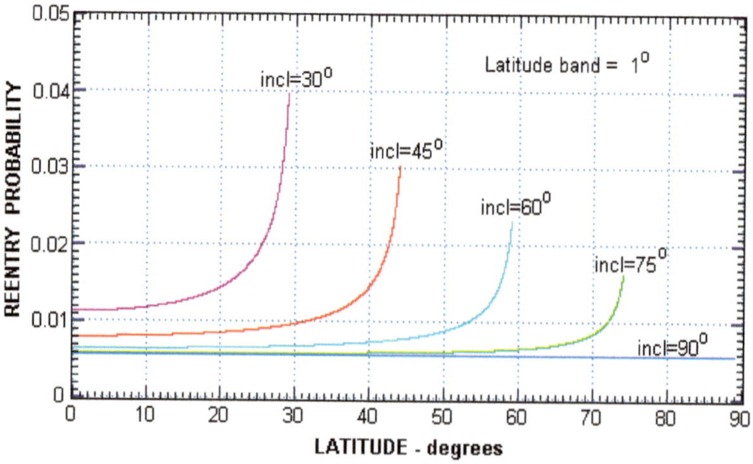

Figure 9.12 Reentry probability as a function of satellite inclination and ground latitude.

Note that the probability of reentry occurring at latitudes near the orbital inclination is greater than it is around the equator. The exact magnitude of the increase ranges from a value of about 4 at an inclination of 30 degrees, to a value of 2.6 at an inclination of 75 degrees. When the inclination is 90 degrees the probability is the same for all latitudinal bands. This is because when $\iota = 90°$ then $\mu = \theta$, and equal increments of time, which are equal increments of θ, are then equal increments of μ(latitude). The latitude independent probability in this case is $2\delta\mu/2\pi$ which is $1/180 = 0.056$ for a one-degree latitudinal band.

The probability values can also be displayed as a density overlay on a ground projection, as shown in Figure 8.17 in Section 8.12, for an orbital inclination of about 55 degrees.

As a final comment we would note that the reentry process spreads debris over a large area, typically measuring at least 20-km cross track by 1,000 km along track. In light of this we need to be careful how we interpret the above probabilities.

References

[1] Hawkins, G. S., *The Physics and Astronomy of Meteors, Comets and Meteorites*, New York: McGraw-Hill, 1964.

Selected Bibliography

Hawkins, G. S., *Meteors, Comets and Meteorites*, McGraw-Hill, 1964.

Opik, E., *Physics of Meteor Flight in the Atmosphere*, Interscience Publishers, 1958.

McKinley, D. W. R., *Meteor Science and Engineering*, McGraw-Hill, 1961.

CHAPTER 10

Space Debris Dynamics

10.1 Population Dynamics

The orbital space debris population is a function of debris sources and debris sinks.

Just launching a satellite into space creates dozens of space debris fragments. The last stage of the rocket that launches the satellite often remains in orbit. The protective shroud that encloses the satellite payload becomes debris. The explosive separation of the shroud, payload, and rocket body releases a lot of small debris fragments.

In orbit, operations can produce space debris. When a satellite reaches the end of its operational life it becomes a large piece of debris. Deterioration of a satellite through the action of solar UV and other environmental factors releases small pieces of debris. A prime example of this is paint flakes.

Fragmentation events are a major source in the generation of larger debris fragments. These events can be either internal or breakup events, or they may be external or collisional events. Breakup events occur when satellites or rocket bodies explosively fragment through the release of internal stored energy (e.g., batteries, unspent fuel). Accidental and deliberate hypervelocity collisions are not currently very frequent in LEO, but can have devastating consequences when they occur (e.g., the deliberate collision in 2007 accounted for about 20% of all low Earth centimetric and larger orbital space debris for many years).

Although not currently important, eventually collisional cascading will account for the majority of orbital space debris creation, at least in LEO [2]. This is equivalent to a chain reaction where one collision produces sufficient debris to cause several other collisions, and so on. Eventually the situation becomes so dominated by collisions that newly launched satellites will be destroyed within a few months of launch. This is why it is important to try and model the evolution of space debris in low Earth orbit, so that the impact of various mitigation actions can be determined.

We shall now examine various sources and sinks in more detail (see Figure 10.1).

10.1.1 Space Debris from Rocket Fuel

Solid rocket fuel is a profligate producer of space debris. Slag particles from the exhaust may range from microns to centimeters (Figure 10.2). Solid fuel rockets are ideal for the base stages of rockets that never reach space, but should not be used in rockets that will remain in space.

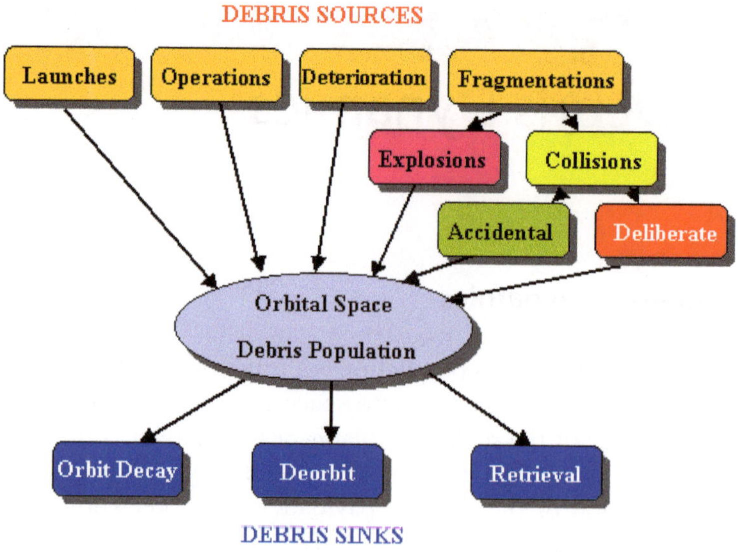

Figure 10.1 Sources and sinks of the orbital space debris population.

Figure 10.2 Slag produced in the exhaust of solid rockets. This piece is almost 50 mm (2 in) in length, large enough to produce severe damage in a hypervelocity impact. (*Source:* NASA ODPO image.)

10.1.2 Fragmentations: Explosions

Fragmentation explosions currently account for the majority of artificial orbital space debris. These may be deliberate or accidental. The Soviets used to routinely destroy satellites at end of life to preserve secrecy. Another Russian spacecraft, Cosmos 2421, underwent a severe fragmentation producing over 250 trackable fragments.

Accidental fragmentation typically occurs due to the release of stored energy, either from unspent fuel or batteries. Hypergolic fuels are notorious. In Delta rockets the separation wall between fuel (hydrazine) and oxidizer (nitrogen tetroxide) would erode and spontaneous reaction would take place, fragmenting the upper-stage motor and casing into a major debris source.

For the first 40 years of LEO history there were approximately 250 known fragmentations.

10.1.3 Accidental Collisions

Only four accidental collisions (involving the larger orbital population) have been recorded to 2020. The current low probability of collision makes this a minor source of space debris at the present time.

The French *Cerise* satellite was improbably unlucky to have its stabilization cable severed by a piece of debris. The collision between the defunct Russian satellite *Cosmos 2251* and the operational Motorola *Iridium 33* satellite produced a large cloud of debris in LEO, and was responsible for a revision of procedures and substantial increase in conjunction assessment in US military space operations.

10.1.4 Deliberate Collisions: ASAT Tests

Antisatellite tests usually result in significant space debris production. Russia, the United States, China, and India have all conducted ASAT tests.

10.1.4.1 Soviet ASAT Tests

The Soviets used in-orbit satellites to close on the target satellite. The antisatellite satellite would then be exploded, showering the target with debris. Approximately 20 ASAT tests of this nature were believed to have been conducted in the Soviet era. No Russian ASAT test was conducted over the next three decades, although a senior Russian military general indicated that tests may soon resume due to the recent Chinese and US tests, and in fact a direct ascent ASAT test occurred in November 2021, which destroyed the defunct *Cosmos 1408* satellite.

10.1.4.2 USA ASAT Tests

The first four US ASAT tests were conducted by firing a missile from an F-15 fighter in a steep climb. The *Solwind* (P-78) satellite was destroyed on September 3, 1985. The last of these tests using an F-15 was conducted in October 1986. The program was terminated by Congress.

The last and most recent US ASAT test was conducted by an SM-3 missile launched from a US Navy *Aegis* class destroyer. The satellite destroyed was the

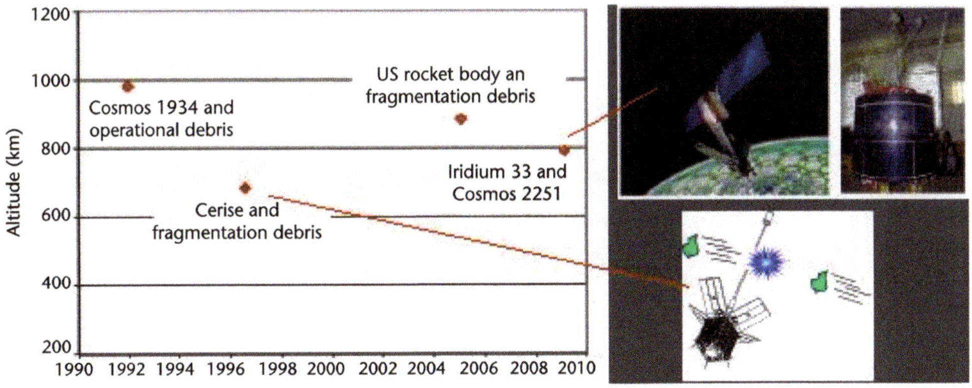

Figure 10.3 Graph showing height of accidental collisions and date of occurrence.

failed intelligence USA-193 craft in February 2008. The reason given was reentry debris mitigation, but may have also been secrecy related (Figure 10.4).

10.1.4.3 Chinese ASAT Test

On January 12, 2007 China launched a ground-based missile and destroyed an old meteorological satellite, creating the worst debris cloud in the history of space flight.

The object population in low Earth orbit before and after this collision is shown in the NASA ODPO plots of Figure 10.5 and 10.6.

(a) (b)

Figure 10.4 (a) A missile released by an F-15 aircraft in a steep climb to destroy the *Solwind* (P-78) satellite (US Air Force image), and (b) a shipborne missile fired to destroy a nonfunctional military satellite (*Source:* US Navy image).

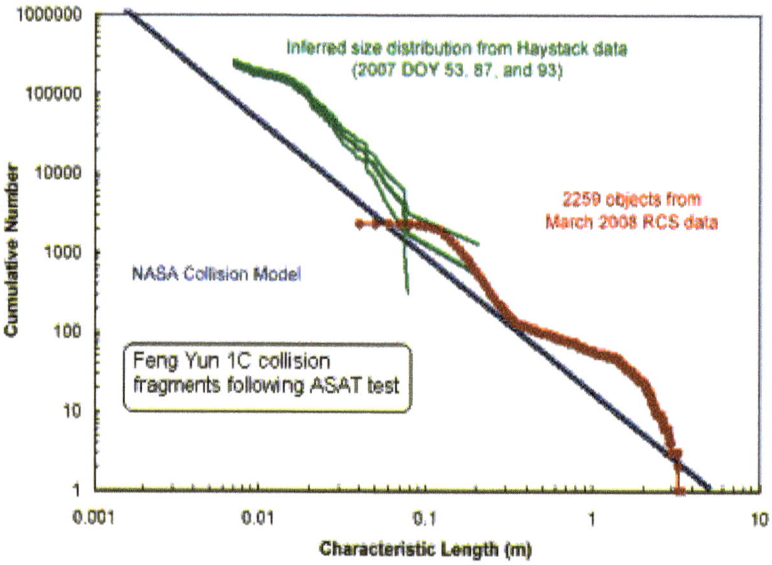

Figure 10.5 FY-1C ASAT fragments. (*From:* NASA ODPO.)

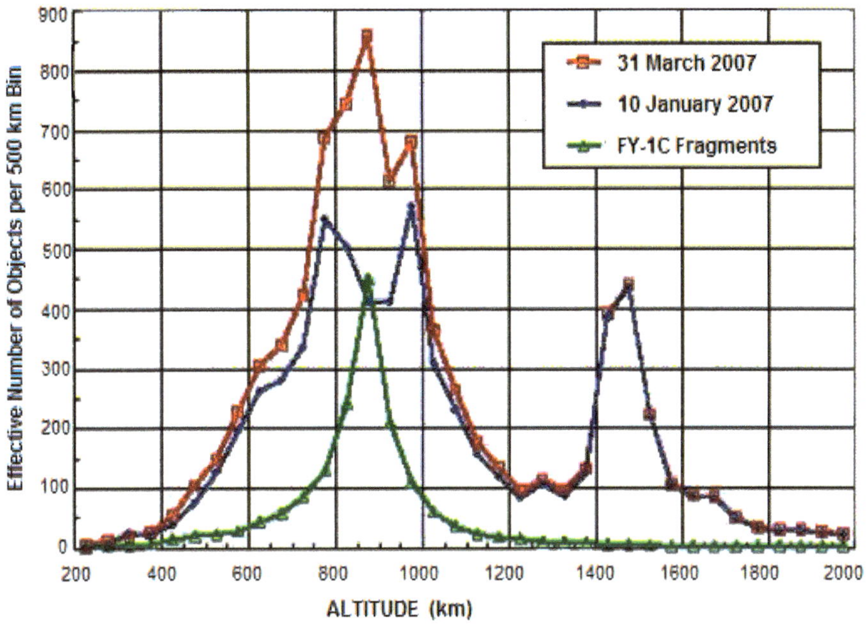

Figure 10.6 FY-1C debris (green) created during a Chinese ASAT test. (*Source:* NASA ODPO.)

10.1.4.4 Indian ASAT Test

In March 2019, India tested an antisatellite weapon in which it hit the Indian *Microsat-R* that was orbiting at an altitude of 283 km. This target satellite was placed in orbit only two months earlier. The kinetic kill interceptor was a ballistic missile defense missile. This test made India, the fourth country after the United States, Russia, and China to have tested an ASAT weapon. Indian officials stated they had started work on the interceptor following the Chinese ASAT test in 2007.

10.1.5 Orbital Decay: A Debris Sink

A small but important amount of the Earth's atmosphere extends to an altitude of 1000 km or more. This will cause aerodynamic drag on all satellites with the approximate decay times given in Table 8.1. The lower altitude figures in this table are very dependent on solar activity. This table assumes an average mass-to-area ratio of around 100 kg/m^2 (the range encountered in larger space objects is usually between 50 and 200), and a drag coefficient $C_D \sim 2$.

10.1.6 Debris Models and Simulations

United States and European researchers have developed computer models to describe the current orbital object population and allow this to be propagated into the future under various assumptions.

The inferences we draw from these models is that low Earth orbit, in the absence of deliberate debris reduction measures, will probably become unusable in a few hundred years. Exactly how many years will depend greatly on the number of continuing fragmentations and deliberate collisions we see in the near future. A war in which space assets are targeted would be catastrophic. And we must remember that it is not only space activities that will be affected; astronomy and other ground-based activities will be affected as well.

Several decades ago, the intrepid explorer Thor Heyrdahl told us that the Earth's oceans were a finite resource, and that we couldn't continue to dump waste cargo into them without experiencing problems in the future. It is sobering to realize that low Earth orbit, with its staggering volume of one trillion cubic kilometers, is likewise a finite resource that we must use wisely.

10.2 Models

The large space agencies have debris propagation models to determine the evolution of the LEO space debris population. The main NASA model is called LEGEND [5]. A team at the University of Pisa in Italy is actively working on debris evolution models [3, 4].

10.2.1 A Simple Model

A simple model developed by Farinella and Cordelli at the University of Pisa was published in 1991 [3], and shows the essential details of the collisional cascade process. Figure 10.7 gives a visual outline of the processes involved.

The model breaks the orbital population into two groups:

1. Large satellite bodies with typical masses on the order of hundreds of kilograms and cross-sectional areas on the order of a few square meters;
2. Much smaller debris pieces with sizes around a centimeter and masses of a few grams.

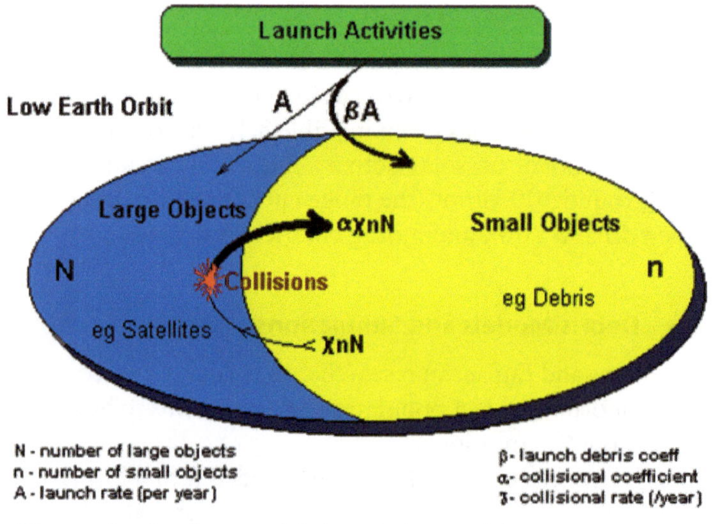

Figure 10.7 Diagram for simple space debris evolution model.

It is assumed that the only significant collisions that occur are between the large satellites and the small debris pieces. Although satellite-satellite collisions are possible, the smaller number of satellites makes the probability of this type of collision small. Debris-debris collisions are also of smaller probability because collision rate is dependent on size. If a body presents a small cross-section to collision then such an event is less likely.

Collision rate is proportional to both the number of satellites and the number of debris fragments. The proportionality constant for this event is signified by χ. For each such collision it is assumed that α debris fragments are generated.

On launch it is assumed that one satellite and β pieces of debris are placed in orbit.

Putting this all together, we have a model that is specified by two first-order coupled differential equations, with four variable coefficients and two initial or boundary conditions:

$$\frac{dN}{dt} = A - \chi nN$$

$$\frac{dn}{dt} = \beta A + a\chi nN$$

where

A = annual satellite injection rate;

n = debris numbers (small bodies);

N = satellite numbers (large bodies);

χ = collision rate;

β = primary debris generation coefficient;

α = secondary debris generation coefficient.

10.2.2 Constants and Coefficients

If we take our start as around the year 2000, then we can assume about $N = 2{,}000$ active satellites and $n = 50{,}000$ debris pieces (greater than a few centimeters in diameter). These are our two initial values or boundary conditions.

The values of the four constants or coefficients will be initially assumed to be:

- The annual insertion rate of satellites into orbit, less the number that decay is taken as around 100 [A = 100/yr];
- Each satellite insertion is assumed to result in an average of 30 primary debris fragments [$\beta = 30$];
- The collision rate is taken as 3×10^{-10} m^{-2} yr^{-1}. See Figure 10.8 for further information [$\chi = 3 \times 10^{-10}$];
- Each collision is assumed to produce about 2,000 debris pieces (of around 1 cm in size), these are considered secondary debris fragments [$\alpha = 2{,}000$].

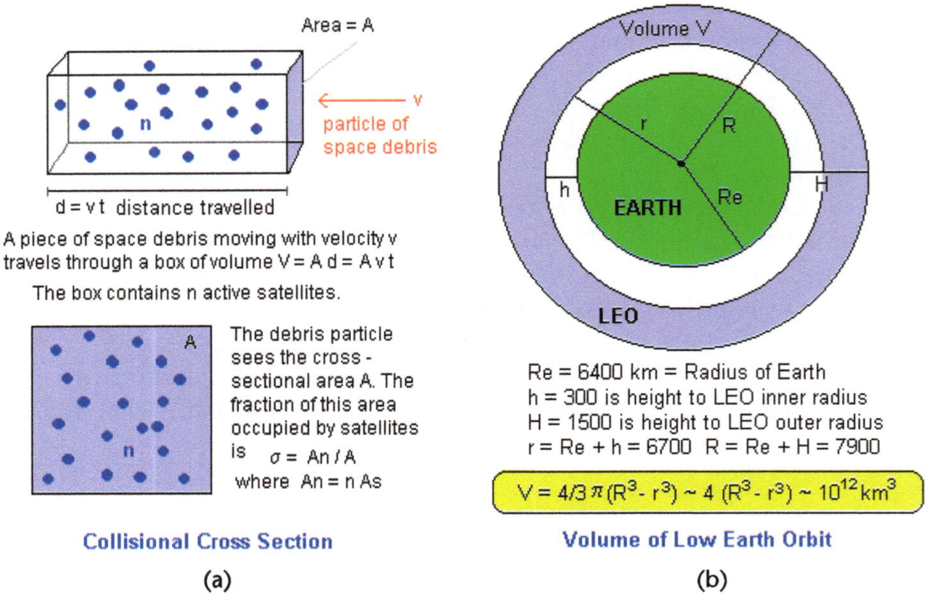

Figure 10.8 Details for collisional cross-section calculations.

10.2.3 Calculating Collision Rate

The collision rate can be estimated by considering the collisional cross-sectional area (Figure 10.8).

Consider a debris particle moving at a velocity v passing through a cross-sectional area A orthogonal to its direction of travel. In a time t it will move a distance $d = vt$, which means that it moves through a volume $V = Ad = Avt$. In this volume there exists n active satellites. These satellites will present a total area to the particle of $A_n = nA_s$, where A_s is the cross-sectional area of a single satellite (~1m^2). The collisional cross-section for the particle is then given as the fractional area that the satellites present:

$$\sigma = \frac{A_s}{A}$$

Now the number of satellites n in volume V is given by:

$$n = \rho_s V$$

where ρ_s is the satellite density in LEO. This is given by:

$$\rho_s = \frac{N}{V_{\text{LEO}}}$$

and the total volume of LEO is about 10^{12} cubic kilometers. This is 10^{21} cubic meters.

Now the collisional coefficient χ, which is the number of collisions per unit time per unit satellite is:

$$\chi = \frac{\sigma}{N}$$

or using the above information:

$$\chi = \frac{A_s v t}{V_{\text{LEO}}} \quad \left[\text{s}^{1}\right]$$

For the original simulation we consider that a satellite cross-sectional area is on the order of one square meter, and for the simulation the unit of time is one year (~30 million seconds). This gives us a value of $\chi \sim 3 \times 10^{-10}$.

10.2.4 Model Output

If the parameters in the model are kept constant, the model can be implemented in a few lines of code, with results over a time span of 500 years being shown in Table 10.1.

A graphical output for the same parameters is shown in Figure 10.9.

Note that the scale is vastly different for satellites and debris fragments.

This simple model can still be usefully used to examine how the space debris population varies under different assumptions. There are four parameters that may be varied in turn, or two together, and the results plotted. Take note of different output metrics, such as time to peak satellite numbers, or tenfold increase in debris numbers. The model becomes even more realistic if you arrange for the parameters to vary with time. For instance, we are currently seeing a considerable increase in satellite launch/injection rate into LEO, but changes in launch technology have made this process more efficient in that fewer debris pieces are produced for each satellite placed into orbit.

Table 10.1 Orbital Debris Evolution Model Output

Year	*Sats*	*Debris*
0	2000	50,000
50	6990	219,371
100	11942	465,719
150	16798	903,299
200	21425	1,799,847
250	25480	3,840,827
300	28079	8,791,400
350	27297	20,506,156
400	21112	43,026,092
450	12199	71,001,504
500	6295	92,959,664

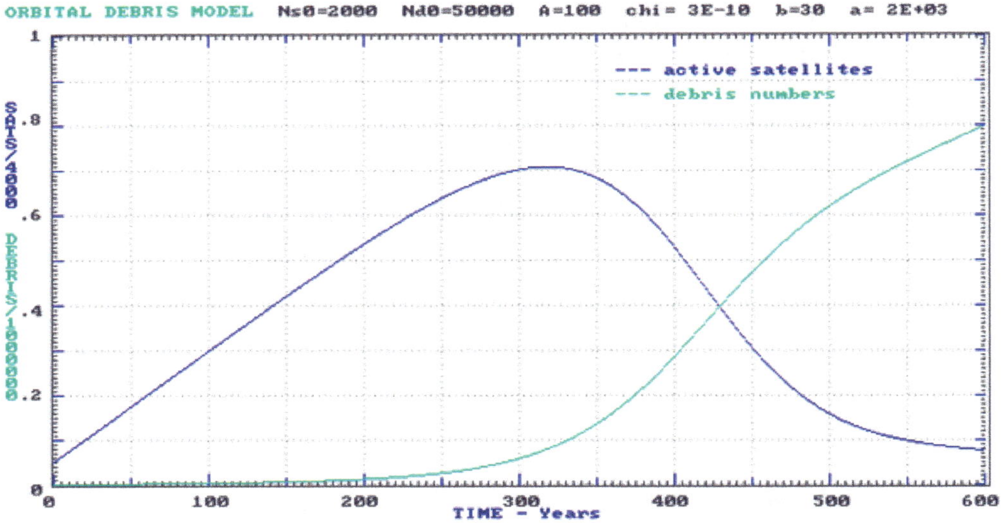

Figure 10.9 Graphical output for space debris evolution model.

10.2.5 Extensions

The model described above has some limitations or physical unrealities that can be rectified relatively easily while still retaining the essential simplicity of the original. The new physics is shown in Figure 10.10.

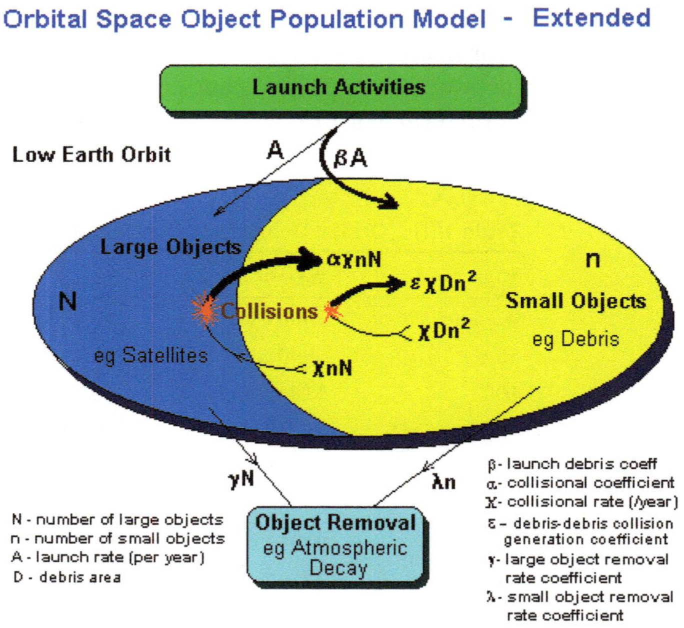

Figure 10.10 Details for extended space debris evolution model, which includes debris-debris collisions.

The new equations for this extended model are given by:

$$\frac{dN}{dt} = A - \chi nN - \gamma N$$

$$\frac{dn}{dt} = \beta A + a\chi nN + \varepsilon\chi Dn^2 - \lambda n$$

Note that collisions are now permitted between debris particles (i.e., debris-satellite and debris-debris). However, the latter class only becomes apparent after a significant fraction of the large population has been depleted.

The extended model also allows for more realistic debris sinks. Large and small objects can be removed separately. While there will be a removal of both types of object through natural causes (e.g., atmospheric drag), which will be proportional to the number of objects present in each population, mitigation effects (say using laser removal on small objects) are more likely to be a constant rate.

In the real world, each collision produces a pseudo-continuum of sizes with the size distribution given by the form: $N (> d) = kd^{-S}$ where N is the number of debris pieces exceeding a diameter of d, and k and s are constants.

After a large number of collisions in any population similar to this (e.g., asteroids and meteoroids) it is found that s tends to a value of 2.

Figure 10.11 shows the empirically determined population size distribution of low Earth orbital debris in 1998 [7]. Even after this short of a time period the exponent is close to −2.

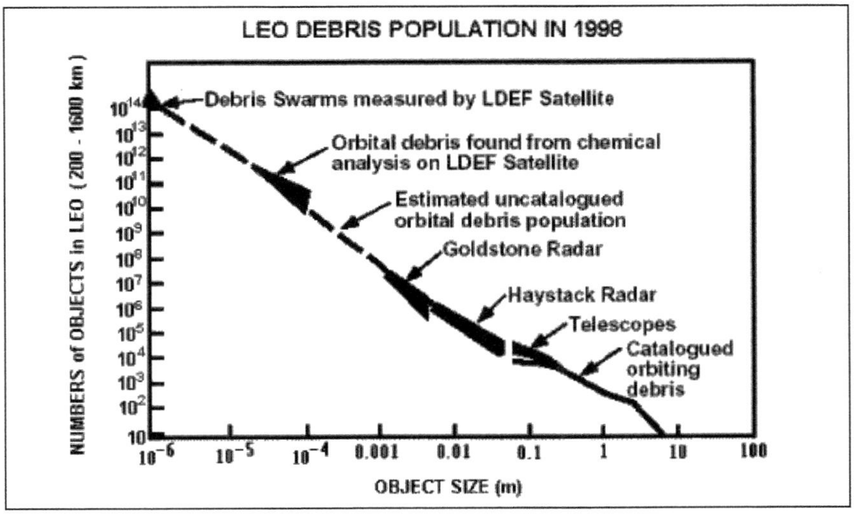

Figure 10.11 The number of debris objects in low Earth orbit (between 200 and 1,600 km altitude) estimated by the NASA Orbital Debris Program Office as of 1998 [7]. This distribution may be approximated by the empirical expression $N(> d) = 800\ d^{-1.9}$ where d is in meters. Many different types of sensors, as indicated, are required to provide the data covering such a large size range.

The output from this version of the extended model is shown in Table 10.2. The constants, coefficients, and boundary conditions are:

Start year = 2000;

Initial number of satellites = 2000;

Initial number of debris fragments = 50,000;

$A = 100$;

$\chi = 3E\text{-}10$;

$\alpha = 2000$;

$\beta = 50$;

$\varepsilon = 1000$ [debris-debris collision generation coefficient];

$\delta = .0001$ [debris removal coefficient];

$D = .0001$ [debris area].

The column labeled microdebris is the predicted number of objects between 100 nm and 1 micron in size assuming an inverse square law size distribution, and a current population of 10-cm-sized debris of 10,000.

The above implementation does not deal with debris removal to the extent that the model allows and could be vastly improved. There is considerable scope to use this model to investigate various mitigation techniques, and some of the events that have occurred in the actual evolution plot of space debris growth to date (see Figure 10.12).

Table 10.2 Extended Orbital Evolution Model Output

Date	*Sats (num)*	*Debris (num)*	*Microdebris (num)*
2000	2000	5.0E+04	1.0E+16
2050	6986	3.3E+05	6.5E+16
2100	11913	7.2E+05	1.4E+17
2150	16689	1.4E+06	2.8E+17
2200	21110	2.8E+06	5.6E+17
2250	24670	6.0E+06	1.2E+18
2300	26181	1.3E+07	2.7E+18
2350	23565	2.9E+07	5.9E+18
2400	16100	5.7E+07	1.1E+19
2450	8206	9.1E+07	1.8E+19
2500	3960	1.3E+08	2.5E+19
2550	2324	1.7E+08	3.5E+19
2600	1525	2.5E+08	5.0E+19
2650	910	4.1E+08	8.2E+19
2700	351	1.1E+09	2.1E+20

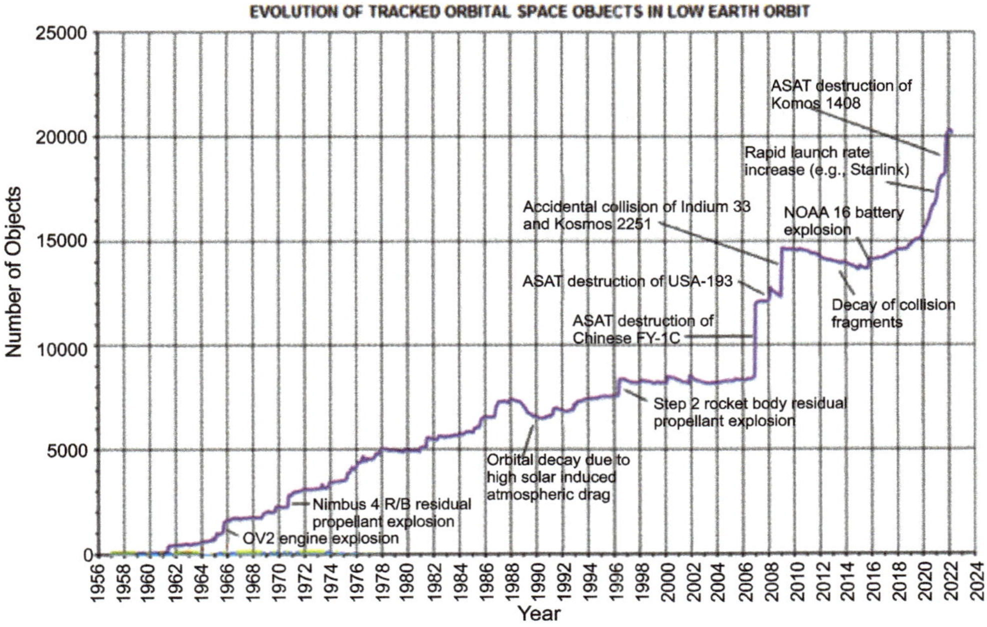

Figure 10.12 NASA ODPO graph of LEO population with time-events labeled.

References

[1] *Orbital Debris: A Technical Assessment*, National Research Council, Figure 1.

[2] Kessler, D. J., and B. G. Cour-Palais, "Collision Frequency of Artificial Satellites: The Creation of a Debris Belt," *Journal of Geophysical Research*, Vol. 83, No. A6, 1978, pp. 2637–2646.

[3] Farinella, P., and A. Cordelli, "The Proliferation of Orbiting Fragments: a Simple Mathematical Model," *Science and Global Security*, Vol. 2, 1991, pp. 365–378.

[4] Rossi, A., et al., "Modelling the Space Debris Evolution: Two New Computer Codes," *Advances in Astronautical Sciences—Spaceflight Mechanics*, 1995, pp. 1217–1231.

[5] Rossi, A., et al., "Long Term Evolution of the Space Debris Population," *Advances in Space Research*, Vol. 19, 1997, pp. 331–340.

[6] Liou, J.- C., et al., "LEGEND: A Three-Dimensional LEO-to-GEO Debris Evolutionary Model," *Advances in Space Research*, Vol. 34, 2004, pp. 981–986.

[7] Portree, D. S. F., and J. P. Loftus, Jr., "Orbital Debris: A Chronology," NASA TP-199-208856, 1999, p.3.

CHAPTER 11

Geostationary and Geosynchronous Orbits

11.1 History

In the English-speaking world, most of us were taught that the concept of the geostationary orbit originated with the famous British author Arthur C. Clarke. However, it has now been recognized that the Slovenian engineer Herman Potočnik in 1929 wrote a work in which he described geostationary and geosynchronous orbits titled *Das Problem der Befahrung des Weltraums*.

However, it was Arthur C. Clarke who popularized and expanded the concept with an article in the British communications magazine *Wireless World* in October 1945, with an article entitled "Extra-Terrestrial Relays—Can Rocket Stations Give Worldwide Radio Coverage?" (Figure 11.1).

The next author of note to advance the concept of geosats was the engineer and scientist John R. Pierce who codeveloped the traveling wave tube, which really made communication satellites possible.

As mentioned in the Preface, the first satellite put into geosynchronous orbit was *Syncom 2* in 1963 (Figure 11.2). The design was led by Harold Rosen who realized the benefits of such an orbit. The first attempt resulted in failure due to an electronics problem, but the second attempt resulted in success although the orbit was not geostationary but had a significant inclination (thus geosynchronous). It was a cylindrical spin-stabilized design and demonstrated the usefulness of this orbit for long-distance communication.

11.2 The Geosynchronous Orbit

11.2.1 The Basic Orbit

A geostationary orbit is one that lies above the terrestrial equator and that has a period exactly equal to the Earth's sidereal rotation period of 23 hours, 56 minutes, and 4 seconds, or 86,164 seconds. This is the time it takes for the Earth to rotate exactly 360 degrees. The is also known as a sidereal day. In a solar day of 24 hours the Earth rotates about 361 degrees. The extra degree is necessary to bring a given line of longitude into lining up with the Sun on successive days. The extra rotation is necessary because the Earth has been carried around its orbit by about a degree.

October 1945 **Wireless World** 305

EXTRA-TERRESTRIAL RELAYS

Can Rocket Stations Give World-wide Radio Coverage?

By ARTHUR C. CLARKE

ALTHOUGH it is possible, by a suitable choice of frequencies and routes, to provide telephony circuits between any two points or regions of the earth for a large part of the time, long-distance communication is greatly hampered by the peculiarities of the ionosphere, and there are even occasions when it may be impossible. A true broadcast service, giving constant field strength at all times over the whole globe would be invaluable, not to say indispensable, in a world society.

Unsatisfactory though the telephony and telegraph position is, that of television is far worse, since ionospheric transmission cannot be employed at all. The service area of a television station, even on a very good site, is only about a hundred miles across. To cover a small country such as Great Britain would require a network of transmitters, connected by coaxial lines, waveguides or VHF relay links. A recent theoretical study[1] has shown that such a system would require repeaters at intervals of fifty miles or less. A system of this kind could provide television coverage, at a very considerable cost, over the whole of a small country. It would be out of the question to provide a large continent with such a service, and only the main centres of population could be included in the network.

The problem is equally serious when an attempt is made to link television services in different parts of the globe. A relay chain several thousand miles long would cost millions, and transoceanic services would still be impossible. Similar considerations apply to the provision of wide-band frequency modulation and other services, such as high-speed facsimile which are by their nature restricted to the ultra-high-frequencies.

Many may consider the solution proposed in this discussion too far-fetched to be taken very seriously. Such an attitude is unreasonable, as everything envisaged here is a logical extension of developments in the last ten years—in particular the perfection of the long-range rocket of which V2 was the prototype. While this article was being written, it was announced that the Germans were considering a similar project, which they believed possible within fifty to a hundred years.

Before proceeding further, it is necessary to discuss briefly certain fundamental laws of rocket propulsion and "astronautics." A rocket which achieved a sufficiently great speed in flight outside the earh's atmosphere would never return. This "orbital" velocity is 8 km per sec. (5 miles per sec), and a rocket which attained it would become an artificial satellite, circling the world for ever with no expenditure of power—a second moon, in fact. The German transatlantic rocket A10 would have reached more than half this velocity.

It will be possible in a few more years to build radio controlled rockets which can be steered into such orbits beyond the limits of the atmosphere and left to broadcast scientific information back to the earth. A little later, manned rockets will be able to make similar flights with sufficient excess power to break the orbit and return to earth.

There are an infinite number of possible stable orbits, circular and elliptical, in which a rocket would remain if the initial conditions were correct. The velocity of 8 km/sec. applies only to the closest possible orbit, one just outside the atmosphere, and the period of revolution would be about 90 minutes. As the radius of the orbit increases the velocity decreases, since gravity is diminishing and less centrifugal force is needed to balance it. Fig. 1 shows this graphically. The moon, of course, is a particular case and would lie on the curves of Fig. 1 if they were produced. The proposed German space-stations would have a period of about four and a half hours.

It will be observed that one orbit, with a radius of 42,000 km, has a period of exactly 24 hours. A body in such an orbit, if its plane coincided with that of the

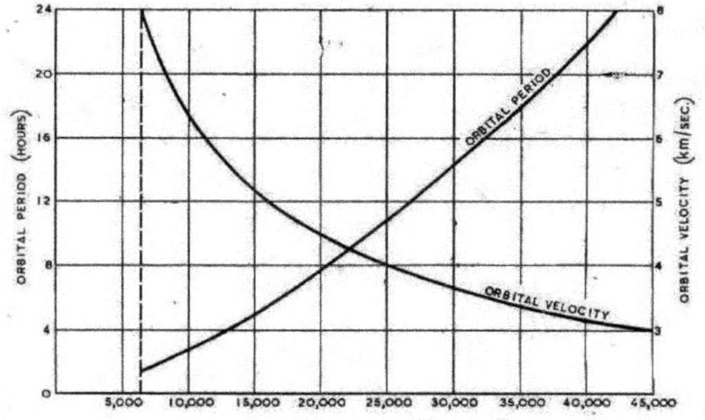

Fig. 1. Variation of orbital period and velocity with distance from the centre of the earth.

Figure 11.1 The first page of the historical *Wireless World* (British) 1945 article by Arthur C. Clarke [1], where he details the geostationary orbit.

Figure 11.2 The first geosynchronous satellite, *Syncom 2*. (*Source:* NASA.)

To compute the height of geostationary orbit we simply equate the sidereal rotation period to the formula for the period of a circular Earth orbit. Thus we have:

$$\omega = \text{sqr}\left(\frac{GM}{r^3}\right) = \frac{2\pi}{86{,}164} \text{ rad/s}$$

This gives:

$$r = \sqrt[3]{\frac{GM}{(2\pi/86{,}164)^2}} = 42{,}164 \text{ km}$$

This is the orbital radius of a geostationary orbit. The height is given by subtracting the Earth's equatorial radius of 6,378 km giving us an orbital height of 35,786, or approximately 36,000 km.

Even if a satellite is placed at the exact distance required with an orbital inclination of zero degrees, perturbative forces due to solar and lunar gravity, the Earth's nonspherical shape, and other factors will soon take it away from exact stationarity and turn it into a geosynchronous orbit, one that has the right period but with other orbital parameters, specifically inclination and eccentricity not exactly equal to one.

The two big advantages of the geosynchronous orbit is that is has a view of around one-third of the Earth's surface area, and that a ground antenna does not have to move to keep the satellite in view. The first factor is what made Arthur C. Clarke indicate that only three ground stations would be necessary for complete global communications coverage. This is not strictly true as a geostationary satellite cannot see the Polar Regions of the Earth north of 81°N or south of −81°S. And even geosat communications stations north or south of this limit generally require much larger antennae to compensate for losses at very low elevation angles. For instance, the Australian Antarctic base at Mawson uses a 20m dish, whereas a 5m might suffice on the Australian continent.

When the inclination of a geosat increases from zero its ground track becomes a figure eight, and when this deviation becomes large enough the ground antenna must follow it with a regular daily motion.

Formula and graphs for locating geostationary satellites are presented and discussed in Chapter 5 (Orbits).

11.2.2 Perturbations

Perturbing forces such as solar radiation pressure, Earth nonsphericity, and lunar and solar gravitational forces, all conspire to remove a satellite from a true geostationary orbit. Ground-directed maneuvers are required to counteract these forces and maintain the satellite in what is really a geosynchronous orbit, an orbit that is close to but not exactly geostationary. These corrections are termed *station keeping*, and are carried out to keep the satellite in a virtual box around the true geostationary point. This box usually has dimensions of less than one degree of arc, as viewed from the ground.

Station keeping has several objectives. If the satellite is used to broadcast TV or data, it means that small receiving antennae do not need to be moved to track the satellite. It also helps minimize interference between adjacent satellites using the same frequency band and it is one step toward minimizing collisions between active satellites.

Sometimes, if the station-keeping fuel on board a geosat becomes low, the inclination of the satellite (the angle that its orbital plane makes with the Earth's equator) is allowed to increase from zero. Such satellites show a figure eight motion in the sky. The satellite sweeps around the figure eight within a period of 24 hours (Figure 11.3).

Thus a ground observer is likely to see a combination of secular drift, periodic motions (e.g., figure eight), and abrupt correctional changes (due to thruster firings) in the motion of a geosynchronous satellite. However, these are usually small, and for many locational purposes it is often adequate to compute look angles assuming the simplicity of geostationarity.

When a geosat is no longer controlled from the ground, its orbit will change in several ways. The most notable initial change is an oscillation in inclination between 0 and 15 degrees, with a period of around 55 years. It will also change in eccentricity and undergo a migration to one of two more stable gravitational wells at 75 and 255 degrees longitude.

11.2.3 Getting to Geosynchronous Orbit

A geosat is not usually injected into geosynchronous orbit directly. In one scheme, a four-step process is used. It is first put into a near-circular temporary low Earth orbit. At this stage all systems may be subject to a preliminary test. Then rockets are fired, placing the satellite into a highly elliptical orbit, which has a perigee equal to the low Earth altitude and an apogee at geosynchronous orbit. At a later apogee the rockets are again fired and the perigee is raised to geosynchronous orbit height.

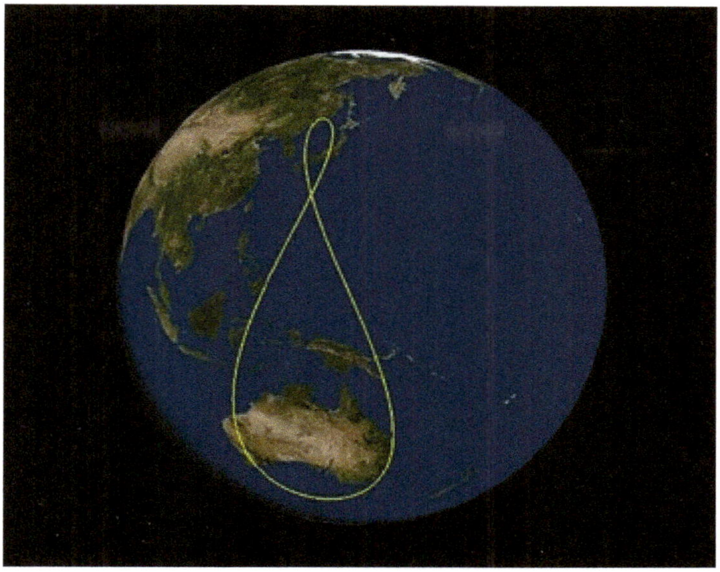

Figure 11.3 Ground track of a geosynchronous orbit with large inclination.

This type of orbital maneuver is very common and is used for planetary transfers. The elliptical transfer orbit is known as the Hohman transfer orbit (Figure 11.4) and is the most efficient way to change from one orbital altitude to another. Even when in a geosynchronous orbit, the satellite will rarely be in its designated location around the orbit. In a fourth maneuver, it must then slowly be flown around the belt until it reaches its designated place.

Sometimes it may require more than one transfer orbit to successively raise the orbit to the required final height (Figure 11.5).

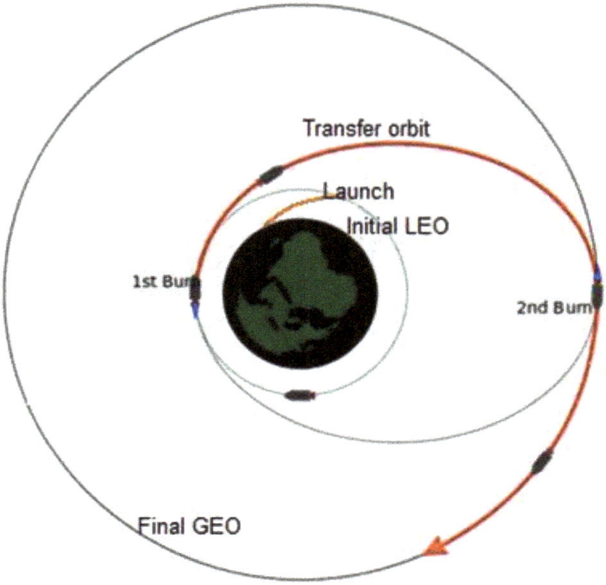

Figure 11.4 A transfer orbit used to get from LEO to GEO.

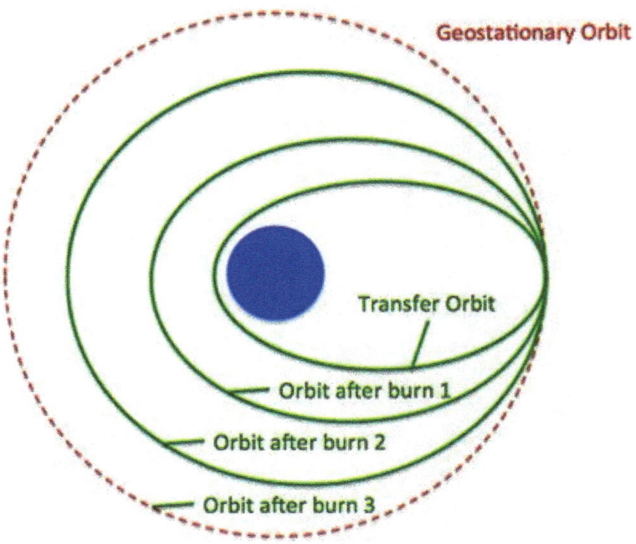

Figure 11.5 Multiple engine firings are often required to achieve geosynchronous orbit.

11.3 Geosat Classification and Operations

There are many ways in which a geosynchronous object may be specified. Two basic classifications may be according to the user, either civilian or military, and a subclassification of the owner and launching state. A second classification may be according to the purpose or use of the satellite, which would include whether or not the object was an active or defunct satellite or a piece of debris. Other classifications may concern the size and form of the object. All of these are of importance for SSA in the examination of an unknown RSO in near-geosynchronous orbit.

Due to the distance of geosynchronous objects, it is difficult to use conventional means to classify the object. Radar can usually only observe the larger geosats due to the inverse fourth-power amplitude dependence of the returned signal (the radar echo). Although optical telescopes can certainly observe geosat objects down to quite faint magnitudes, they cannot resolve such objects (although special interferometric systems may give some information in this regard). We are thus left with optical observations that can determine optical magnitude, object color, and polarization as the main ground-based tools to analyze geosynchronous objects.

For this reason, some states have resorted to space-based means to perform high-resolution observations of other states' orbital geosynchronous assets. These military satellites (milsats) are often equipped with sufficient amounts of fuel to enable them to be flown around geosynchronous orbit and observe any satellites along the way. Such maneuvers are termed rendezvous and proximity operations. RPOs are currently a hot topic in SSA. China, Russia, and the United States have all been actively involved in RPOs. There is also research into the detection of RPO satellites and how they might be separated from geosats, which frequently perform housekeeping maneuvers, which may include large orbital changes to move to a different orbital slot to keep up with changing customer requirements.

At the end of life of a geosat, guidelines specify that it should be removed out of geosynchronous to a supersynchronous orbit that is at least 300 km above GEO. This requires an operator to maintain a sufficient amount of stations-keeping fuel to conduct such a maneuver. This was not done in the early days of geosynchronous operations (Australia's first two GEO commsats were not so moved), and it is also impossible if a satellites fails before its nominal end of life (either due to communication failure or a station keeping system failure).

11.4 Imaging Geosats

Imaging a geosat basically depends on its location and its brightness. It must obviously be above the horizon of the observer. Apart from this is must be bright enough to be within the imaging capabilities of the imaging instrument. Most of the time a telescope is necessary to give sufficient collect power to observe the light from a geosat. However, geosats show considerable variation in brightness and some may even be visible to the naked eye around the equinoxes. The brightness of a geospace object depends on several factors. These factors include:

1. The size (surface area) of the satellite;
2. The orientation or attitude of the satellite;
3. The materials out of which the satellite surfaces are made;
4. Degradation of the surface materials due to the space environment;
5. The changing position of the Sun with respect to the satellite and observer;
6. The distance between the satellite and the observer.

Some of these factors change with time, and can lead to variations of brightness on time scales from minutes to years throughout the year and throughout the observing night.

The size of an active satellite is dependent on its task, the amount of equipment it must carry, the amount of power it must develop to power its equipment, and the size of external antennas or sensors. However, objects other than active satellites inhabit the geosynchronous orbit. At the end of its life, current guidelines specify that a geosat should be moved to a supersynchronous orbit at least ~300 km above geosynchronous orbit. As previously mentioned, this has not always been done in the past, and such injection maneuvers are not always successful Thus, quite large debris objects are found in geosynchronous orbit. During the lifetime of a geosat a fragmentation event may occur, which leads to many small pieces in orbit. Such an event may be due to a collision but is more likely the result of an internal explosion due to stored energy.

The end result is that there are geosynchronous space objects with sizes from millimeters and smaller, to a few tens of meters. This size diversity results in brightnesses that range from below detectability by even the largest telescopes (>mag 25) to incredibly bright objects (mag –4) during times of a high-intensity glint. Observers must consider the object and the time of observation to evaluate the potential range of brightness that they can expect. It may require a change of filter to record a satellite during a time interval when the brightness changes by more than a few magnitudes, to avoid saturating a CCD camera. The mean brightness of a satellite will determine the exposure required for a good reducible image.

As the satellite is further away from an observer's central meridian, the range of possible solar phase angles will increase, and the distance to the satellite will also increase. This latter change has been shown to be negligible and does not normally need to be corrected, as it is generally smaller than the photometric noise, and other factors. However, the possibility of larger phase angle changes will need to be considered.

The time of year during which satellites are observed is an important consideration for many purposes. Light curves in general are more stable around the solstices, but show considerable change around the equinoxes. This is important if filter spectrometry is to be conducted. If relative spectral differences are important to a project, it is imperative that the different spectral images (using different filters) be obtained over a short enough time frame that the total brightness of a satellite shows no significant change. Determination of a light curve to show the (time) spectrum of temporal fluctuations is a good preliminary project to determination if and over what time period filter spectroscopy is possible.

Temporal fluctuations will also be totally different for an active stabilized satellite, and an uncontrolled and drifting piece of space debris that may be spinning and/or tumbling, or a combination of various motions about several axes. A light curve taken over a short period of time is quite useful in determining such motions. This requires many images to be taken at short intervals.

In summary, the brightness of a geosat is an important parameter (Figure 11.6). It is sometimes the only observable parameter, except for the two sky angles that

Figure 11.6 Image of *Chinasat 10* taken by Arie Verveer with the ASA Skynet telescope at Meckering, Australia. The geosat is a point image because the telescope tracking was stopped. It is also much brighter than the nearby stars that appear as trails. To measure the brightness of the satellite, it must be compared with nearby stars in a tracked image exposed for the same time.

characterize its position in the sky. Changes in brightness, both temporally and spectrally, can be used to determine the nature (size, shape, and surface materials), and even the type of an unresolved geosynchronous space object. Degradational changes in surface materials may also be studied, and these may sometimes be used to give an estimate of age (time in orbit).

It can thus be seen that geosat brightness is very important in SSA for the identification of such objects. This is discussed in the next sections.

11.5 The Brightness of Geosats

11.5.1 Overview

Objects in geosynchronous orbit shine by reflected sunlight. They generate no light of their own, except in the far infrared, where they are visible through their own thermal radiation, particularly if they carry high-power active systems.

The total solar flux at Earth's orbit is about 1,370 watts per square meter, and most of this is in the visible spectrum. At the Earth's orbital radius this flux gives the Sun an astronomical magnitude of −27, the brightest object in the sky.

The brightness of a geosynchronous satellite as seen from the surface of the Earth depends upon several factors. First, on how much sunlight the satellite intercepts, and if it is of an irregular shape, how it is oriented with respect to the Sun. Second, it depends on how much of the illuminated surface can be seen by the terrestrial observer. Third, it depends on the nature of the surface of the satellite. Some materials reflect or scatter more light than others. Not only is the type of material important, but the roughness of the surface is also an important factor. In the case of an irregularly shaped satellite with different parts made of different materials, the problem of brightness becomes quite involved. Aging of material surfaces with time will change their reflectivity, often referred to as their albedo.

Other factors that may affect the brightness are the distance of the satellite to the observer, whether the reflection is diffuse, as is usually the case, or whether it is specular (mirror-type reflection), as happens at certain times of year, around the equinoxes, producing what are called glints. It is also necessary to consider if the satellite is in eclipse, when the Sun is hidden from the satellite by the Earth. Again, this will occur at certain times around the equinoxes.

11.5.2 A Simple Geosat Brightness Model

Consider Figure 11.7. A spherical uniform satellite at a distance r from the observer is 50% illuminated by the Sun (as seen by the observer). If the satellite has a diameter d then the area of the disc projected to the observer illuminated by the Sun is $A_i = \pi d^2/8$.

Now the solar flux from the Sun is L_s (~1380 W/m^{-2}), so the intercepted power is $P_i = L_s A_i$.

If we define the albedo (α) of the body as the fraction of intercepted power that is diffusely reradiated (i.e., radiated in all directions), then the reradiated power is $P_r = \alpha L_s A_i$.

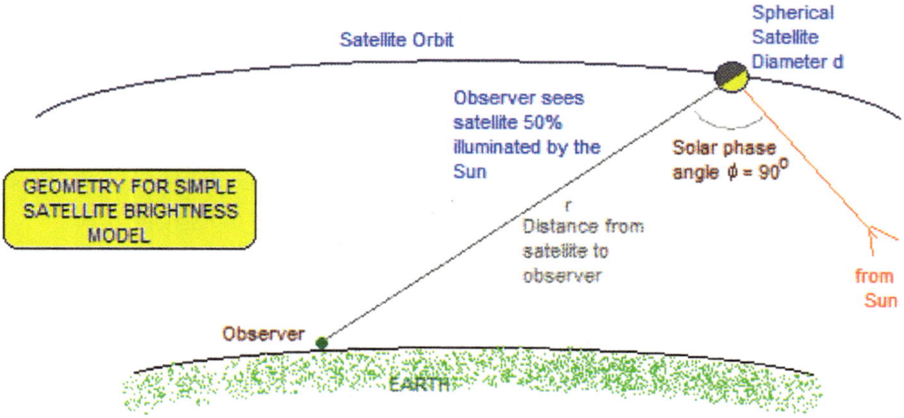

Figure 11.7 Simple satellite brightness model geometry.

The flux observed by the ground observer then becomes

$$L_o = \frac{P_r}{4\pi r^2} = \frac{\alpha L_s A_i}{4\pi r^2}$$

Substituting in the relevant values gives a ratio

$$\frac{L_o}{L_s} = \frac{\alpha d^2}{32 r^2}$$

Now the formula that defines the astronomical magnitude of an object is

$$m_s - m_o = 2.5 \log \frac{L_o}{L_s}$$

where in our case m_s is the magnitude of the Sun, m_o is the magnitude of the satellites as observed on the ground, L_o and L_s are the luminous flux of the satellite and the Sun, respectively (as defined above). The logarithmic function is to the base 10.

This then enables us to calculate the observed magnitude of the satellite as

$$m_o = m_s - 2.5 \log\left[\frac{\alpha d^2}{32 r^2}\right]$$

or

$$m_o = 2.5 \log\left[\left(\frac{32}{\alpha}\right)\left(\frac{r}{d}\right)^2\right] - 27$$

For a satellite of diameter 5m, a typical albedo of 0.1 and a distance of 40,000 km, we calculate a magnitude of 13.8. This is quite comparable to the observed magnitude of a medium size geosat.

The magnitude of a medium-sized communication satellite is around 14 when observed away from the equinoxes. It can rise to around magnitude 9 when observed near the equinoxes (March 22 and September 22), and during these times specular reflections or glints may produce a much brighter apparent magnitude for a short period of time.

11.5.3 Satellite Size, Shape, and Effective Area

The physical size of a satellite affects its apparent size seen by a ground observer (Figure 11.8) according to the formula:

$$\theta = \frac{d}{r} \text{ (radians)}$$

If we use the angular units of arcseconds then

$$\theta'' \sim 200{,}000\frac{d}{r}$$

where d and r are in the same units.

Most ground-based telescopes cannot resolve much more than 1" due to atmospheric turbulence. Using the above formula, a geosat must be over 200m in diameter to be resolved. No geosats of this size currently exist. This is why current telescopic observations of geosats appear as points.

A perfectly spherical satellite is totally symmetric. That is, it has an infinite number of axes about which any amount of angular rotation will produce the same view. This means that its size, and thus its effective area, are the same from any point of view. When considering its brightness as a satellite, it does not matter if it is rotating or not.

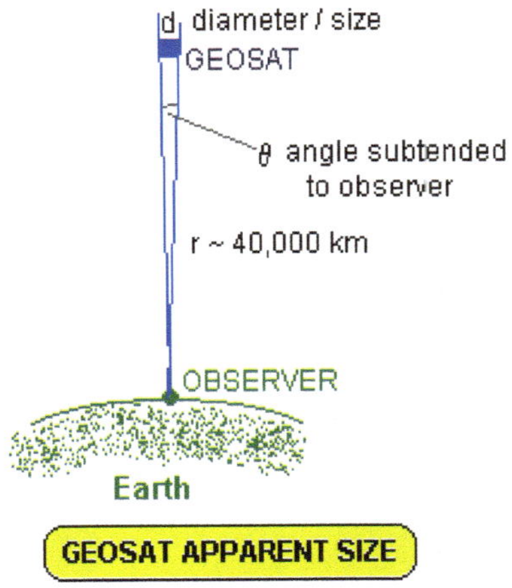

Figure 11.8 Current telescopic observations of geosynchronous space objects see them only as point sources. They cannot resolve detail or the size of the object through astrometry. Brightness therefore plays an indispensable role in estimating size and other properties of such objects. (Note: An exception to this might lie in the future development of optical interferometry, which has much greater resolving power.)

Real satellites are quite different. Two common geosat shapes are cylindrical and the so-called box-wing.

Figure 11.9 shows these two different shapes. It is apparent that the effective area for illumination by the Sun, and the effective area seen by a ground observer can show large changes as the position of the Sun, and position of the observer, change. The cylindrical satellites are normally spun about their longitudinal axis for stability, whereas the box-wing type of satellite uses three-axis stabilization of the box, and the solar panels can be rotated to intercept maximum sunlight. Cylindrical satellites have a higher degree of symmetry than do box-wings and so can be expected to display a simpler light curve, both when stabilized as active satellites, and when unstabilized after the end of life.

To compute the effective illuminated area that a physical satellite displays to an observer requires not only a knowledge of the solar phase angle, but also of the actual satellite 3D shape. This is not generally known, or is too complex for detailed calculations. Instead, statistical properties of the light curve of a satellite can be used to infer its general shape.

The largest geosynchronous satellite is believed to be a US signals intelligence satellite with an unfurled skeleton parabolic dish 100m in diameter. Because it is only a skeleton dish, it has a low effective area for illumination and thus is not a particularly bright satellite. The largest commercial dish in orbit is believed to be just under 20m in diameter when fully unfurled.

Most communication satellites in GEO orbit have dimensions in the meters to tens of meters range. A cylindrical satellite will typically have a diameter of less than 5m and an axial length when fully deployed of less than 15m. A box-wing satellite will typically be less than 25m in length (from one solar panel tip to the other). The two Australian NBN *Sky Muster* satellites are among the largest box-wing satellites in GEO orbit. Their dimensions are quoted as 26m long, 12m wide, and 9m tall. This, as can be seen from the diagram, is by no means all solid satellite, and the effective illuminated area to the observer is very dependent upon the box and the wing orientations.

On the lower limit, small debris particles will undoubtedly exist that are submicron in size, way below the detectability limit of any ground-based sensor.

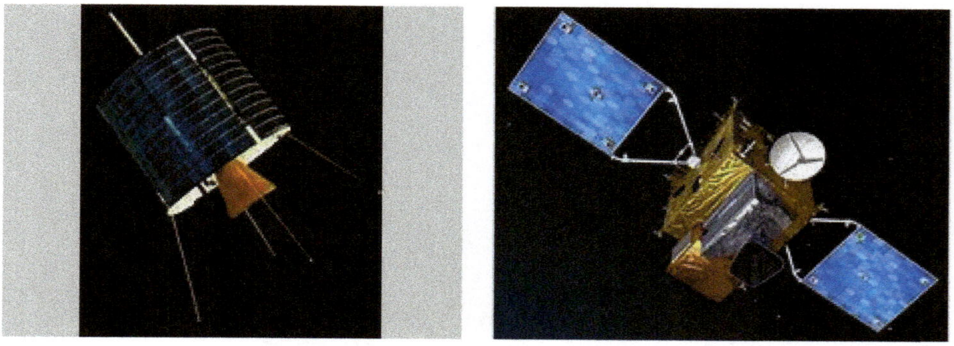

(a) (b)

Figure 11.9 Two different shapes of geosynchronous satellites: (a) cylindrical, and (b) box-wing. (*Source:* NASA.)

11.5.4 Albedo

The albedo, or reflective ability, of a satellite depends on the nature of the materials that make up its surface, and the condition of that surface. A smooth surface will have a higher albedo than a rough surface.

Albedo (α) ranges in value from 0 to 1. A value of 0 means that all the light incident on the surface is absorbed and none is scattered. The absorbed energy goes into heating the satellite and will eventually be reradiated, but in the far infrared part of the spectrum.

An albedo of 1 means that all of the energy incident upon the surface will be scattered back out to space. In diffuse scattering, the energy is reradiated in all directions (i.e., isotropically over a hemisphere). Note that this is different from specular reflection where the incident energy is also sent back into space but in one very specific direction that obeys the law of reflection (reflected ray angle = incident ray angle). See Section 11.5.9 on glints for a diagram and discussion explaining this further.

It is difficult to achieve albedos either close to 1 or nearly zero. However, some military satellites employ special coatings to reduce their albedo in the visible spectrum close to zero so that they are very difficult to observe. However, even these satellites cannot reduce their emission in the thermal infrared spectrum (8 to 14 μm), unless they are powered down. In this condition, however, they can do no useful work. They need to be powered up to perform their function and in this condition they are readily observed by an infrared telescope. Such telescopes must be located on top of high mountains or in space to avoid atmospheric absorption of thermal infrared radiation.

The albedo of most surfaces is generally less than 0.5, sometimes significantly less. Where the albedo of a space object is unknown it is common to assume a value of $\alpha = 0.1$.

As material surfaces age in the space environment, particularly under the influence of strong ultraviolet light from Sun and particle radiation, their albedos will change, usually showing a decrease. At any one time the albedo of a surface will also vary with wavelength, and it should be noted that the albedo for diffuse scattering (α_d) may be quite different than the albedo observed in specular reflection (α_s). See Section 11.5.9 on glints.

11.5.5 The Effect of Distance

From Section 5.2.3 we have a formula for the distance from a ground observer to a geostationary satellite as

$$r = \sqrt{\left[R^2 + (R + h)^2 - 2R(R + h)\cos(g)\right]}$$

where $R = 6{,}378$ km is the equatorial radius of the Earth, h is the height of geostationary orbit above the equator ($R + h = 42{,}165$ km), and g is the great circle angle between the observer and the point on the equator directly beneath the satellite. This can be computed from the equation

$$\cos(g) = \cos(\delta\text{Lon})\cos(\text{Lat})$$

where Lat is the latitude of the observer and δLon is the difference between the longitude of the observer and the longitude of the satellite.

Because h is considerably larger than R, the variation in the distance or range to a geosat is not all that great, irrespective of where the geosat may be in the sky. Table 11.1 lists various parameters, including range, of geosats at 5-degree longitude intervals visible from a midlatitude site.

The slant range is the relevant parameter here. It varies from around 41,200 km at low elevation angles to 36,900 km when the satellite lies on the central meridian (and is closest to the observer). At an elevation angle of 15 degrees, which is the lower limit for many telescopes, the range is around 40,200 km.

Thus for telescopic observations we have a variation in luminous flux from

$$L_{\text{max}} = \frac{L_{\text{ref}}}{r_{\text{min}}^2} \text{ to } L_{\text{min}} = \frac{L_{\text{ref}}}{r_{\text{max}}^2}$$

where L_{ref} is an arbitrarily chosen reference flux (actually the flux at unit distance from the satellite).

Table 11.1 Geosat Look Angles for a Specified Observer Site

Geostationary Satellite Look Angles
Observer LAT = –31.4 LON = 115.7 Zadko@Gingin

SAT-LONG (deg)	*ELEVANGLE (deg)*	*AZANGLE (deg)*	*HA (hrs)*	*DEC (deg)*	*SLANT (km)*	*GND (km)*
41.0	4.3	278.1	5.471	4.62	41199	8570
51.0	12.9	283.8	4.784	4.73	40278	7637
61.0	21.5	290.2	4.081	4.83	39410	6728
71.0	29.8	297.7	3.362	4.93	38628	5859
81.0	37.7	306.9	2.628	5.02	37960	5056
91.0	44.7	318.5	1.881	5.09	37433	4356
101.0	50.1	333.2	1.124	5.14	37071	3820
111.0	53.1	351.0	0.361	5.16	36888	3526
121.0	53.0	10.1	–0.405	5.16	36894	3535
131.0	49.9	27.7	–1.168	5.13	37087	3845
141.0	44.3	42.2	–1.924	5.08	37460	4393
151.0	37.3	53.7	–2.671	5.01	37995	5100
161.0	29.3	62.7	–3.404	4.92	38670	5908
171.0	21.0	70.2	–4.122	4.83	39458	6780
–179.0	12.4	76.5	–4.824	4.72	40330	7691
–169.0	3.8	82.2	–5.510	4.61	41254	8625

Using the definition of astronomical magnitude

$$m_1 - m_2 = 2.5\log\frac{L_2}{L_1}$$

we can compute the difference in magnitude due to the variation in range to the geosat as

$$\begin{aligned}\Delta m &= 2.5\log\left[\frac{L_{\max}}{L_{\min}}\right] \\ &= 2.5\log\left(\frac{r_{\max}}{r_{\min}}\right)^2 \\ &= 5\log\left(\frac{r_{\max}}{r_{\min}}\right) \\ &= 5\times 0.037 \\ &\sim 0.2 \text{ magnitudes}\end{aligned}$$

This value is very small and will generally be lost in the noise of variations due to other causes. It may in general be neglected. If, however, it is desired to correct for distance, we would suggest that a distance-normalized flux be computed with respect to the flux observed on the equator directly below the satellite (i.e., at a range of $h = 35{,}787$ km). This is of course the brightest magnitude that the satellite will have from any point on the Earth's surface.

The magnitude correction will then be given by

$$m_{\text{corrected}} = m_{\text{observed}} - 5\log\left(\frac{r}{35{,}787}\right)$$

where r (km) is the range to the satellite with an observed magnitude of m_{observed}.

11.5.6 Satellite Orientation or Attitude

An active satellite must be kept in a fixed orientation or attitude so that its antennae of sensors continually point to the same point on the Earth. At the same time its solar panels need to collect as much sunlight as possible.

The two different types of satellites discussed previously, cylindrical and box-wing, tackle their attitude control in totally different ways.

Cylindrical satellites are rotated about their axis of symmetry to achieve a stable attitude in space. Often an end platform on which are mounted the antennae is derotated from the main cylinder. Box-wing satellites use reaction wheels to achieve stability. In each case the main objective is to maintain precise ground pointing for the antennae, and/or any sensors imaging the ground. In the case of a cylindrical satellite, the orientation of the solar panels are fixed and so their interception of maximum solar energy is compromised. This means that more solar panels must be used to supply the required power for satellite operation. In the case of the box-wing satellite, the box is stabilized to ensure accurate antennae and sensor pointing. The wings, or solar panels, are then rotated to achieve maximal interception of solar energy.

The accurate ground pointing required of all active communication satellites means that, for cylindrical satellites, the effective area of the satellite seen by a ground observer is constant. The only factor that then affects brightness is the position of the Sun relative to the satellite. This will be true for the box in a box-wing satellite, but it will not be true for the solar panels. These are rotated to try and ensure maximal illumination by the Sun, but in doing so they present vastly different areas to the observer due to foreshortening. It is important to remember this when decoding a satellite light curve.

At the end of a satellite's life, it will eventually drift into a spinning and tumbling regime that will cause much more rapid changes in the observed light curves. As noted before, an unstabilized box-wing satellite is likely to produce a more complex light curve than is a cylindrical satellite. This can be significant when trying to identify unknown or lost satellites.

11.5.7 Solar Phase Angle

11.5.7.1 Computation of Solar Phase Angle

With reference to Figure 11.10, we can write the unit vectors in the directions of the Sun and the geosat in terms of the unit vectors along the coordinate system axes (boldface denotes a vector quantity with three components along the x, y, and z axes).

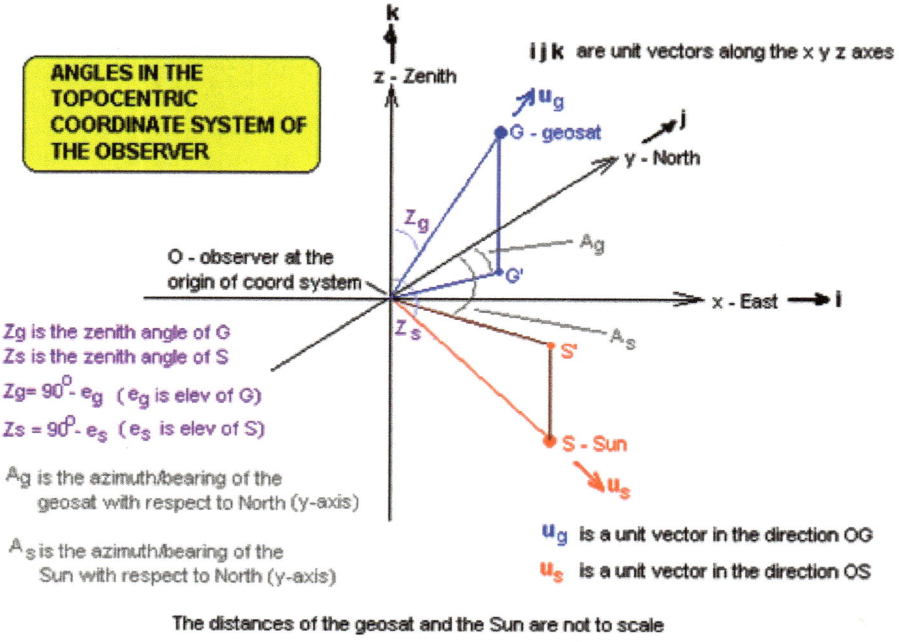

Figure 11.10 Angles and vectors of the geosat and the Sun in the observer's topocentric coordinate system.

$$\begin{aligned}\mathbf{u}_g &= u_{gx}\mathbf{i} + u_{gy}\mathbf{j} + u_{gz}\mathbf{k}\\ &= \sin(A_g)\mathbf{i} + \cos(A_g)\mathbf{j} + \cos(z_g)\mathbf{k}\\ \mathbf{u}_s &= us\mathbf{i} + u_{sy}\mathbf{j} + u_{sz}\mathbf{k}\\ &= \sin(A_s)\mathbf{i} + \cos(A_s)\mathbf{j} + \cos(z_s)\mathbf{k}\end{aligned}$$

From Figure 11.11 it can be seen that the solar phase angle ϕ can be computed by a dot (scalar) product between two unit vectors:

$$\begin{aligned}\cos(\phi) &= -\mathbf{u}_g \cdot \mathbf{u}_s\\ &= -\sin(A_g)\sin(A_s) - \cos(A_g)\cos(A_s) - \cos(z_g)\cos(z_s)\end{aligned}$$

The phase angle can thus be found with a knowledge of the satellite elevation and azimuth together with the solar elevation and azimuth. The satellite elevation and azimuth can be found in Section 5.6. Observer latitude and longitude and satellite longitude are the required inputs. Solar elevation and azimuth can be found using the information in Section 5.7. The required inputs are observer latitude and longitude and date and time.

In this analysis we have assumed that the direction to the Sun from the satellite is the same as the direction to the Sun from the Earth-based observer. This is accurate to better than 0.02 degrees. The maximum angle that the observer-satellite line subtends at the Sun is $\theta = h/r_s = 36{,}000/150{,}000{,}000 = 2.4 \times 10^{-4}$ radians = 0.014 degrees. This is more accurate than the satellite and solar algorithms used to give azimuth and elevation angles.

11.5.7.2 Variation of Solar Phase Angle

There are constraints on the possible values of the solar phase angle that are due to the fact that the observer must be in the dark (night) to observe the satellite, and to the presence of the Earth, which causes eclipses.

Two cases are shown in Figure 11.12. These represent the extreme case for satellite position. When the satellite is in the observer's central meridian (maximum elevation) the maximum solar phase angle is governed by the fact that the observer must be in the dark. Now sunset occurs when the solar elevation (at the observer's

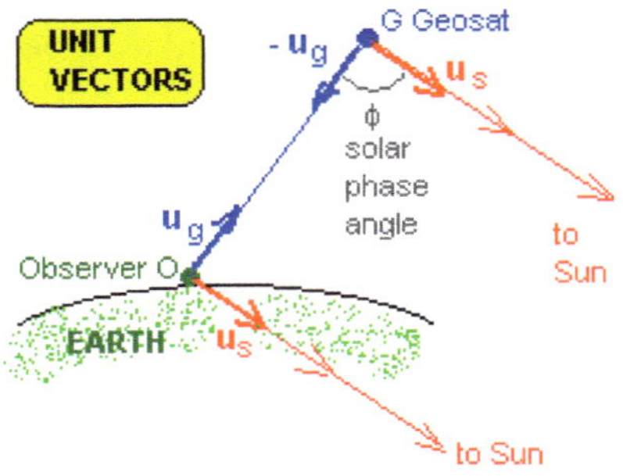

Figure 11.11 The unit vectors needed to compute the solar phase angle.

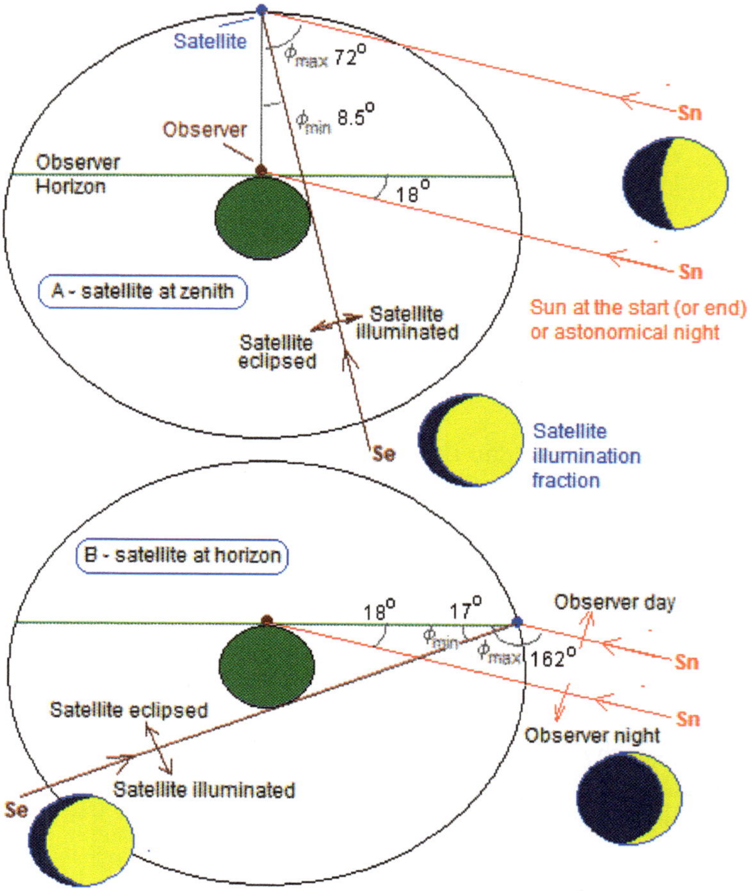

Figure 11.12 Diagrams showing the constraints on the minimum and maximum values of the solar phase angle for the extreme positions of a satellite.

location) is zero (actually a little below zero so that the whole of the solar disc is below the horizon). However, it does not get fully dark (what is called the start of astronomical night) until the solar elevation reaches −18 degrees.

For normal viewing we will be restricted to astronomical night. However, for very bright satellites we may be able to observe before this time in the evening, and after this time in the morning, in what is known as astronomical twilight, when the solar elevation lies between −12 and −18 degrees.

If we restrict ourselves to astronomical nighttime, we can see from the diagram that for an overhead satellite the solar phase angle will be restricted to between 8.5 and 72 degrees. The lower number is due to the presence of the Earth (which subtends 17 degrees from geosynchronous orbit). The upper limit is to ensure that the sky is sufficiently dark for observations to be made. In both these cases, the fractional area of the disc seen by the observer that is illuminated is well in excess of one half.

For a satellite on the observer's horizon, the solar phase angle can range from a minimum of 17 degrees when the satellite is in the ecliptic plane (8.5 if it is not), and 162 degrees. The portion of the disc seen by the observer as illuminated can thus range over a much larger fraction. In the worst case scenario, with a phase angle of 162 degrees, the illuminated portion will be seen as only a small crescent.

Observation down to the horizon is rarely possible due to the large mass of air the light from the satellite must pass through, and this will cause several magnitudes extinction (reduction) in the brightness and will also cause significant twinkling or scintillation of the satellite. Many observatories have a lower elevation limit of somewhere between 10 and 30 degrees.

11.5.7.3 Including Phase Angle in the Simple Model

In Figure 11.13 the plan view shows the portion of the spherical satellite that is illuminated by the Sun at a phase angle ϕ. We are looking down on the satellite from above the Sun-satellite-observer plane. The satellite is then rotated in the second half of the diagram so that we are seeing it as a disc normal to the line of sight of the observer. This is, of course, exactly how the observer would see it.

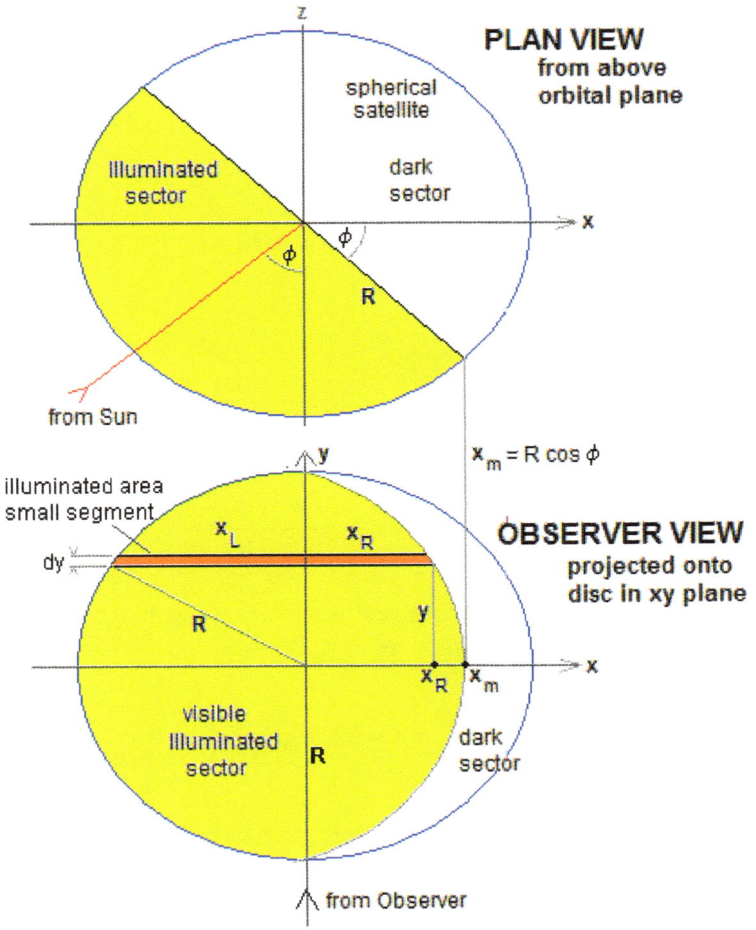

Figure 11.13 The geometry necessary to calculate the fractional area of a spherical satellite illuminated at various solar phase angles.

The left-hand quadrants are fully illuminated, but the right-hand quadrants have a crescent-shaped area that is not illuminated. The x coordinate of the maximum illuminated segment is $x_m = R\cos\phi$. This is the distance projected from the top diagram; R is the radius of the sphere.

To compute the area that is illuminated we need to first consider a small horizontal segment of area that has it located a distance y above the x axis, and has a height dy. The length of this segment is the sum of the left and right lengths x_L and x_R. These have the values:

$$x_R = \frac{x_m}{R}\sqrt{R^2 - y^2} = \cos(\phi)\sqrt{R^2 - y^2}$$
$$x_L = \sqrt{R^2 - y^2}$$

The small segment of illuminate area is then given by

$$dA_{\text{illum}} = (1 + \cos\phi)\sqrt{R^2 - y^2}$$

Integrating this gives the total area

$$A_{\text{illum}} = (1 + \cos\phi)\int_{-R}^{R}\sqrt{R^2 - y^2}\,dy$$

Reference to a good calculus textbook will show that this definite integral equates to $\pi R^2/2$. If we then divide by the disc area πR^2 we have the required expression for the illuminated fraction as

$$f_{\text{illum}} = 0.5(1 + \cos\phi)$$

Referring back to our simple model of Section 11.5.2, we note that the disc illuminated can now be more fully written as

$$A_{\text{i}} = \frac{\pi d^2(1 + \cos\phi)}{8}$$

Following through this gives us an observed magnitude of the satellite that includes the solar phase angle as

$$m_{\text{o}} = 2.5\log\left[\frac{32(r/d)^2}{\alpha(1 + \cos\phi)}\right]$$

Appropriate values of phase angle can be substituted in this equation to show what effect the phase angle has on magnitude. This of course assumes that the albedo does not vary significantly with phase angle. For some surfaces (particularly solar panels), this assumption breaks down as the phase angel approaches zero and diffuse scattering turns to specular reflection (see Section 11.5.9 on glints).

The other thing to remember is that no real satellite is a sphere, and changing phase angles may result in large changes in intercepted light, let alone the fraction of the intercepted light that can be seen by the observer.

11.5.8 Eclipses

Most of the time a geosynchronous satellite will see full sunlight. However, around the equinoxes, when the solar declination is close to zero, it will pass into the Earth's shadow and suffer an eclipse (Figures 11.14 and 11.15). When this occurs it will not be visible from the ground as there is then no sunlight to reflect off it.

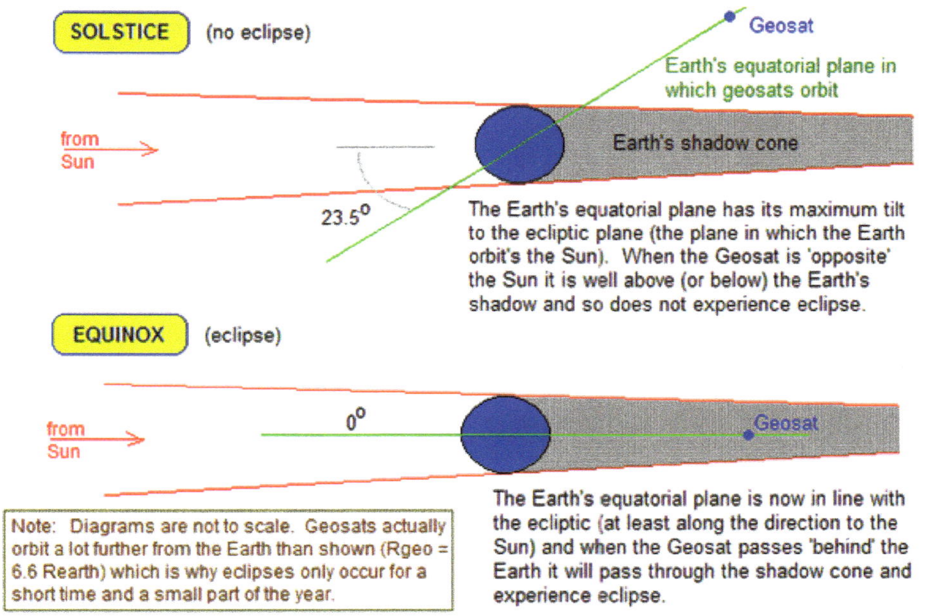

Figure 11.14 Orbital circumstances that lead to geosat eclipse around equinoxes.

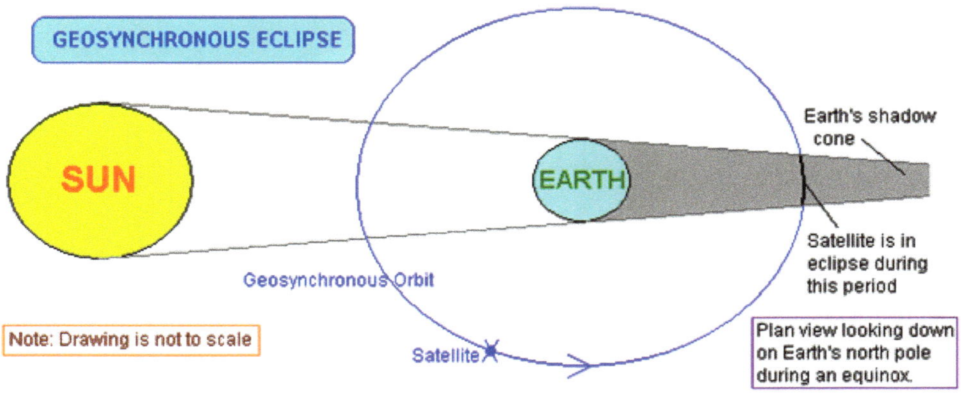

Figure 11.15 Plan view of a geosynchronous satellite eclipse during an equinox.

During the eclipse phase of the geosynchronous orbit the satellite's solar cells generate no power and all equipment must then be powered from storage batteries. If storage capacity is insufficient then some systems must be shut down.

The eclipse season lasts for a little over a month and is centered on each equinox (Figure 11.16). The duration of the eclipse is minimal at the start of the season, as the satellite just grazes the Earth's shadow. The deepest eclipse is when the satellite passes through a full diameter of the Earth's shadow. At this time the eclipse will last for around an hour. Altogether, the total time that a geosat spends in eclipse is only about 0.8%.

The time of deepest eclipse occurs when the satellite is directly opposite the Sun on a line from the Sun to the Earth. This will occur when the longitude of the satellite is 180 degrees different from the apparent longitude of the Sun. This can also be stated as saying that the eclipse will occur when the satellite subpoint is near local midnight. This happens around a UT time of

$$T_{\text{eclipse}} = \frac{24 - \lambda_s}{15} \text{ hours}$$

where λ_s is the satellite easterly longitude measured in degrees.

11.5.9 Glints

Most of the time sunlight will reflect, or more correctly be scattered, from the surface of the satellite in what is known as diffuse scattering. That is, for each incoming ray of sunlight the reradiated light will occur in an isotropic fashion (i.e., at all angles). The total power of the scattered light is only a fraction (given by the albedo) of the incident light.

On occasion, however, specular or mirror-type reflection (Figure 11.17) may occur from a particularly shiny object on the satellite, such as the solar panels. Such a reflection is called a glint and causes a significant transient brightening of

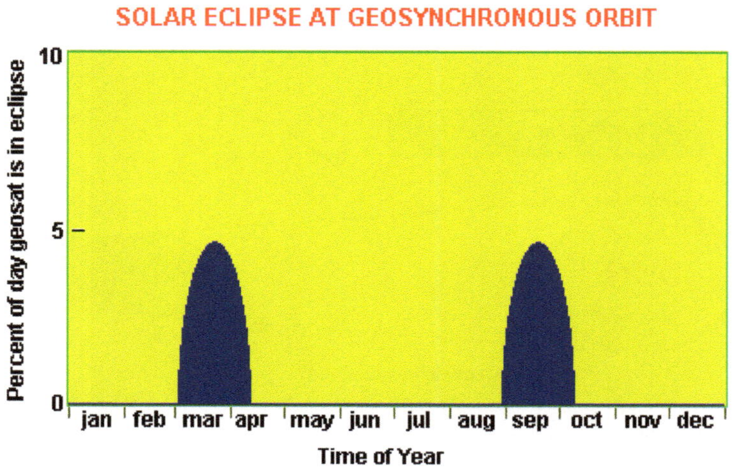

Figure 11.16 Graph showing eclipse seasons for geosynchronous satellites and the amount of time they spend in eclipse each day. Averaged over a year, the total time in eclipse is only 0.8%.

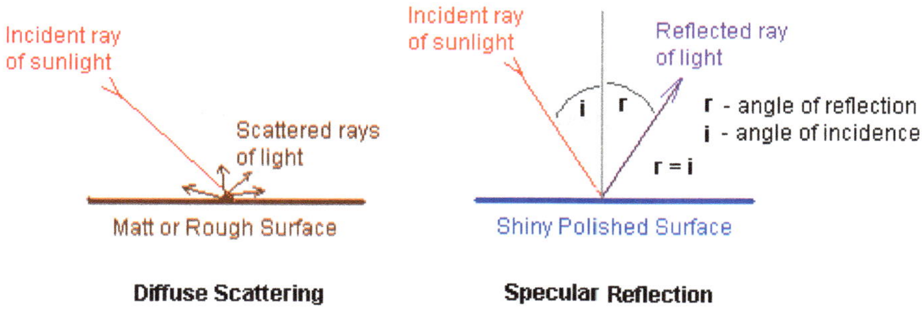

Figure 11.17 The difference between diffuse scattering of light and specular reflection.

the satellite. This will only occur when the reflected angle of the light equals the incident angle. Even with a stabilized satellite (where the attitude of the satellite remains constant in the rotating Earth reference frame) the Sun is still moving and so such a specular reflection will only exist, as far as an observer is concerned, for a brief period of time.

In perfect mirror-type specular reflection, the sunlight is simply bent through an angle (equal to $180° - [i + r]$) without losing any significant intensity. This makes the reflected ray even brighter than diffuse scattering with an albedo near unity. This is because scattering creates a new light source that follows its own inverse square law dispersion, whereas specular reflection simply adds the distance traveled by the reflected ray to the distance traveled by the incident ray, with the new distance being under the same inverse square law as the incident ray (Figure 11.18).

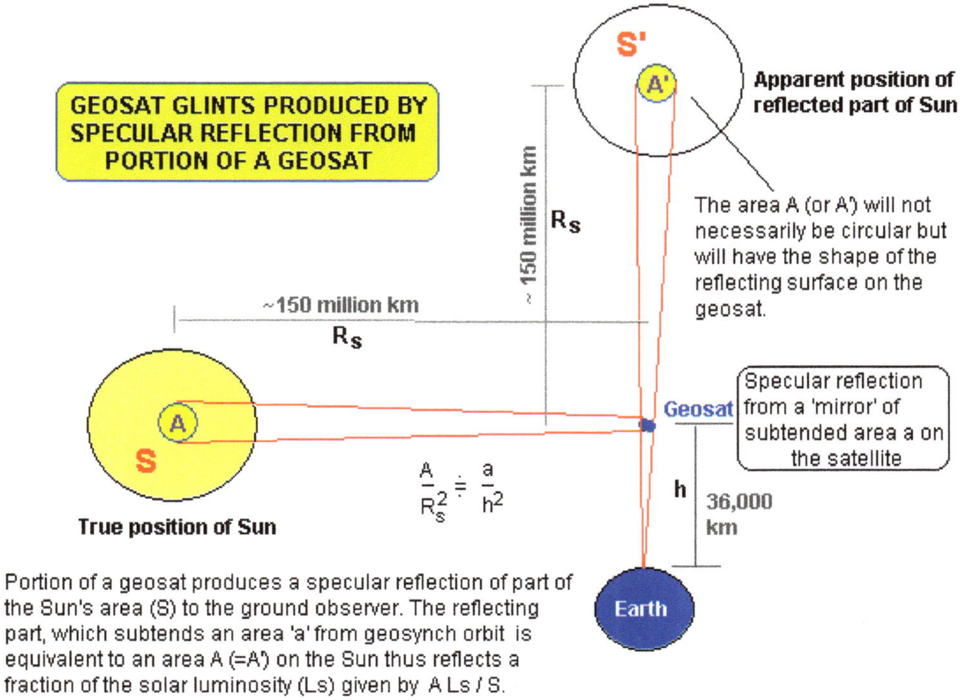

Figure 11.18 Production of a glint by a mirror-type geosat surface.

So scattering tends to result in inverse fourth law dependence, whereas specular reflection is only inverse square law.

With reference to Figure 11.18, one square meter (i.e., $a = 1$) of specularly reflecting panel on the geosat will reflect an area of the solar surface given by

$$\begin{aligned} A &= \frac{aR_s^2}{h^2} \\ &= \frac{\left(150 \times 10^9\right)^2}{\left(36 \times 10^6\right)^2} \\ &= 1.74 \times 10^7 \text{ square meters} \end{aligned}$$

The area S of the subtended solar disc is given by

$$S = \pi r_s^2$$

where $r_s = 6.96 \times 10^8$ m is the radius of the solar disc.

Thus, $S = 1.52 \times 10^{18}$ square meters, and so the fractional area of the Sun reflected is

$$\frac{A}{S} = 1.14 \times 10^{-11}$$

This reflects a flux of AL_s/S at the Earth's surface where $L_s = 1{,}380$ W/m^{-2}. The magnitude of the glint is then given by

$$\begin{aligned} m_g &= 2.5\log\left(\frac{S}{A}\right) - 27 \\ &= 0.4 \text{ magnitude} \end{aligned}$$

Thus, a 1-meter square perfect specular reflector at geosynchronous orbit will produce a glint equal in brightness to the star Procyon.

Although the size of the specular reflector will be a small fraction of the size of the Sun, the portion of the Sun that it reflects may come from anywhere over the solar surface. As this has an angular diameter of 0.5 degrees, the patch of light reflected to the Earth will also spread over 0.5 degrees. At the Earth's surface this will result in a patch size of $\Delta = r_g\theta_s$ where r_g is the distance from geosat to observer (~40,000 km) and θ_s is the solar angular diameter of 0.5 degrees (= 8.7×10^{-3}). This gives $\Delta = 350$ km. This patch of light will move over the observer in about 2 minutes (Figure 11.19).

Glints are most likely when the solar phase angle is zero, because the solar panels will be pointing directly at both the Sun and the observer. In practice, this does not occur because the satellite will then be in solar eclipse (see Section 11.5.8). The minimum solar phase angle is usually 8.5 degrees (see Section 11.5.7). Although there have been reports of glints reaching a magnitude of −3 to −4, this is usually due to offset solar panel pointing and occurs outside of an eclipse. Generally

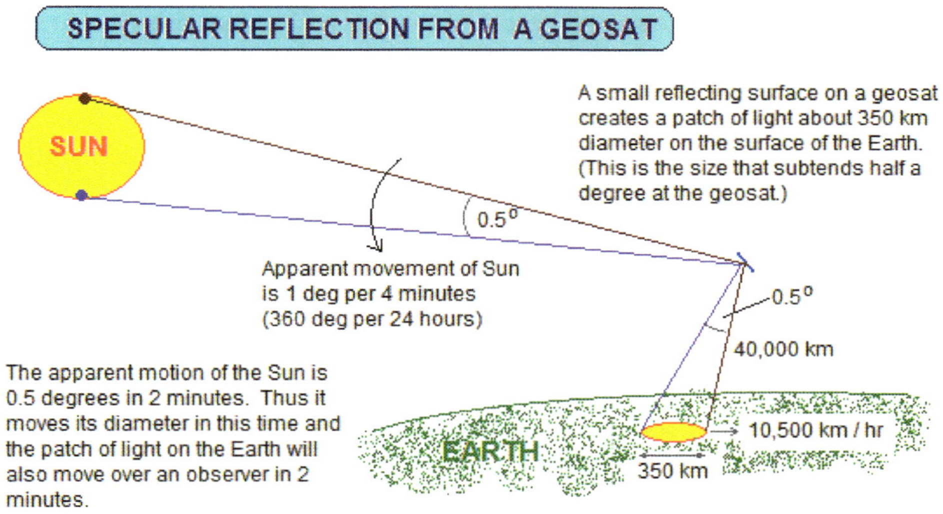

Figure 11.19 A perfect reflector in geosynchronous orbit will produce a patch of light on the Earth. This patch will move past an observer in about 2 minutes, the time it takes for the Sun to move its own diameter in the sky.

observations show a maximum brightness of +2 and the peak being spread over a much longer time of up to an hour. This is thought to be due to multiple reflections from smaller sections of solar panel, which have relative tilts to each other of a few degrees, rather than from a larger smooth continuous surface.

The larger glints are generally attributable to reflections from the solar panels. However, smaller peaks in the light curve are often observed (faint glints) and these come from specular reflection from other parts of the spacecraft. The magnitude of the glint is not only proportional to the size of the reflecting object (A_s) but also due to its less than perfect reflectivity. We can define a specular albedo (α_s) to account for this reduced magnitude. This can be quite different than the albedo for diffuse scattering. The amplitude of the glint is then proportional to A_s and α_s.

The glint season for geosats is around the equinoxes, the same time as the eclipse season. However, because of the fact that zero phase angle does not occur (due to eclipse), the exact time of maximum observed glint will be shifted from the exact equinox. In the Northern Hemisphere the glint season will be prior to the vernal equinox and after the autumnal equinox, whereas in the Southern Hemisphere it will be after the vernal and prior to the autumnal equinox. In both hemispheres it will occur before the March equinox and after the September equinox.

11.6 Geosat Communication

Geosynchronous satellites orbit at an altitude of 36,000 km. This large altitude means we always need a directional antenna for geosat communications. However, we have two points in our favor. Because the orbit is at a radius 6.6 times larger than the Earth's radius, the variation in range from the zenith to the horizon is not very great (about 17%). We can assume a mean range of 38,000 km in our calculations. The second benefit of the geosynchronous orbit is that such a satellite stays relatively in the same place in the sky and thus it does not need to be tracked. This makes possible direct-to-home TV broadcasts at low cost (because the antenna can be fixed in look position).

Geosynchronous downlinks vary in power and bandwidth. Bandwidths can be very narrow in the case of beacons, to very large in the case of TV transmissions, which might be 36-MHz wide. Powers can vary from a few watts to hundreds of watts. In Figure 11.20 we look at how SNR varies with ERP and bandwidth. We will assume a low noise receiving system noise temperature of 50K.

11.7 The Geosat Environment

The geosynchronous space is quite difference from the low Earth orbit environment. The two are separated by the van Allen radiation belts. This shields LEO from a lot of particulate radiation. The geosynchronous orbit lies at the outer edge of the outer radiation belt, and thus is also a lower radiation environment than that experienced by global positioning satellites in middle Earth orbit, but higher than that experienced by satellites in low Earth orbit. Also, because the outer belt is mainly due to energetic electrons it is more easily shielded against.

However, excess electron flux injected into the Earth from the Sun due to CMEs can cause charging problems on the outer surfaces and solar panels. This is particularly so when the pressure of the solar wind on the magnetopause of the Earth's magnetosphere causes it to compress and move inside geosynchronous orbit. GEO is then exposed directly to the interplanetary medium, and any satellite that relies on the Earth's magnetic field for stabilization, or other function, will have a problem. For this reason space weather agencies sometimes offer a magnetopause model.

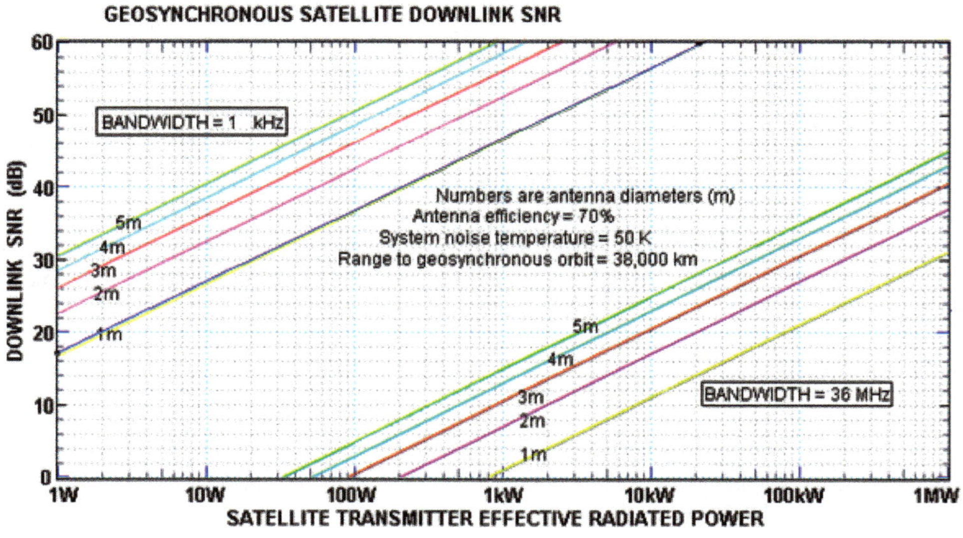

Figure 11.20 Geosynchronous communication parameters.

11.7.1 Magnetopause Model

Geostationary satellites are usually protected to some extent from the effects of solar energetic particles by the Earth's magnetic field. But, when the magnetopause, the abrupt boundary separating the solar wind from the Earth's magnetic field, is pushed towards Earth (inside GEO orbits) by a major coronal mass ejection, some satellites will lose this protection and suffer damage from an increased flux of high-energy particles.

Damage of this sort is sometimes severe. Take for example the Equator-S mission [2]:

> "Unfortunately, the mission was damaged by several malfunctions. On December 17, 1997, the primary master processor unit failed, and on April 30, 1998, the second unit experienced the same problem. This caused the premature end of the mission. Failure analysis attributed both events to a latch-up in one the RAM chips used by the processors. It is not unlikely that deep dielectric charging by the high energetic electron fluxes in December 1997, as well as in April 1998, was the cause of the latch-ups."

The stand-off distance and shape of the magnetopause are controlled by the dynamic and static pressure of the solar wind, the pressure of the Earth's magnetic field, and the orientation of the Interplanetary Magnetic Field (IMF) [3]. The solar wind and IMF field data used for this model originate from the NOAA/NASA DSCOVR (Deep Space Climate Observatory) spacecraft orbiting the L1 point (a point between the Sun and Earth where gravitational forces balance).

The graphical presentation (Figure 11.21) is in the equatorial plane as viewed from above the Earth's North Pole. Geosynchronous orbit is shown at a distance of 6.6 Earth radii (Re) from the center of the planet, with four longitude spokes. The locations of several geosynchronous satellites are indicated, with the legend in the lower left identifying the satellites. All the satellites are actually at the same orbital radius, though the display adjusts these in some cases for clarity. More information on this model can be found in the chapter on Space Weather (Chapter 14).

11.7.2 Electron Fluence and Deep Dielectric Discharge

A satellite may become charged due to different physical phenomenon. The photoelectric effect (sunlight) may cause charging. If the satellite is orbiting in a high electron flux, it may acquire charge from the space environment. Satellites in geosynchronous orbits are particularly susceptible to this phenomenon due to their location in the outer regions of the outer radiation belt. This belt is predominantly populated by electrons.

If the charge on a satellite reaches a critical value, then a discharge can occur that may cause damage to one or more of the satellite systems. Of particular concern is the phenomenon termed deep dielectric discharge (DDD). In a flux of high-energy electrons, the electrons penetrate the spacecraft's outer surface and bury themselves in dielectric materials, such as circuit boards and the insulation in coaxial cables.

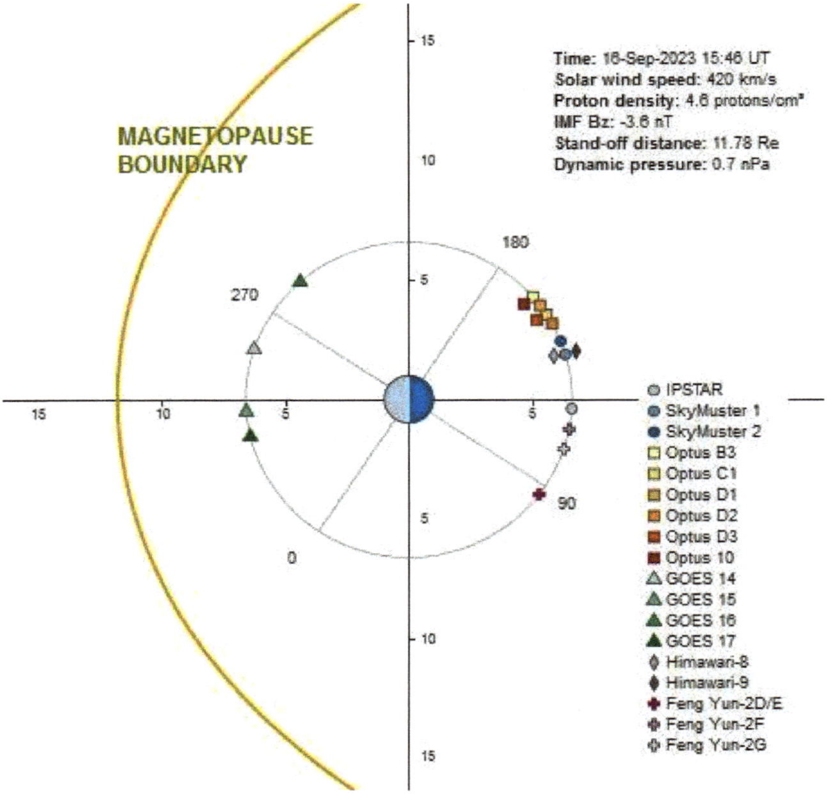

Figure 11.21 Magnetopause model output. (*Source:* SWS/BOM.)

The build-up of charge will continue until the dielectric strength of the material is exceeded, when a sudden electrical discharge will occur. This miniature lightning stroke can cause permanent damage in the associated or nearby circuitry.

Exact data on the DDD phenomenon is hard to come by. US Air Force satellite operators are issued warnings when the daily electron fluence >2 MeV (electrons/sq cm/steradian/day) at geosynchronous orbit meets either of the following conditions:

- Greater than 3×10^8 per day for 3 consecutive days;
- Greater than 10^9 for a single day.

An article in *Space Weather* [4] provides a little more qualitative information. They provide the experimental results for a DDD event as a function of electron fluence >300 keV from the *SCATHA* satellite, near geosynchronous orbit. Table 11.2, which is drawn from the results of Vampola (stated as a private communication) [4], provides data points for electron energies greater than 300 keV.

Now, the GOES electron fluence, which is reported on the BoM/SWS website daily, is for the electron fluence >2 MeV. For the above graph to be useful, we must find a relation between the >300 keV and the 2 MeV electron populations. The AE-8 electron radiation belt model is the one most commonly used for examining

Table 11.2 Data Used to Develop the Deep Dielectric Discharge Model

Daily Fluence (electrons/cc)	*Discharge Probability*
3×10^8	0.01
1×10^9	0.02
3×10^9	0.04
1×10^{10}	0.12
3×10^{10}	0.14
1×10^{11}	0.23
3×10^{11}	0.44
1×10^{12}	0.51

radiation belt electron fluxes at this time. An examination of this model shows that the relation between the two fluxes is not a simple one, and varies with position in the radiation belt. However, at an L value of 6.6, which is where the geosynchronous orbit lies, the flux of >2 MeV electrons is typically two orders of magnitude below the >300 keV flux.

Thus for this model the fluence axis is shifted by two orders of magnitude and then a Lorentzian growth curve is fitted to the points to produce the model algorithm. The graphical model is shown in Figure 11.22.

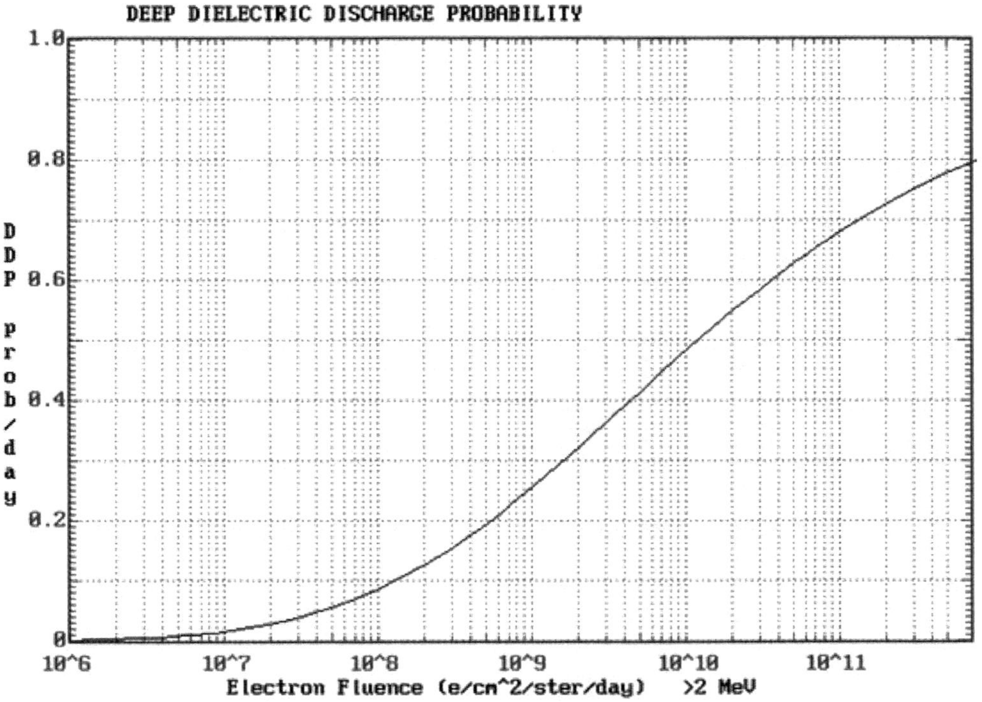

Figure 11.22 Graphical output for deep dielectric discharge model.

This curve has the form:

$$\text{prob}\{DDD\} = \frac{z^{4.5}}{1200 + z^{4.5}}$$

where $z = \log_{10}(\text{Fluence}) - 5$.

This is the probability of a deep dielectric discharge during the day in which the electron fluence (>2 MeV) is specified.

The occurrence of a DDD on a particular satellite will depend on the actual configuration of, and the electronics in, the satellite. Engineering techniques employed in the construction of a satellite can make it more or less susceptible to DDD. For this reason, it is probably best to term the probability computed in this model the normalized (or standardized) daily probability of a DDD occurring given a specified electron fluence.

Implementation of the model requires the fluence to be input from the GOES value (available from the SEC website). The formula above is then employed to calculate a single daily normalized probability of DDD for a satellite in geosynchronous orbit (where the fluence was measured).

11.7.3 The GEO Space Debris Environment

The GEO space environment is also different with respect to space debris. The orbital speed is lower and all satellites are moving in the one direction. The total length of the geosynchronous orbit is about 260,000 km. With a spacing of 10 km, the orbit could potentially accommodate 26,000 satellites. As of early 2024, there are currently less than 600 active satellites in GEO, or 1% of this theoretical capacity.

Congestion in the geosynchronous orbit is mainly a result of the need to allow a box around each satellite for station keeping, and in the case of communications satellites by the consideration of the beamwidth of the user antenna. In this case the satellites should be spaced far enough apart so that adjacent satellites using the same frequency allocations do not cause mutual interference. In the case of C-band transmissions (~4 GHz) the spacing should be around 1 degree, but for Ku band (~12 GHz) around one-half a degree. In some parts of the geosynchronous orbit satellites are reaching these limits, which is why more and higher frequency allocations are being sought by satellite operators. Laser communication, with an enormous potential downlink bandwidth, are at the forefront of current research.

Earth observation satellites are not so constrained by the above limit as their control and data links use much larger ground dishes. These satellites can thus be placed closer together, with their spacing only determined by the volume required for station-keeping operations.

Even with impending congestion for some satellite types, it was initially thought that space debris would never be a concern at GEO. After all, the satellites are all going around the Earth in the same direction (as opposed to the random directions of satellites in LEO) and their orbital speed is a lot less than in LEO (~3 versus ~8 km/s). It would thus seem reasonable to assume that the collision rate would be a lot less, and so the debris density would be consequently a lot lower.

However, a wake-up call occurred around the year 2000 when a Swiss research group managed to image a significant population of small debris objects in and around GEO. Thomas Shildnecht, and others in Switzerland [5], proved that the GEO orbital debris population had been severely underestimated. Obviously, fragmentations from unspent fuel sources and energy storage (batteries) had occurred, but had not been noticed. This is not surprising considering the size of such debris. They are simply a lot harder to observe because of their distance.

Debris observations in geosynchronous orbit are continuing to this day. The examination of the smaller-sized population is of course inhibited by the much larger distances involved compared to LEO, and the subsequent low brightness of these objects falling below the observing magnitude threshold of even large telescopes.

11.8 The Asteroid Hazard to Geosats

The occasional near-Earth asteroid will come so close that is passes by inside geosynchronous orbit. It is thus natural to consider the possibility of a collision between an NEA and a geosat.

Table 11.3 lists some NEA close approaches in the years 2011 to 2013, and Figure 11.23 shows the actual path of asteroid *2021D14*.

The following asteroid-geosat hazard model approximates the geosynchronous belt by a spherical shell (Figure 11.24).

The volume of the spherical shell is given by:

$$V_{geo} = 4\pi r_g^2 \delta$$

The number density of geosats in this shell is:

$$\rho_{geo} = \frac{N_{geo}}{V_{geo}} = \frac{N_{geo}}{4\pi r_g^2 \delta}$$

The asteroid travel distance in the shell is:

$$a = 2x = \frac{2\delta}{\cos\theta} = \frac{2\delta r_g}{\sqrt{r_g^2 - d_{min}^2}}$$

Table 11.3 Some Asteroid Close Approaches Over a 3-Year Period

Asteroid	*Date*	*Miss (km)*	*Size (m)*
2011CQ1	2011-02-04	11,900	3
2011MD	2011-06-27	18,700	10
2012KT42	2012-05-29	20,800	7
2012DA14	2013-02-15	27,700	45

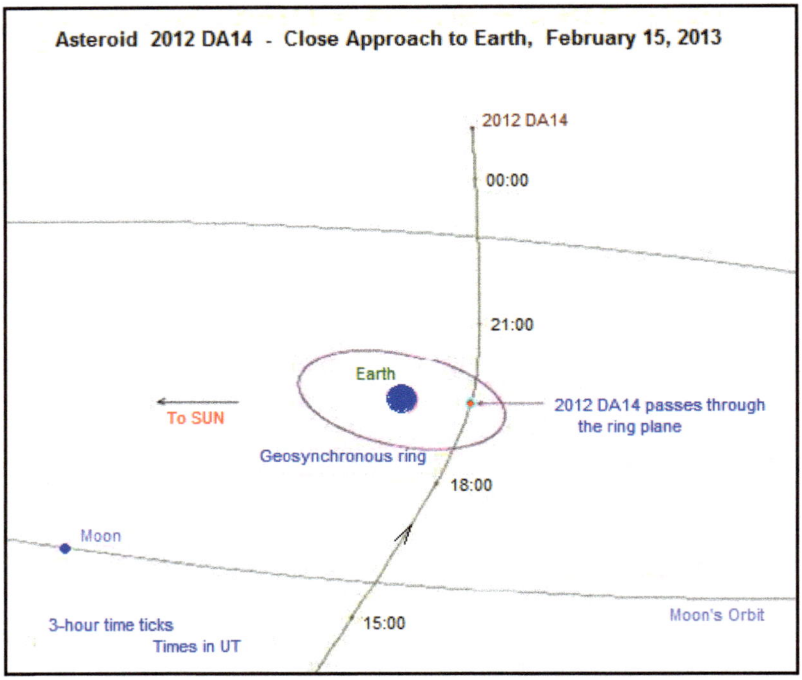

Figure 11.23 Path followed by asteroid *2012 DA14*. (*Source:* NASA/JPL.)

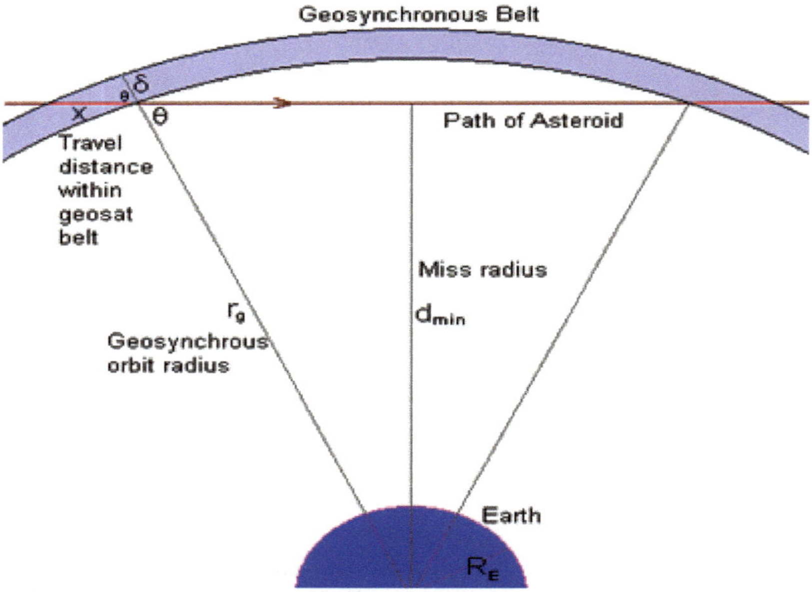

Figure 11.24 Diagram for asteroid collision probability.

The volume traversed by the asteroid in the shell:

$$V_a = \alpha^2 a = \frac{2\alpha^2 \delta r_g}{\sqrt{r_g^2 - d_{\min}^2}}$$

The expected number of geosats hit by the passage of one asteroid is:

$$N_{\text{hits}} = \alpha^2 a \rho_{\text{geo}} = \frac{\alpha^2 N_{\text{geo}}}{2\pi r_g \sqrt{r_g^2 - d_{\min}^2}}$$

(Note that the shell thickness does not appear here.)

Assuming an active geosat population of 1,000, an asteroid diameter of 50m, and an Earth miss distance of 35,000 km, then the expected number of satellites hit in a single asteroid pass is $N_{\text{hits}} = 3 \times 10^{-10}$.

To have a reasonable expectation of one satellite hit by an asteroid we would need 3 billion asteroids to traverse the satellite belt. It is estimated that there exists only around 8 million near-Earth asteroids (NEA)/potentially hazardous (PHA) with a diameter of 50m or greater (as of 2025).

Even if we assume these pass within the geosat region every year, it would take approximately 100 years to accumulate the required number of passes. In fact, we probably need a minimum of 100 years for this many near misses to occur. So we might expect one geosat hit in 10,000 years! We might reasonably conclude that asteroids do not present a significant hazard to geosynchronous satellites.

11.9 Geosynchronous Orbital Debris

Space debris in low Earth orbit has been well discussed. It is in the LEO regime that the problem of space debris is most acute. However, geosynchronous orbit has problems of its own, and some of these have largely gone unnoticed, because the increased distance to GEO makes detection of smaller objects very difficult with ground-based sensors [5].

One of the main differences about geosynchronous orbit is the absence of atmospheric drag and decay as a sink for the debris population. The four main perturbing forces on an object in or near geosynchronous orbit are gravitational forces due to the Earth's oblateness, the Moon, the Sun, and solar radiation pressure.

The standard procedure to dispose of a geosat at the end of its life is to boost it into a supersynchronous orbit; that is, an orbit above geosynchronous orbit. It is now agreed that these graveyard orbits should be at least 300 km above the active geosynchronous belt.

To solve the sensing problem, the US Air Force has recently launched a satellite into GEO that can maneuver around the orbital belt, allowing both observation of the overall environment, and observation of individual satellites that may experience problems.

References

[1] Clarke, Arthur C., "Extraterrestrial Relays," *Wireless World*, 1945.

[2] Haerendel, G., R. B. Torbert, and H. Höfner, "Introduction: The Equator-S Mission," *Annales Geophysicae*, 1999, Vol. 17, No. 12, pp. 1499–1502.

[3] Shue, J. H., et al, "A New Functional Form to Study the Solar Wind Control of the Magnetopause Size and Shape," *Journal of Geophysical Research: Space Physics*, Vol. 102, May 1997, pp. 9497–9511.

[4] Baker, D. N., S. G. Kanekal, and J. B. Blake, "Characterizing the Earth's Outer Van Allen Zone Using a Radiation Belt Content," *Space Weather Journal*, Vol. 2, 2004.

[5] Schildknecht, T., et al., "Investigation of Properties and Characteristics of High-Area-to-Mass-Ratio Objects Based on Examples of Optical Observation Data of Space Debris Objects in GEO-like Orbits," *Adv. Space Res.*, 34,901, 2004.

Selected Bibliography

Inglis, A. F., and A. C. Luther, *Satellite Technology: An Introduction*, Focal Press, 1997.

Pocha, J. J., *An Introduction to Mission Design for Geostationary Satellites*, Microcosm Press and Springer, 1987.

Soop, E. M., *Handbook of Geostationary Orbits*, Microcosm Press and Springer, 1983.

CHAPTER 12
Electromagnetic Awareness in SSA

Satellites or space debris can be detected by the light they reflect from the Sun, at least when they are illuminated by the Sun. When the object is in the Earth's shadow, or when the observer is in sunlight, this detection method is usually not possible. Most satellites use some form of radio communication with the Earth, which can be used to detect and track a satellite. Space debris and uncommunicative satellites can be detected by radar, although this becomes more difficult at the higher altitudes due to the inverse fourth power of distance-limiting radar detection. A knowledge of satellite communication frequencies used by various countries is essential to SSA.

12.1 Satellite Beacons

12.1.1 Introduction

A *beacon* can be defined as both a guiding or warning signal, and a source of light or inspiration.

As mentioned in Chapter 2, the first artificial satellite, *Sputnik-1* (Figure 12.1), launched into orbit on October 4, 1957, carried a radio beacon transmitting on the frequency of 20.005 MHz. This could be received by a sensitive short wave receiver on the ground, and many people heard the beep-beep-beep sound of this satellite as it orbited around the Earth every 90 minutes or so. It served as an announcement of its presence to the world and allowed it to be tracked by ground stations. It is also thought that the duration and spacing of the beeps carried information about

Figure 12.1 Model of *Sputnik-1*, the first orbital satellite. (*Source:* Moscow Museum.)

the conditions in the satellite. This was the first instance of telemetry (measurement from afar) from space.

Scientists quickly discovered that the signals from such a beacon could be used to gather information about the state of the Earth's upper atmosphere, in particular the region of the atmosphere containing charged particles called the ionosphere.

Satellite beacons have been used for a variety of purposes, ranging from tracking and telemetry through propaganda and advertisement, to ground navigation and monitoring. This chapter will explore in more detail some of these purposes.

12.1.2 History

Following *Sputnik-1* came *Sputnik-2* (Figure 12.2). This carried a live dog (Laika) into space. Not only was a beacon needed to track the spacecraft and telemeter information about its health, but it was also needed to relay information about the dog's condition. The cartoon shown as Figure 12.2 was published by a Russian newspaper of the time illustrating the essential details.

Following the Russian successes, the Americans launched satellites in the Explorer series and then in the Vanguard series. The early US satellites carried beacons that transmitted on 108 MHz. Slightly later satellites used frequencies in the newly created space band at 136 MHz.

Satellite beacons today are considerably more complex, generally transmit on higher frequencies, and are controlled by digital logic circuitry (e.g., computers).

12.1.3 Types of Beacons

Satellite beacons may be classified according to their function or according to the frequency they use, or a mixture of both. Table 12.1 lists some major categories.

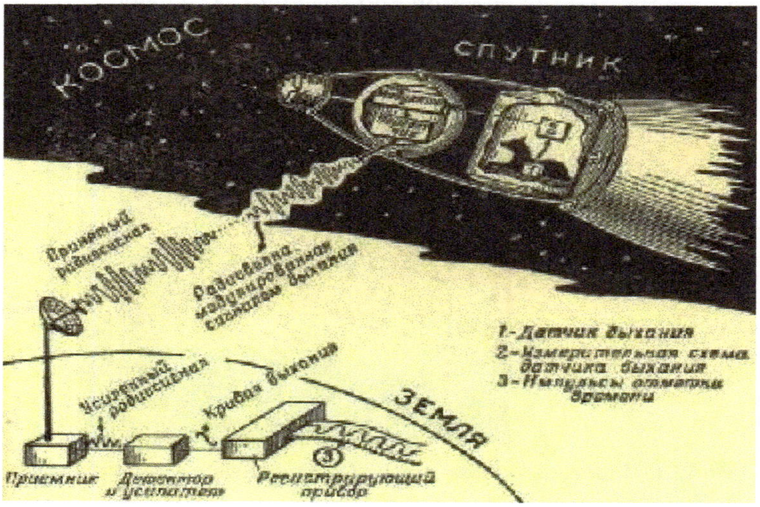

Figure 12.2 Russian press information about *Sputnik-2*. (*Source:* PRAVDA.)

Table 12.1 Satellite Beacon Details

Beacon Function	*Band*	*Notes*
Early Russian tracking and telemetry	HF	Example: 20.005 MHz
Early US tracking and telemetry	VHF	Example: 108 / 136 MHz
Geosat tracking and telemetry	Ku / Ka bands	Example: 14/26 GHz
Deep space tracking and telemetry	S / X bands	Example: 2 / 8 GHz
Ionospheric studies	VHF / UHF	GPS now commonly used
Tropospheric studies (rain attenuation)	Ku / Ka bands	14 / 26 GHz
Navigation	L-band	Example: GNSS / GPS
Optical	Visible light	Mostly experimental

12.1.4 Beacons for Ionospheric Research

The ionosphere is a region of the Earth's upper atmosphere (altitude about 80 to 1,000 km) ionized (to form a weak plasma) by UV and X-ray radiation from the Sun [1]. This region affects radio waves passing through it, producing refraction, absorption, scintillation, and other effects. The refractive index of the ionosphere is given by:

$$n = \sqrt{\left[1 - \left(\frac{f_p}{f}\right)^2\right]}$$

where f is the radiowave frequency and f_p is the plasma frequency given by:

$$f_p^2 = \frac{N_e e^2}{4\pi^2 \varepsilon_0 m_e}$$

where N_e is the ionospheric electron density, e and m_e are the charge and mass of the electron, respectively, and ε_0 is the permittivity of free space.

As long as f is greater than f_p the radio wave can pass through the plasma and transionospheric propagation is possible.

Satellite beacons used for ionospheric research normally transmit in the VHF (30 to 300 MHz) or UHF (300–3000 MHz) bands. They should be above the ionospheric critical frequencies (typically less than 30 MHz) so that the signal will penetrate the ionosphere, but they should not be at too high a frequency or sensitivity to ionospheric parameters is reduced (most ionospheric characteristics show a variation according to the inverse square of the frequency $-1/f^2$).

A block diagram of a satellite radio beacon transmitter ideally suited for ionospheric studies is shown in Figure 12.3.

The beacon employs three widely spaced phase-locked frequencies. The frequency range covered allows for the examination of ionospheric irregularity structures over a wide spectral range for the derivation of a spectral power law in observed ionospheric scintillation. Two widely separated frequencies allows for accurate determination of ionospheric total electron content (TEC). The highest frequency

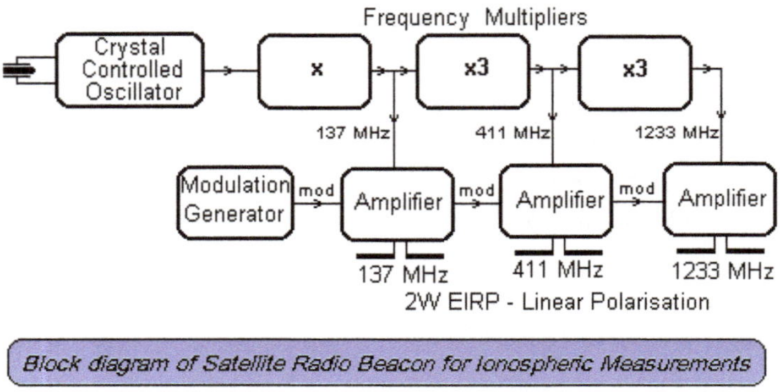

Figure 12.3 Three-frequency satellite beacon schematic diagram.

essentially acts as a reference to the low frequency, which shows a high sensitivity to TEC. The phase-locked relationship between the frequencies allows differential phase comparison. The relatively high transmitter output (2W) allows the beacon to be used in GEO as well as in low Earth orbit. The presence of a modulating signal (usually a pseudorandom binary code) allows for the measurement of differential time delay between frequencies.

The most common ionospheric measurements made using satellite beacons are (1) Faraday rotation, (2) delay due the ionospheric line of sight electron content, and (3) signal scintillation.

12.1.5 Measuring Faraday Rotation

Instruments used to determine the Faraday rotation angle of a satellite radio beacon signal, and monitor its change with time, are called polarimeters (Figure 12.4). The basic principle is to rotate a linearly polarized antenna until a maximum or minimum signal is obtained. The antenna can be rotated mechanically, or several stationary antennas can be employed in a switched arrangement to electronically

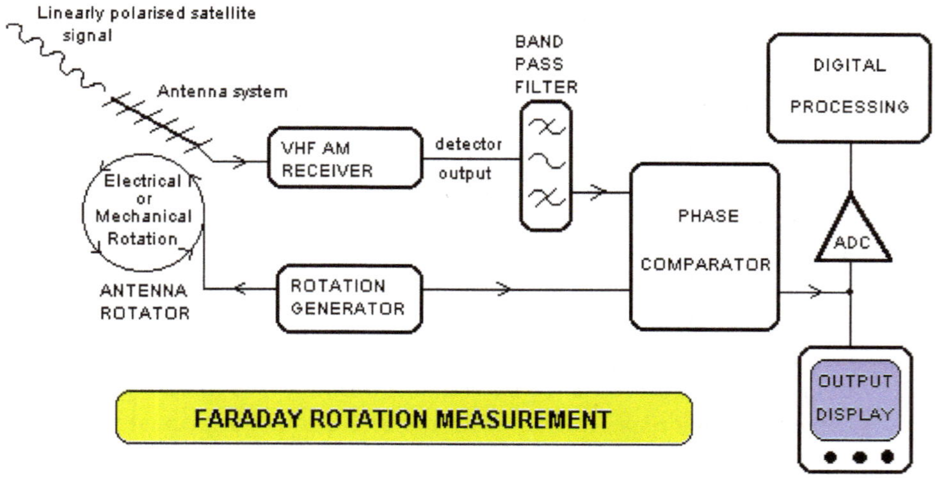

Figure 12.4 Setup for measuring ionospheric Faraday rotation.

simulate rotation. The output of the system is a time history of Faraday rotation angle. If required, this may be converted into TEC using a model of the Earth's magnetic field.

12.1.6 Measuring Total Electron Content

Differential time measurements, sometimes called differential Doppler measurements, may be performed in one of two ways (Figures 12.5 and 12.6). Two of the frequencies (the lowest and the highest) may be coherently downconverted and then divided to produce a common reference frequency. A phase comparison is then performed at this common frequency to determine the phase difference between the two frequencies. An ambiguity can exist in the output of the comparator, especially if the reference frequency is high. This ambiguity, which must be resolved by other data or knowledge, is $2n\pi$ in-phase, where n is the unknown integer to be determined. The output data stream is a time history of TEC.

The second type of differential time measurement is often referred to as group path delay. It relies on a modulation frequency common to both signals. In the ideal case of binary modulation, the zero crossing of the fastest signal (highest frequency) would be used to start a time counter. The zero crossing of the lower frequency would then be used to stop the counter. The time interval measured is a direct measure of TEC. In practice, the modulation frequencies of each signal may require downconversion (or division) to effectively expand the time measurement required. In this way, time differences of milliseconds, rather than microseconds or even nanoseconds can be made, with a resultant increase in accuracy.

12.1.7 Measuring Ionospheric Scintillation

Amplitude scintillation measurements can be made with simple equipment (Figure 12.7). A narrowband low noise AM receiver connected to a suitable display to show the signal amplitude/intensity is the simplest configuration. The main requirement

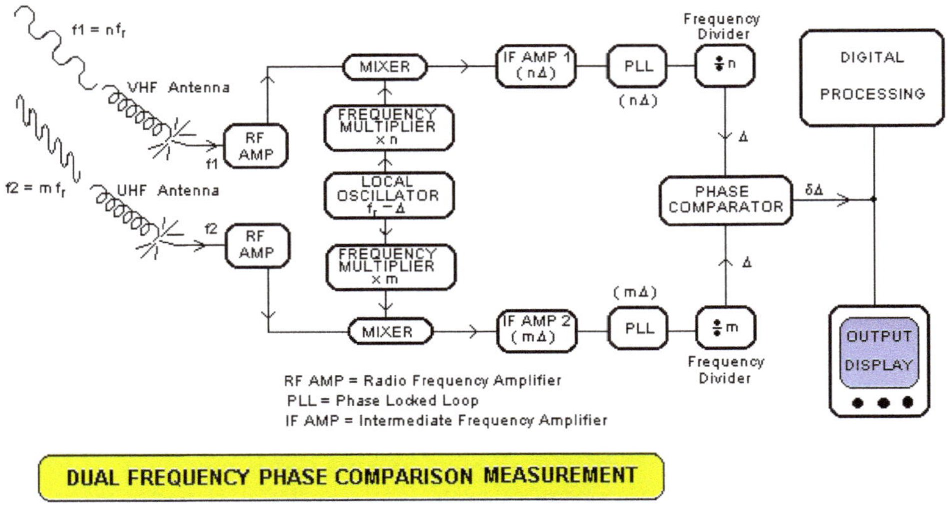

Figure 12.5 Dual-frequency measurement of ionospheric total electron content.

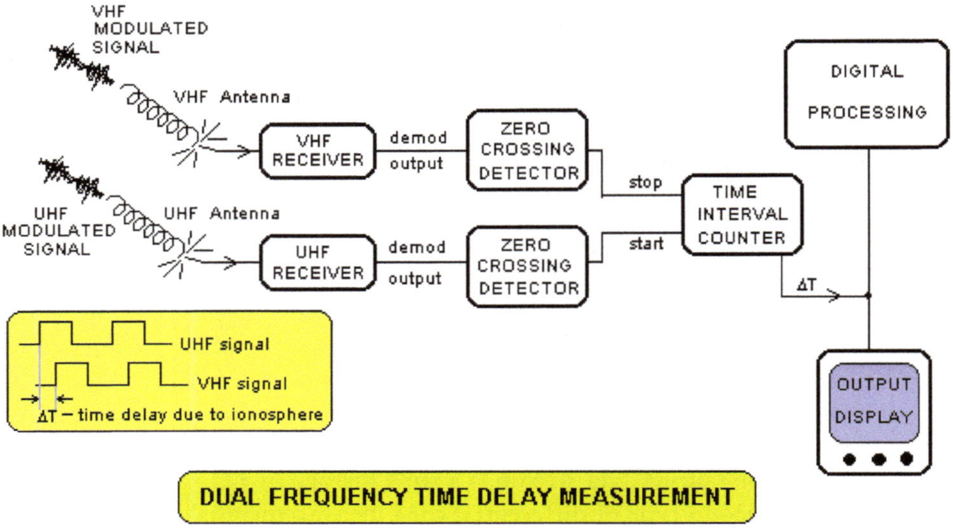

Figure 12.6 An alternative method of measuring ionospheric total electron content.

for this type of system is a high signal-to-noise ratio, so that signal scintillations do not saturate the receiver or fall below the noise level. A high SNR can be achieved with a high beacon power, a low noise receiver, and/or a large receiving antenna (high gain).

12.1.8 Ionospheric Satellite Beacon Status

In the early days of space research, there was a multitude of VHF beacons on both geosats and satellites in low Earth orbit for ionospheric research. Today, in the twenty-first century there are very few such beacons, especially phase coherent

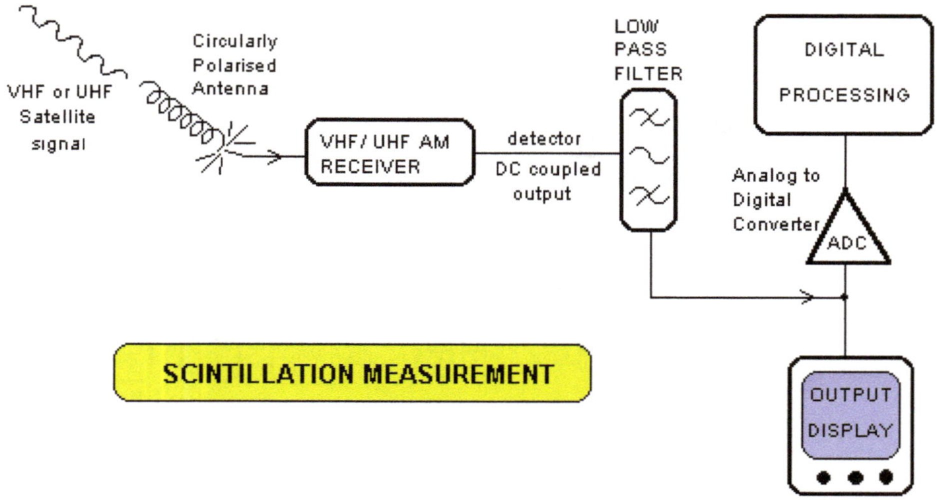

Figure 12.7 The measurement of satellite signal scintillations caused by ionospheric irregularities.

multifrequency VHF beacons. Ionospheric researchers have thus turned to the readily available and prolific GPS beacons to continue ionospheric research.

While these monitors are ideal for some purposes, the high frequencies used (L-band) do not have adequate resolution/sensitivity to support lower frequency systems and applications (e.g., low frequency radio astronomy).

12.1.9 Beacons for Tropospheric Research

Satellite beacons (Figure 12.8) above X-band are affected little by the ionosphere. However, as the frequency is increased, the beacon starts to be affected more and more by the troposphere, the first 10 km of the Earth's atmosphere in which our weather occurs.

The refractive index of the troposphere to microwave signals is given by:

$$n = \frac{N}{10^6 + 1}$$

where the refractive modulus N is given by:

$$N = 0.776\frac{P}{T} + 3,730\frac{P_{wv}}{T^2}$$

and where P is the atmospheric pressure in Pascal (Pa), P_{wv} is the water vapor partial pressure (Pa), and T is the temperature in Kelvin. This formula is accurate to $N <$ 0.1 over the frequency range 1 to 50 GHz.

Note that standard atmospheric surface pressure is 101.3 kPa (= 1013 millibar), and 20°C = 293K.

As the pressure, temperature, and water vapor change with altitude, geographic location, and time, the troposphere thus produces bending (refraction), attenuation (absorption), and scintillation (variable amplitude) of high microwave band satellite-to-ground signals.

At VHF and UHF frequencies very long periodic fadings (5 to 30 min) have been observed on 130 and 400 MHz transmissions of geostationary satellites, and these have been correlated with cold fronts, ground and elevated temperature inversions.

12.1.10 Beacons for Telemetry and Tracking

Satellite beacons are often used to telemeter data concerning the health of the spacecraft and its environment. Such data may include the status of the various satellite subsystems, the power output of the solar panels, and the current drawn by the various pieces of equipment, the satellite temperature, orientation (gyro status), and the external magnetic field, solar flux, and radiation environment, for example.

In the early satellites, data was transmitted in analog form using various multiplexing schemes. Tones, frequency, and time division multiplexing on a single carrier or several subcarriers were used. Nowadays, data is almost always sent in digital form and often in packets using standard protocols. Various error-correcting schemes are used to mitigate noise on low-signal levels.

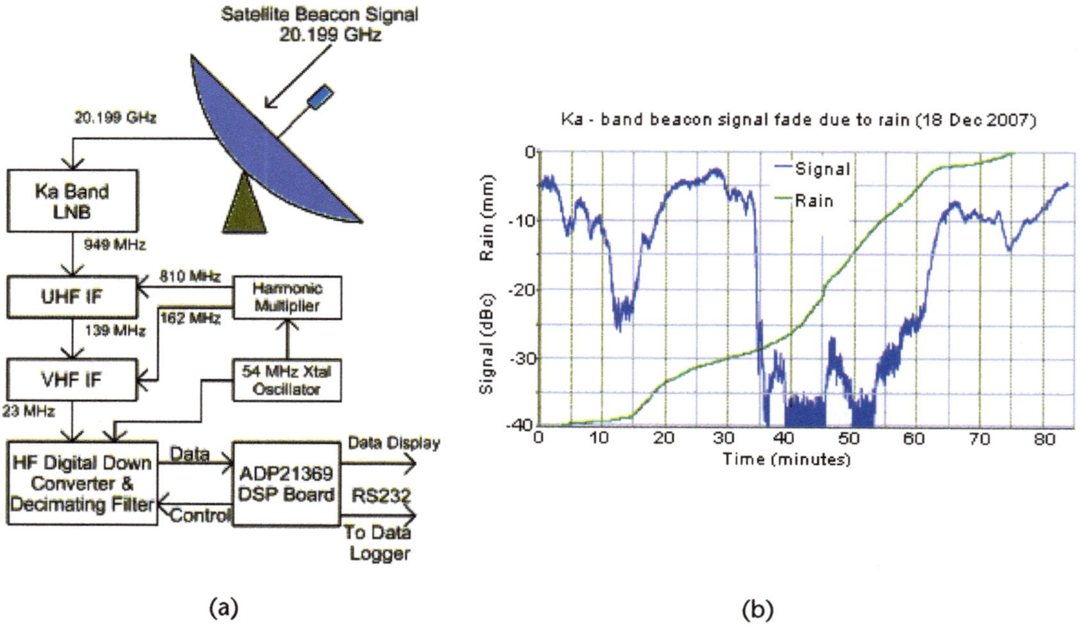

Figure 12.8 (a) Ka band beacon receiver developed at James Cook University, and (b) rain fade detected by the receiver. (Kikkert & Kenny JCU.)

Although hardware decoder designs are available, software is increasingly used to decode satellite telemetry. This often requires only a simple preprocessor before input to the decoding computer. A simple infinite clipping processor schematic is shown in Figure 12.9, which acts as an interface between a satellite receiver and a computer serial port. Often times the receiver output is fed directly to the sound card of a computer.

With the advent of smallsats and particularly low-cost cubesats, and the ability of some launchers to place dozens of these into orbit at one time, there is now a plethora of these satellites in orbit transmitting telemetry down to Earth, with

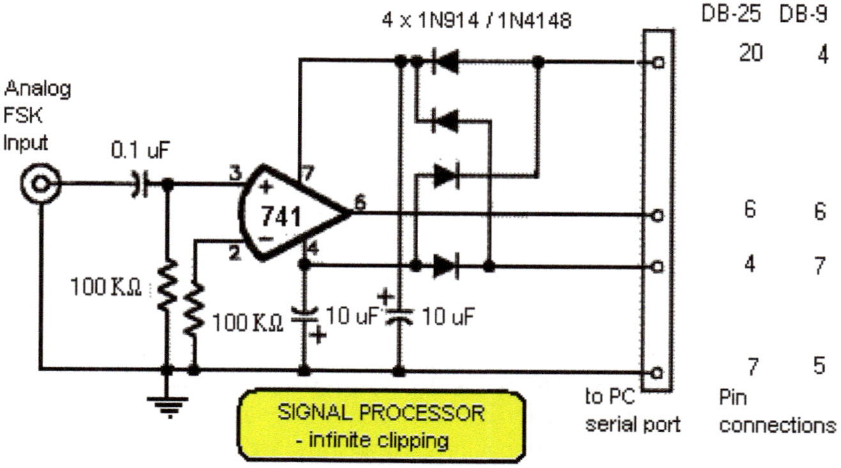

Figure 12.9 Simple signal processor.

an inadequate number of receiving stations around the globe to ingest the data. Many of the satellite owners (often universities with limited funds) are pleading for people to receive and send them the data from their vehicles. There is thus a great opportunity for individual participation in this field.

The FunCubeSat organization in the United Kingdom is making available a small receiving dongle that can be connected between an antenna and a USB port of a PC to enable reception and decoding of such satellite telemetry [2].

Individual cubesat (and other smallsat) websites often have details of their telemetry format.

Commercial geosynchronous communication satellites use a beacon for tracking purposes. These are often K-band beacons as the higher frequencies (and thus shorter wavelengths) allow for more accurate radiometric measurements (range and velocity).

12.1.11 Optical Beacons

Most beacons carried by satellites have transmitted radio frequencies. However, an unusual satellite built by Japanese engineering students at Fukuoka Institute of Technology in Japan carried a high-intensity LED modulated by Morse code to make visual patterns across the sky as seen by a ground observer.

A few other satellites have carried xenon flashtubes for optical beacon test purposes.

12.1.12 An Ionospheric Beacon Ground Station

Satellite radio beacons offer a good opportunity to introduce students to the elementary concepts of space systems: acquisition, tracking, downlink signal calculations, data decoding and display, as well as a study of the near-space environment. Figure 12.10 outlines a system to monitor ionospheric TEC from Faraday rotation

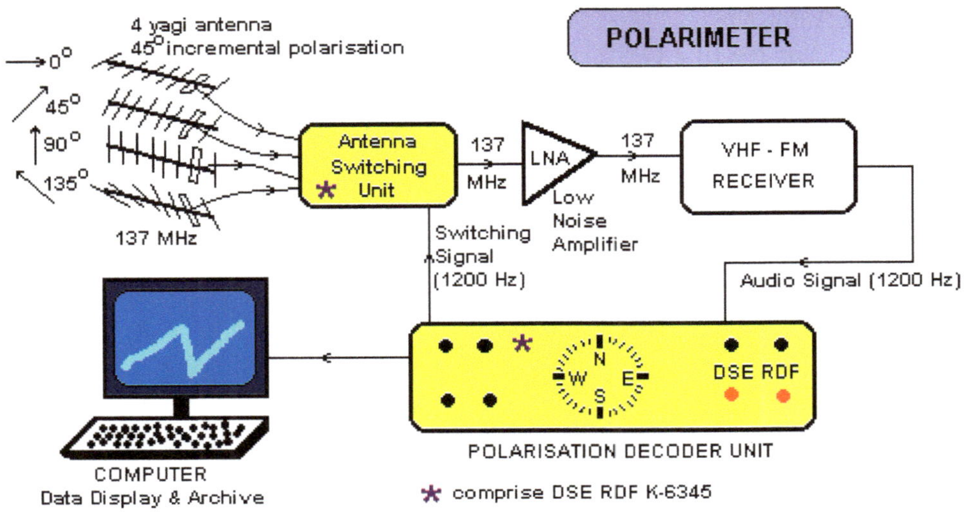

Figure 12.10 A satellite receiving system to measure satellite signal polarization changes.

measurements. A knowledge of both is important in satellite communications, tracking, and radio astronomy.

All components of the system are essentially off-the-shelf. The two nonstandard items are the antenna switching unit and the polarization decoder unit. These were originally available as one package in the form of a radio direction finding adaptor from Dick Smith Electronics (catalog K-6345). In its intended application, these two devices are used with an FM receiver to simulate a rotating antenna. The rotation results in a sinusoidal Doppler shift or frequency variation. Details and schematics were given in the *Electronics Australia* magazine for February 1986.

The decoder unit performs a phase comparison of the received signal variation with the reference switching frequency to locate the direction of the incident signal. This process is exactly that required to measure Faraday rotation. In this instance the four switched antennae have linear polarizations offset in steps of 45 degrees. (With a little ingenuity and vector addition of signals, two orthogonal antennae are sufficient.) A single antenna with a continuously rotating plane of polarization is thus simulated. Comparison of the signal, suitably amplified by the receiver, from this virtual rotating antenna with the reference signal indicates the required plane of polarization.

The output indication is by 32-front panel LEDs. A 5-bit digital representation of this is readily available within the device (small circuit variations could increase this to 6 bits for greater accuracy). This digital signal could be fed to a PC for display and archive purposes. Regular diurnal variations in the plasma density of the ionosphere will be readily apparent. On some days ionospheric storms may be noted.

12.2 Communication Delay

Space communications are carried by electromagnetic radiation, usually by radio waves, but increasingly by light. The speed of propagation of these carriers is about 300,000 km/s, the speed of light (and all other electromagnetic radiation). While this is an incredibly high speed, it does cause problems in space communication (see Table 12.2).

Only the first two circuits listed have a delay that is barely noticeable in a two-way voice conversation. On telephone calls between continents that are routed via a geosynchronous satellite, the time between when one person stops speaking and

Table 12.2 Communications Delay in Various Terrestrial and Space Circuits (One Way)

Circuit	*Distance*	*Delay Time*
HF link (UK-NZ)	~20,000 km	0.07s (67 ms)
Submarine cable(UK-NZ)	~20,000 km	0.07s (67 ms)
Geosat Link (US-Aus)	~80,000 km	0.3s
Earth-Mars	55–378 million km	3–21 minutes
Earth-Jupiter	590–970 million km	33–53 minutes
Earth-Pluto	~5,800 million km	5 hours
Earth-Nearest Star	~40 trillion km	4 years

then hears the other person reply is half a second. This can cause immense confusion if the other person starts speaking before the first has finished. It can take several sentences before the confusion is finally sorted out. Two-way digital communication between machines at high-data rates suffers from this problem even more acutely, and protocols must be established that prevent conflicts arising.

Communication with a future lunar base will be worse, and for voice communications will necessitate an "over to you" simplex radio-communications-type approach. Two-way interactive communication with any station beyond the moon is basically impossible. Although there are no manned bases currently on other planets, this delay is presently of great concern to people who send remotely controlled spacecraft to Mars. There is no possibility of detecting an incipient vehicular crash in time to do anything about it. The vehicle must thus be given a very large degree of autonomous control (e.g., for navigation).

12.3 Radio Frequencies for Space Communication

12.3.1 Introduction

To be useful, satellites and spacecraft must communicate, sometimes to relay communications between two points, sometimes to transmit data they have collected. Although there have been some experiments in optical communications using lasers, most satellite communication is accomplished by radio, one part of the electromagnetic spectrum. Radio frequencies must be shared with terrestrial radio services, and international frequency assignment is essential to avoid interference between all the different uses made of the radio spectrum.

The International Telecommunications Union (ITU) is the global body that assigns radio frequency allocations. In doing this they divide the world into three regions, regions I, II, and III. Australian lies in region III. The general frequency assignments may be found at the ITU website [3], and Australian allocations may be seen at the ACMA website [4]. The ITU unfortunately charges horrendous prices for any of its publications. However, the Australian Communications and Media Authority (ACMA) makes all spectral allocation information available for free download. This includes a book, an attractive wall poster, and a simplified spectrum graphic.

This section discusses the frequencies that are used for space communications.

12.3.2 The Electromagnetic Spectrum

There are four, and only four, known forces in the universe (although so-called dark energy hints at another, if it exists). These are, in order of strength, the nuclear strong force, the electromagnetic force, the nuclear weak force, and the gravitational force. The two nuclear forces exert their influence over only very short (nuclear) distances, and apart from holding all matter together, do not directly influence us in everyday life. It is gravity and particularly electromagnetism that are of direct concern to us in our daily interactions.

Gravity springs from the property of matter we call mass, while electromagnetic effects derive from the property we call charge (Figure 12.11). When a charge is

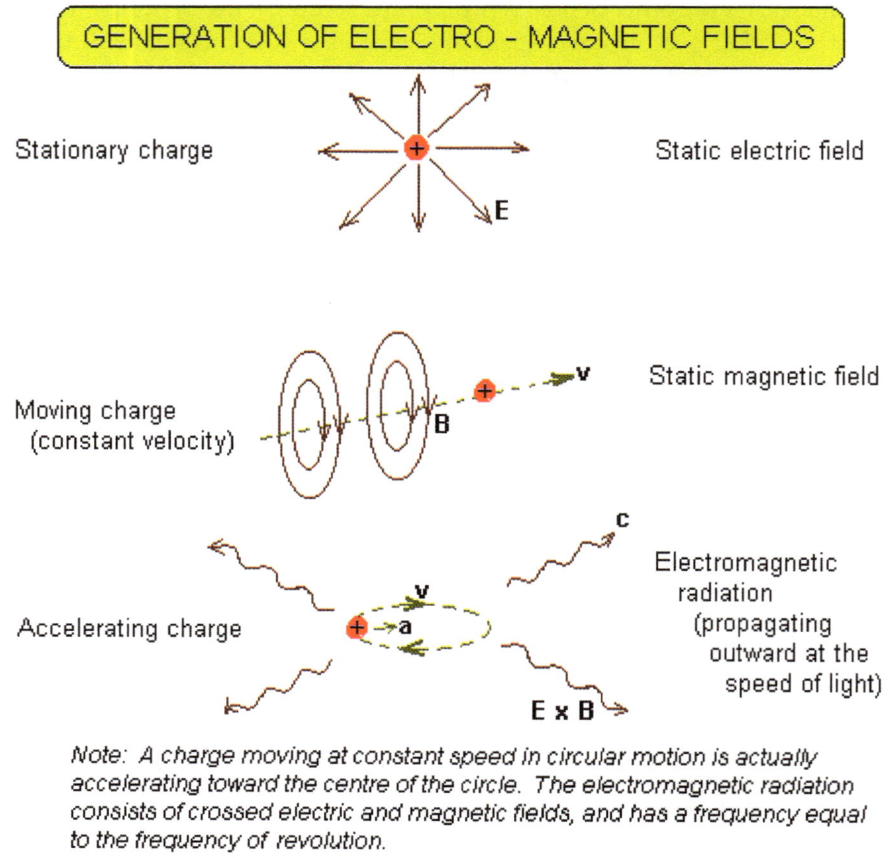

Figure 12.11 The different fields around a charge, dependent upon its motion.

stationary, it has around it an electrostatic field. If it moves with a constant velocity it produces a magnetic field, and when it accelerates or decelerates it generates electromagnetic radiation.

Electromagnetic radiation is a coupled oscillation of electric and magnetic fields that propagates through space with a velocity of about 3×10^8 meters per second. The properties of this electromagnetic radiation vary markedly depending on the frequency of the oscillation. This gives rise to what we know as the electromagnetic spectrum.

Figure 12.12 shows the major divisions of the electromagnetic spectrum. An electromagnetic wave may be characterized by its frequency f (the number of times

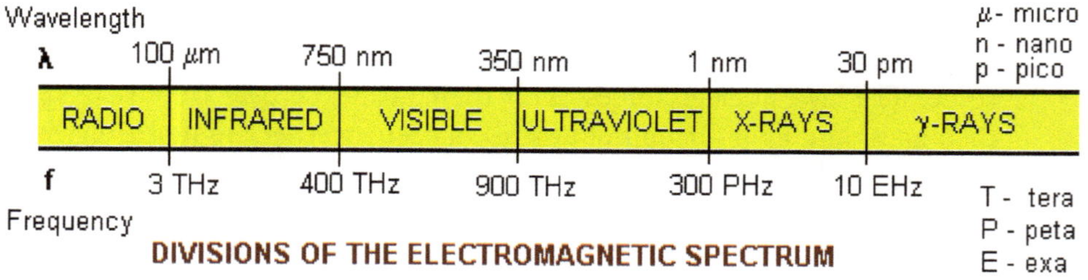

Figure 12.12 The entire electromagnetic spectrum.

per second the signal undergoes a complete oscillation at a specified point in space), or its wavelength λ (the distance between successive extremal values of the wave at a specified time).

Figure 12.12 shows a basic division of the entire electromagnetic spectrum. Note:

1. For clarity the bands are not shown with uniform frequency or wavelength spacing;
2. The visible spectrum occupies only a very small part of the total EM spectrum;
3. Bands also have subdivisions (this is particularly true of the radio spectrum);
4. The band divisions are not as sharp as shown, but rather fuzzy, merging into one another;
5. In the frequency scale $T = 10^{12}$, $P = 10^{15}$, $E = 10^{18}$;
6. In the wavelength scale $\mu = 10^{-6}$, $n = 10^{-9}$, $p = 10^{-12}$.

12.3.3 The Radio Spectrum

The radio spectrum is a subset of the electromagnetic spectrum. It extends from frequencies below 1 Hz up to around 3,000 GHz or 3 THz, where it gives way to the infrared spectrum. Different frequencies have different uses because of different propagation, generation, and general properties. The radio spectrum is divided into many different bands (Figure 12.13).

This figure shows the usually accepted division of the radio spectrum. The left-hand column lists the frequency (f), the center column is the band designator, and the right column is the wavelength (λ). The relationship between frequency and wavelength is given by the expression:

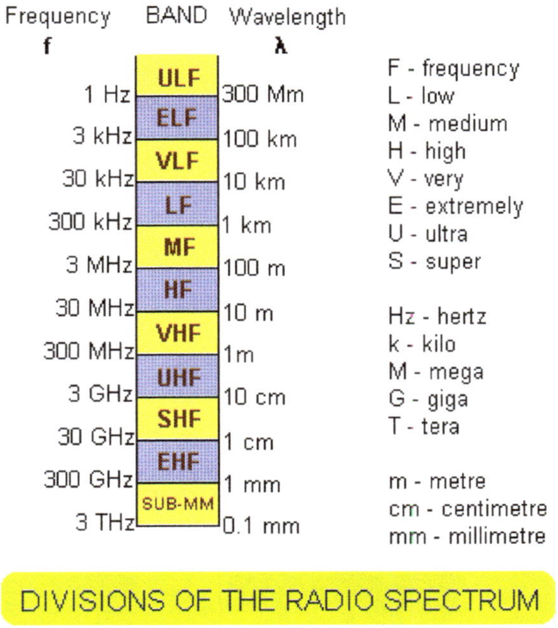

Figure 12.13 The radio spectrum.

$$f = \frac{c}{\lambda}$$

where $c = 3 \times 10^8$ meters per second and is the speed of light (and other EM radiation) in free space.

In the above relation, frequency is given in hertz (Hz) when wavelength is specified in meters (m).

Note that the designated band qualifiers are not in the same order going toward lower frequencies as they are going toward higher frequencies. Also note that the division between ELF and ULF is not universally agreed. It can be placed anywhere from 1 to 100 Hz. The one shown (1 Hz) is that employed by geophysicists.

There is no lower limit to the ULF band and magnetic signals with periods of years can be identified.

Microwave is a term that was historically applied to signals with wavelengths less than 1 ft (30 cm), and this region has been subdivided into letter bands. However, there are several schemes of designation for microwave bands (Figure 12.14). Two of these, which we shall call traditional and new, are shown. Despite the efforts of many engineers to have the new division adopted, the traditional scheme seems to be firmly entrenched among space communicators.

12.3.4 Windows to Space

Not all of the electromagnetic spectrum can pass through the Earth's atmosphere. Obviously, visible light can—we can see the stars at night, at least when there are no clouds. However, ultraviolet and higher frequencies are mostly absorbed by different components of the atmosphere.

There are, in fact, only two main windows of the EM spectrum that are open to space (Figure 12.15). One is the visible spectrum, as mentioned above, and the other is the radio spectrum. However, not all of the radio spectrum is useable for space communication. The available window spans from about 30 MHz to 30 GHz, although these are not absolute end frequencies.

Traditional	f range(GHz)	New	f range(GHz)
L	1 - 2	D	1 - 2
S	2 - 4	E	2 - 3
C	4 - 8	F	3 - 4
X	8 - 12	G	4 - 6
Ku	12 -18	H	6 - 8
K	18 - 27	I	8 - 10
Ka	27 - 40	J	10 - 20
V	40 - 75	K	20 - 40
W	75 - 110	L	40 - 60
mm	110 - 300	M	60 - 110
sub-mm	> 300		

MICROWAVE RADIO BANDS

Figure 12.14 The microwave radio spectrum.

Below 30 MHz, the ionosphere, at altitudes from around 100 to 500 km, absorbs and reflects signals. Above 30 GHz, the lower atmosphere or troposphere, below 10 km, absorbs radio signals due to oxygen and water vapor. Even between 20 and 30 GHz, there are some absorption bands that must be avoided.

12.3.5 Historical Space Frequencies

The first satellite to orbit the Earth was *Sputnik 1*, launched by the Soviets in October 1957. It carried two radio beacons on frequencies of 20.005 and 40.01 MHz.

The Soviets continued to use frequencies around 20 MHz, and even some around 15 MHz for many subsequent missions.

The first satellite launched by the United States was *Explorer 1*, which carried beacons on 108.00 and 108.03 MHz. This lay just above the terrestrial FM broadcast band (from 88 to 108 MHz) and just inside the civil aviation band, which extends from 108 to 136 MHz. This frequency had been specified by an international committee for the International Geophysical Year (IGY–1957/8) as the one to be used for all scientific satellites launched in pursuit of IGY objectives. The Soviets had chosen to ignore this recommendation and use the much lower frequencies previously mentioned.

12.3.6 Space Communication Bands

The following is a list of some of the more heavily used frequency bands for space communication. Specific frequencies may be found in the links provided at the end of this section.

VHF Band

- *136–138 MHz:* This band was used heavily by many different types of satellites in the past. Today (2012), most activity is restricted to 137–138 MHz (which is the current allocation), and consists of meteorological satellites transmitting data and low-resolution images, together with low data rate mobile satellite downlinks (e.g., Orbcomm).

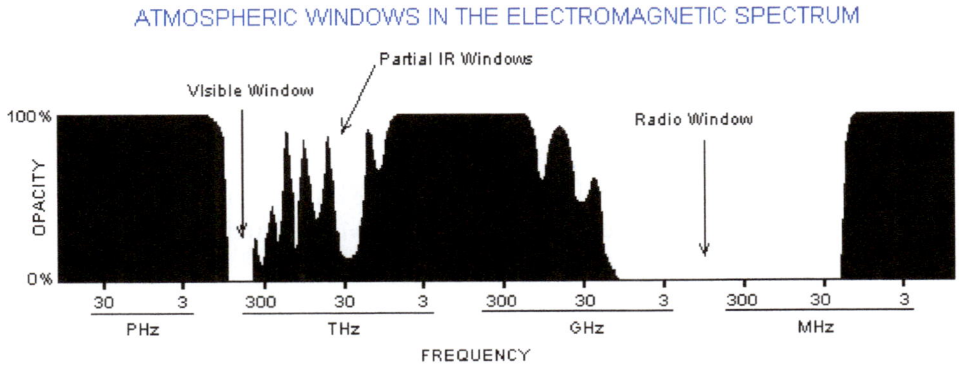

Figure 12.15 Atmospheric transparency.

- *144–146 MHz:* One of the most popular bands for amateur satellite activity. Most of the links are found in the upper half of the band (145–146 MHz).
- *148–150 MHz:* This tends to be used for uplinks of the satellites that downlink in the 137–138 MHz band.
- *149.95–150.05 MHz:* This is used by satellites providing positioning, time and frequency services, by ionospheric research, and other satellites. Before the advent of GPS it was home to large constellations of US and Russian satellites that provided positioning information (mainly to marine vessels) by use of the Doppler effect. Many satellites transmitting on this band also transmit a signal on 400 MHz.
- *240–270 MHz:* Military satellites, communications. This band lies in the wider frequency allocation (225–380 MHz) assigned for military aviation.

UHF Band

- *399.9–403 MHz:* This band includes navigation, positioning, time and frequency standard, mobile communication, and meteorological satellites. Around 400 MHz is a companion band for satellites transmitting on 150 MHz.
- *432–438 MHz:* This range includes a popular amateur satellite band as well as a few Earth resources satellites.
- *460–470 MHz:* Meteorological and environmental satellites, includes uplink frequencies for remote environmental data sensors.
- *900 MHz:* Iran has used frequencies in this band.

L Band

- *1.2–1.8 GHz:* This frequency includes a very diverse range of satellites, and encompasses many suballocations. This range includes the GPS and other GNSS (Global Navigation Satellite Systems: Russian *Glonass*, European *Galileo*, Chinese *Beidou*). It also hosts SARSAT/COSPAS search and rescue satellites that are carried on board US and Russian meteorological satellites. It also includes a mobile satellite communication band.
- *1.67–1.71 GHz:* This is one of the primary bands for high-resolution meteorological satellite downlinks of data and imagery.

S Band

- *2.025–2.3 GHz:* Space operations and research, including deep space links from beyond Earth's orbit. This encompasses the Unified S-band (USB) plan that is used by many spacecraft, and which was also used by the Apollo lunar missions. It also includes military space links, including the US Defense Meteorological Satellite Program (DMSP). Many Earth resources (remote sensing) satellites downlink in this band.
- *2.5–2.67 GHz:* Fixed (point-to-point) communication and broadcast satellites, although the broadcast allocation is only used in some Asian and Middle Eastern countries.

C Band

- *3.4–4.2 GHz:* Fixed satellite service (FSS) and broadcast satellite service (BSS) downlinks. International TV broadcast uses this allocation heavily.
- *5.9–6.4 GHz:* This is the FSS/BSS uplink for the 3.4–4.2 GHz downlink band.

X band

- *8–9 GHz:* This is used heavily for space research, deep space operations, environmental and military communication satellites. Many satellites/spacecraft carry complementary S- and X-band transmitters.

Ku band

- *10.7–11.7 GHz:* Fixed satellite services.
- *11.7–12.2 GHz:* Broadcast satellite service downlinks. This band is used for domestic TV programs.
- *14.5–14.8 GHz:* The uplink for the previous Ku downlink band.
- *17.3–18.1 GHz:* An alternate Ku-band BSS uplink.

Ka band

- *23–27 GHz:* A region that will be used increasingly by a variety of fixed link, broadcast, environmental, and space operations satellites in the future as more bandwidth is required than can be provided in the lower bands. The disadvantage of this band is the increased absorption due to water vapor and rain. Not very useful for tropical regions of the Earth.

12.3.7 Specific Space Communication Frequencies

12.3.7.1 Russia

Russian manned spacecraft use 143.625 and 121.5 MHz FM for voice communications. Other frequencies used on manned missions include 166 and 923 MHz. The Russian ISS module uses the band from 628–632 MHz.

12.3.7.2 China

China uses 180 MHz for weather satellite downlinks, and possibly for manned missions. Meteorological satellites also use 480 MHz for downlink.

12.3.7.3 North Korea

North Korea made five attempts to launch their own satellite, of which the last two were successful. Named *Kwangmyongsong* (Bright Star), *KMS-4* was launched in 2012, and *KMS-5* in 2016. They have stated that a slogan (Juche Korea) was transmitted in Morse code on 27 MHz and two revolutionary hymns on 470

MHz. A reconnaissance satellite (*Malligyong-1*) launched in November 2023 and is believed to use 8 GHz for imagery downlink and a UHF frequency of 468.738 with FM modulation.

12.3.7.4 Amateur Satellites

Many satellites have been launched which use the amateur radio bands for downlinking data, telemetry, and imagery, and for providing relay communications and store and forward communications. Most of the amateur radio frequency bands have a satellite allocation subband. The most popular bands are in the 144–146 and 435–438 MHz range. The Russians have often used the HF bands at 21 and 29 MHz for amateur communications.

A very popular frequency for many amateur satellites is 145.825 MHz. Amateur Radio on the International Space Station (ARISS) typically uses frequencies in the 144–146 band.

12.3.7.5 NASA Deep Space Network

Three NASA ground stations in the Deep Space Network, Goldstone (California), Tidbinbilla (Canberra), and Madrid (Spain) provide data and tracking services for all NASA spacecraft outside of Earth's orbit. These use S-, X- and Ka-bands (Table 12.3).

The earliest DSN spacecraft used S-band (1960s), then in the 1990s moved to X-band, and Ka-band started to be used in the twenty-first century. Many spacecraft have dual frequency capability (S/X and lately X/Ka).

The use of higher frequencies allows larger bandwidths, better tracking capability, and minimizes ionospheric effects. It also requires greater pointing accuracy.

12.3.8 Links

The links below provide detailed information about specific frequencies of various types of satellites:

Commercial satellite communications (broadcast): http://www.lyngsat.com

Meteorological satellites: http://mdkenny.customer.netspace.net.au/metsat_frequencies.html

Amateur satellites: http://www.amsat.org/amsat-new/satellites/frequencies.php

Monitored satellites: http://www.zarya.info/Frequencies/FrequenciesAll.php

Table 12.3 Specific Frequency Allocations for Space Use

Band	*Uplink Frequency (MHz)*	*Downlink Frequency (MHz)*
S	2110–2120	2290–2300
X	7145–7190	8400–8450
Ka	34200–34700	31800–32300

12.4 Space Communication Calculations

12.4.1 Introduction

In designing systems for space communication, it is essential to know the transmitter power that will be required and the antenna sizes, both in space and on the ground, which will allow adequate effective communication with acceptable signal-to-noise ratio or bit error rate.

Fortunately, communication in space is usually line of sight with no intervening absorption and little scattering media involved. It is therefore easy to calculate and predict the actual communication parameters quite accurately.

12.4.2 The Transmitting End

The basic physical principle behind all space communication is the inverse square law. This expresses the fact that all electromagnetic radiation spreads out as it propagates, and has an intensity that is proportional to the reciprocal (or inverse) of the square of the distance from the source. This is expressed mathematically as:

$$I = \frac{k}{r^2}$$

where r is the distance from the emitter and k is the proportionality constant.

This law is illustrated in Figure 12.16.

At a distance r, the beam passes through an area A. At a distance of $2r$, the beam passes through an area of $4A$, and at a distance of $3r$, the beam passes through an area of $9A$. It can be seen that the area increases as the square of the distance. If no absorption of the beam energy occurs, the same power must pass through each area. At a distance of $2r$ each square of area only receives one-quarter the total beam power, and at a distance of $3r$ each unit of area only receives one-ninth of the total beam power. Thus the flux density of the signal (i.e., the power per unit area) is proportional to the reciprocal square of the distance.

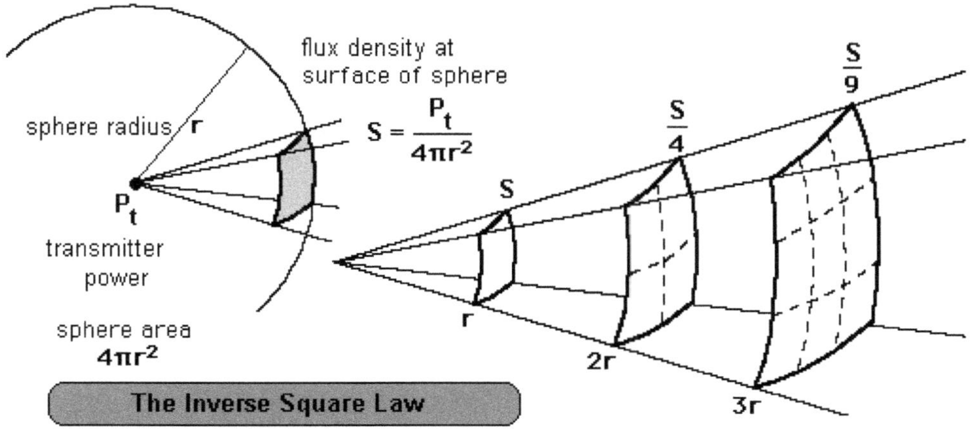

Figure 12.16 Illustrating the inverse square law of signal propagation.

How do we determine the proportionality constant? Imagine we have an isotropic radiator, a transmitter that broadcasts a power of P_t uniformly in all directions. If we also imagine a sphere surrounding the transmitter, with the transmitter at its center, then the power at each point on the sphere will be the same. Now we know that the total surface area of a sphere of radius r is $4\pi r^2$. All the transmitted power must pass through this area, and so the flux density of the signal at the surface of the sphere is

$$S = \frac{P_t}{4\pi r^2}$$

This is also shown in Figure 12.17.

Note: If we consider the matter closely, we see that the inverse square law is a consequence of us living in a three-dimensional universe. In a 2D universe we would have an inverse distance law where $S_{2D} = P_t/(2\pi r)$. That is, in 2D the power falls off as the inverse of the distance, not the inverse square of the distance (this means that the power falls off at a less rapid rate).

Very few transmitting systems are isotropic radiators. Normally they send out signals in a beam (Figure 12.18), which concentrates power along the boresight of the antenna at the expense of reducing the signal in other directions. The beam has a beamwidth, which may be broad (if the antenna gain is low) or narrow (when the antenna gain is high).

Nonisotropic radiators still obey the inverse square law. Even laser beams, which have extremely narrow beams, obey the inverse square law. No device has yet been made whose radiated energy does not diverge with distance from the emitter. For a nonisotropic emitter with an antenna gain G_t we use the effective radiated power (the power at the center of the beam) where:

$$P_{\text{eff}} = G_t P_t, \text{ and thus}$$

$$S = \frac{G_t P_t}{4\pi r^2}$$

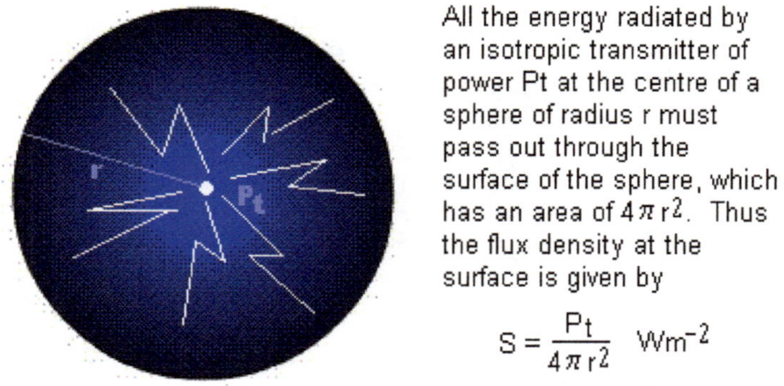

Figure 12.17 Computing signal flux density at a distance r from a transmitter of power P_t.

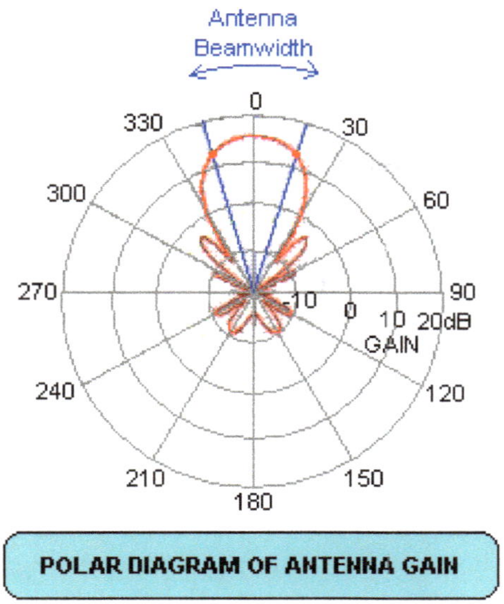

Figure 12.18 Antenna polar diagram showing beamwidth and gain.

12.4.3 The Receiving End

At the receiving end is an antenna that intercepts a portion of the transmitted power. The main parameter associated with a receiving antenna is its effective area. For a parabolic antenna, as shown in Figure 12.19, the effective area is a fraction 'e' of the physical cross section that the antenna presents to the incoming signal. That fraction 'e' is thus termed the efficiency of the antenna. For a different type of antenna, the effective area can be determined as described later.

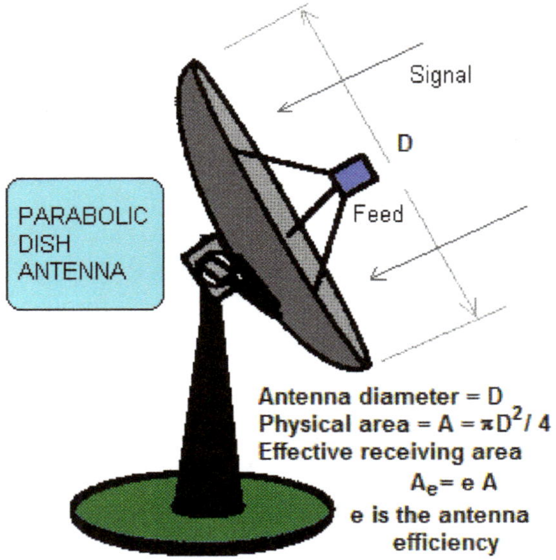

Figure 12.19 The antenna as a collector.

Now the units of flux density are watts per square meter (W/m^{-2}) and the units of area are square meters. The power (in watts) collected by the antenna is thus the flux density times the effective collecting area of the antenna:

$$P_r = SA_e$$

Substituting for S and A_e we find that:

$$P_r = \frac{eD^2 P_{\mathrm{eff}}}{16r^2}$$

Now in any receiving system this signal power has to contend with a noise power. This noise can arise from external sources but it is also unavoidably generated in the receiver by normal electronic processes associated with active devices, such as transistors that are necessary to amplify the signal. Even resistors generate noise, and this noise is proportional to the temperature of the resistor. It is thus usual to talk about the noise temperature of a receiver as the temperature of a matched resistor connected to an ideal (no-noise) receiver that generates the same noise power output as the actual receiver.

This noise power is given mathematically by the formula:

$$P_n = kBT$$

where

k is a constant (called Boltzmann's constant);

B is the bandwidth of the receiver (in Hz);

T is the equivalent noise temperature of the receiving system.

In SI units, k has the value 1.38×10^{-23}.

For a signal to be detectable it must usually be greater than some number 'n' times the noise. That is:

$$P_r > nP_n$$

The value $n = P_r/P_n$ is usually termed the signal-to-noise ratio (although an alternative definition of SNR is SNR = $(P_r + P_n)/P_n$). We will use the first definition. SNR is often expressed in decibels. Thus $n = 2$ implies SNR = 3 dB, $n = 10$ implies SNR = 10 dB, and $n = 100$ implies SNR = 20 dB.

For voice communication the SNR must be greater than 3 dB if the signal is to be understood at all. A value of 10 dB is normally the minimum desirable and an SNR of 20 dB ($n = 100$) is required for perceived noise-free communication. An even higher SNR might be required for a noise-free TV image. However, for some types of modulation, in particular that known as *spread-spectrum* modulation, it is still possible to decode a signal for values of n less than 1 (i.e., negative values of the SNR in decibels).

SNR is calculated from the expression:

$$n = \frac{P_r}{P_n} = \frac{eD^2 G_t P_t}{16r^2 kBT}$$

If the collecting area of an antenna needs to be calculated it can be done using what is known as the reciprocity theorem. This states that the gain G of a receiving antenna is identical to the gain of the same antenna used to transmit a signal. This transmitting gain can be measured on an *antenna range* by comparison with an omnidirectional antenna (or one with a known gain). The effective collecting area of the antenna is then calculated from:

$$A_e = \frac{\lambda^2 G}{4\pi}$$

or

$$eD^2 = \frac{\lambda^2 G}{\pi^2}$$

We now have enough information to calculate communication parameters for a wide range of space communication situations (Figure 12.20).

Note: Engineers often use logarithmic versions of these formula so that quantities can be added and subtracted instead of multiplied and divided. With computers this is no longer really necessary. Many engineering textbooks on communication also give an expression for *free-space* path loss in decibels, which depends on frequency. This is entirely misleading, as we have proven above that free-space path loss is entirely due to the geometry of space and is *independent* of wavelength or frequency. The reason the engineers get confused is because they use the wrong parameter for the receiving antenna. They use gain instead of collecting area, and in the process introduce a wavelength dependence of the gain, which they then wrongly ascribe to the free-space path loss. These equations work, but give a misleading picture of the physical situation.

For some of the following calculations we need to consider the beamwidth of the antenna. For an antenna with an effective linear aperture 'a' the half-power beamwidth is given by the expression:

$$\theta \approx \frac{\lambda}{a}$$

where

θ is the beamwidth in radians;

λ is the operating wavelength in meters;

a is the aperture in meters.

Note that $\lambda(m) = 300 / \text{freq(MHz)}$.

For a parabolic dish antenna $a = D$ (the dish diameter). For other types of antenna we can compute the effective diameter through the gain equation.

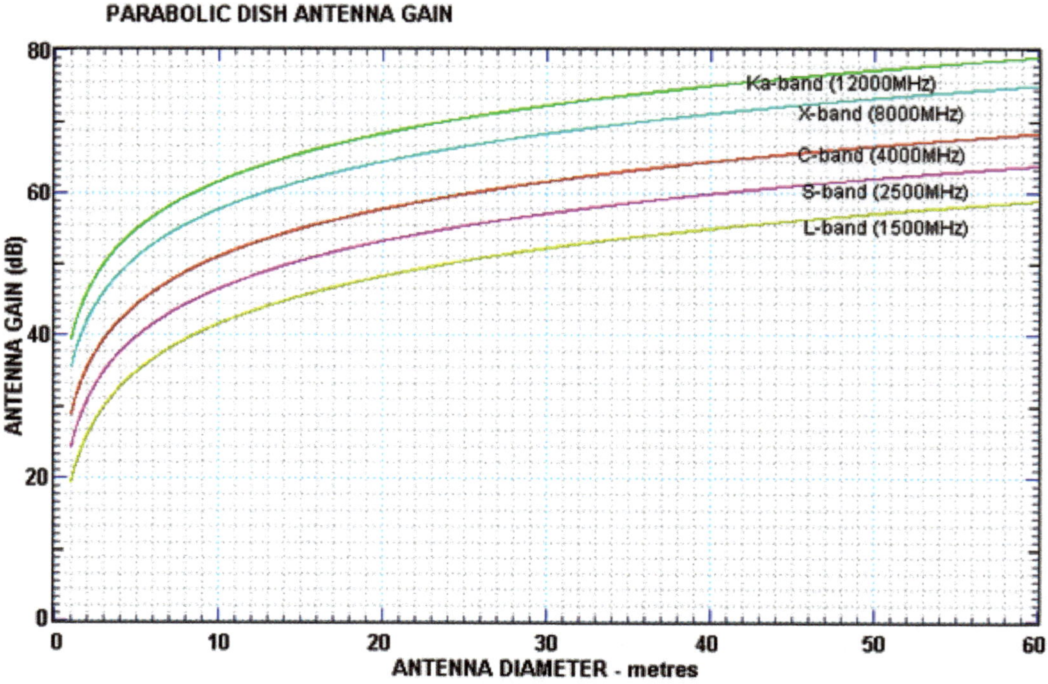

Figure 12.20 Antenna gain for parabolic dish antennae.

The graphs in Figures 12.21 and 12.22 plot antenna beamwidth as a function of antenna diameter.

The following sections investigate the transmitting powers and antenna sizes necessary to give specified SNRs over various distances.

12.4.4 Communicating with the International Space Station

The ISS hosts an amateur radio experiment (ARISS) whereby schools and amateur radio operators can communicate with astronauts. The ISS orbits the Earth at an altitude of around 400 km. This is toward the lower end of the range of orbits called low earth orbit. At this altitude atmospheric drag is significant and the ISS orbit must be periodically boosted to keep it from reentering back to Earth. Thus the orbital altitude may change from about 350 to 450 km.

One of the problems with communications from Earth to LEO is that the satellite moves quickly, and that the distance to the satellite changes by a large factor from when it is overhead to when it is on the horizon (Figure 12.23).

Thus the range to the ISS may vary from ~400 km for an overhead pass to almost 2,300 km when the station is on the horizon, either at acquisition of signal (AOS), or loss of signal (LOS). Since received power varies as the square of the distance, this makes for a 33 times (15 dB) change in signal strength.

The rapid change in direction to the ISS as it passes across the communicator's sky could be a problem for low-cost communication if a directional antenna was required. This is because the directional antenna would have to track the ISS as it

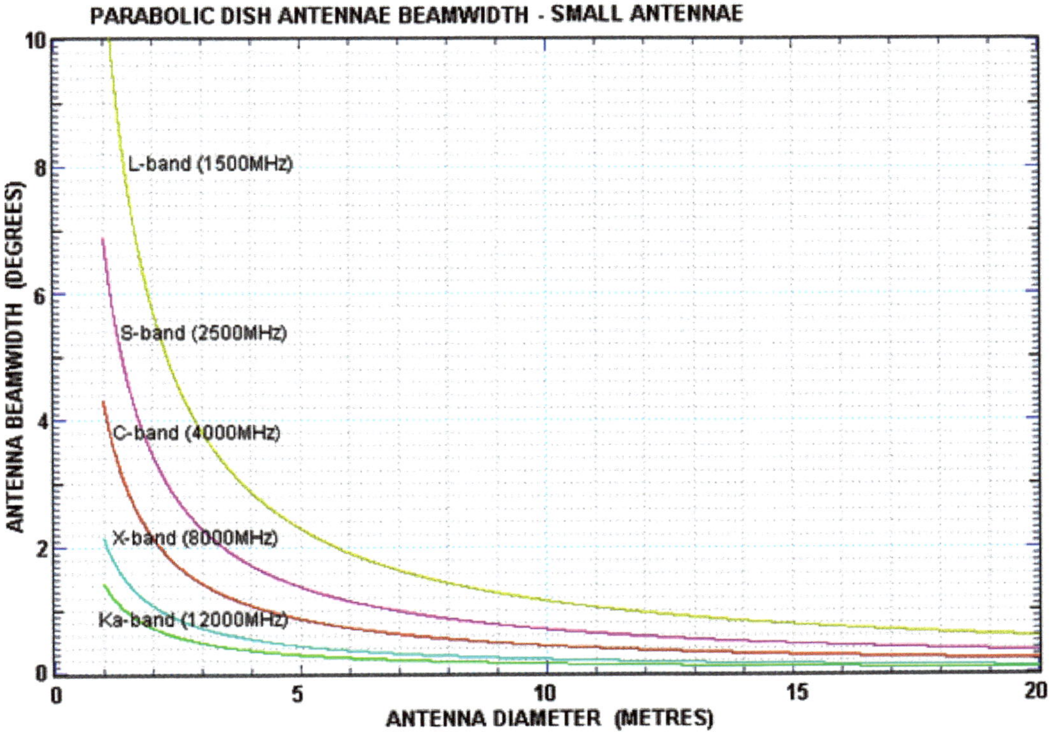

Figure 12.21 Antenna beamwidth for small antennae.

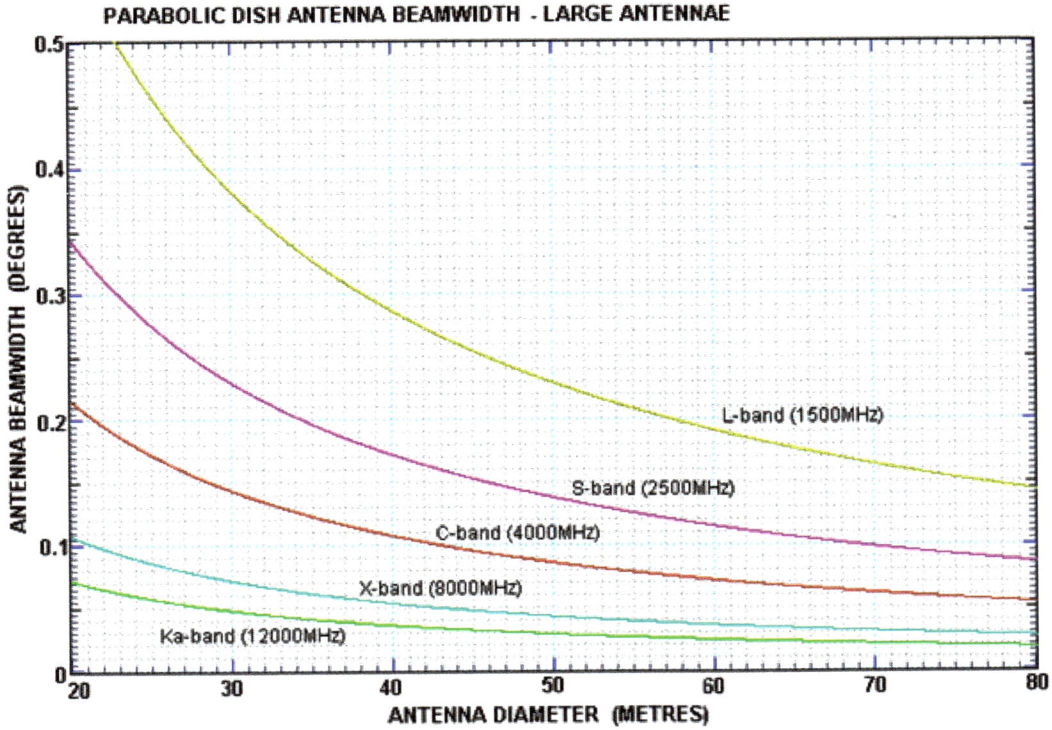

Figure 12.22 Antenna beamwidth for large antennae.

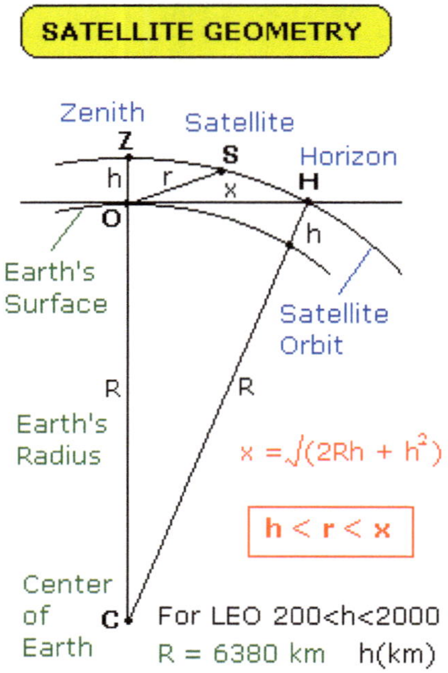

(a)

Satellite Altitude h (km)	Horizon Distance x (km)
200	1610
300	1980
400	2290
500	2570
600	2830
700	3070
800	3290
900	3510
1000	3710
1100	3900
1200	4090
1300	4280
1400	4450
1500	4620
2000	5430

(b)

Figure 12.23 Calculations for satellite observations.

moved. Fortunately, the low frequency used (145.825 MHz) allows the use of hand-held radios with almost omnidirectional whip antennae.

Our problem for computation will thus be to determine the received SNR for the range of distances over which communication must take place. We will do so for three transmit powers, 1W, 3W, and 5W, which are typical of the powers that simple amateur hand-held transceivers might employ. These are effective radiated powers that might result from a transmit power half of that times an antenna gain of two. We also assume a clear line of sight between ground communicator and the ISS. The results are shown in Figure 12.24.

As can be seen, this results in a strong signal for the whole pass of the ISS over the ground station. A 20 dB SNR using narrow-band FM modulation is sufficient to ensure no-noise (full-quieting) communication. Amateurs have long played a significant role in space communications, and have orbited leading edge technology for experimentation and communication.

12.4.5 Low Earth Orbit Communications

A more typical LEO satellite might orbit at around 800-km altitude. From Figure 12.23 this results in ranges out to over 3,000 km during the ground station pass.

Let us consider the case of a polar-orbiting weather satellite at 800-km altitude, and transmitting low-resolution imagery at VHF (137 MHz) and high-resolution imagery at L-band (1700 MHz). We will assume each transmitter radiates with an

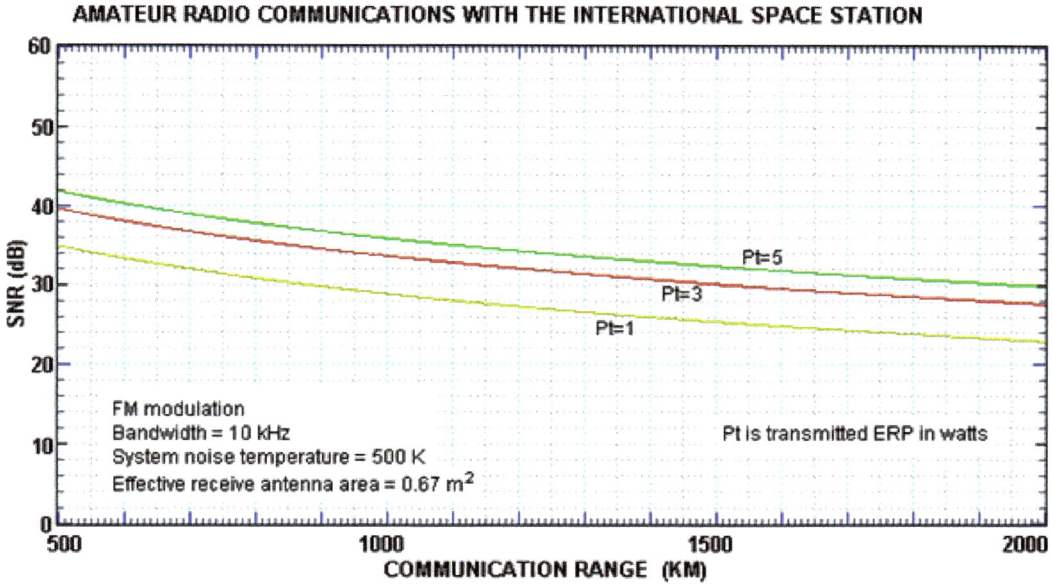

Figure 12.24 Communication with the International Space Station.

effective power of 5W. For the receiver we will need a bandwidth of 70 kHz and a system noise temperature of 300K (Figure 12.25).

For this exercise we will look at the SNR that results from the use of 3 different receive antenna gains. One at 3 dB, which is what we expect from a nontracking omnidirectional antenna, 13 dB, which is the typical gain of a 10-element yagi antenna, and 27-dB gain antenna. The latter can be achieved at L-band with a

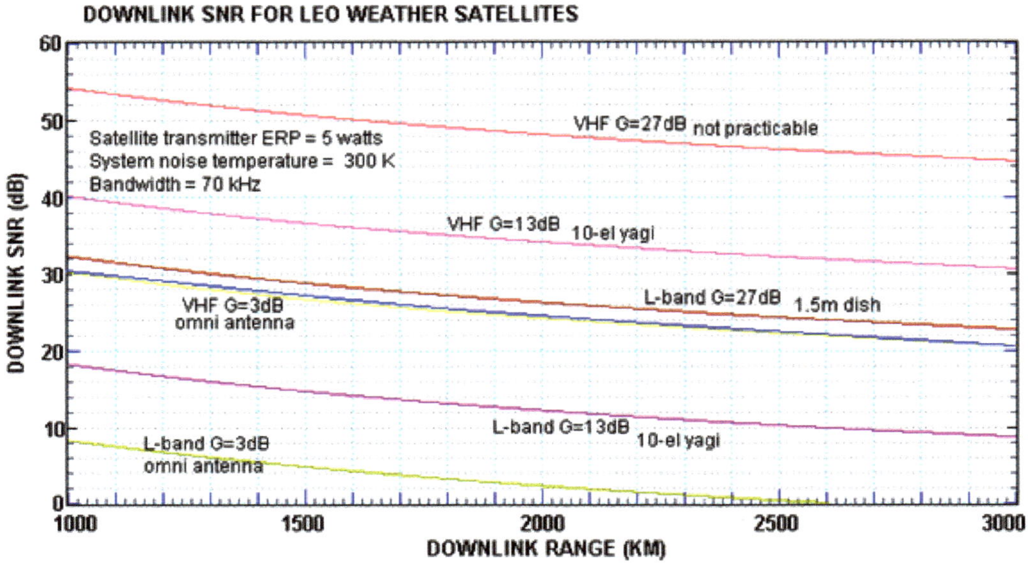

Figure 12.25 Receive signal SNR for meteorological leosats.

1.5m dish, whereas it would require a much larger dish to give this gain at VHF, or alternatively an enormous array of yagis.

We immediately see that a nontracking antenna can be used at the VHF frequency but not at L-band. We really need the 27 dB (1.5m dish) at L-band to get a good quality signal. This means that L-band and above requires a tracking antenna, with all the additional hardware and expense this entails.

In fact, the situation is even worse at L-band than the above figure depicts and that is because we need a bandwidth considerably more than 70 kHz to download the high-resolution imagery. This decreases the SNR. To offset this, the satellite L-band transmitter radiates substantially more power than the VHF one. We are also helped by nature in that cosmic noise, which sets our system noise at VHF, decreases rapidly with frequency and allows the use of a lower system noise temperature at L-band.

12.4.6 Communicating to Geosynchronous Orbit

Geosynchronous satellites orbit at an altitude of 36,000 km. This large altitude means we always need a directional antenna for geosat communications. However, we have two points in our favor. Because the orbit is at a radius 6.6 times larger than the Earth's radius, the variation in range from the zenith to the horizon is not very great (about 17%). We can assume a mean range of 38,000 km in our calculations. The second benefit of the geosynchronous orbit is that such a satellite stays relatively much in the same place in the sky and thus it does not need to be tracked. This makes possible direct-to-home TV broadcasts at low cost (because the antenna can be fixed in look position).

Geosynchronous downlinks vary in power and bandwidth. Bandwidths can be very narrow in the case of beacons, to very large in the case of TV transmissions that might be 36-MHz wide. Powers can vary from a few watts to hundreds of watts. In Figure 12.26 we look at how SNR varies with transmission power (ERP) and bandwidth. We assume a low noise receiving system noise temperature of 50K.

12.4.7 Lunar Communications

Communication with the Moon is simplified in that the distance from the Earth to the Moon is effectively constant, no matter where the Moon is in the sky. In Figure 12.27, we are interested in the terrestrial antenna size we would need to receive data from a scientific package left on the lunar surface, together with a 1W transmitter with a nondirectional antenna.

Although the Earth always appears at the same point in the sky from a fixed position on the Moon, and thus a high-gain antenna pointed to the Earth does not have to track, it must be pointed in the right direction in the first place. This might not be a problem for a sophisticated lander, but it might impose an impossible weight penalty on a simple package.

We can see (Figure 12.27) that even for a low data rate (1 kHz bandwidth) we need at least a 15m dish to get an acceptable SNR (10 dB). Real-time TV (with a required bandwidth in excess of 100 kHz) is simply not practicable in this situation, even with a very large antenna.

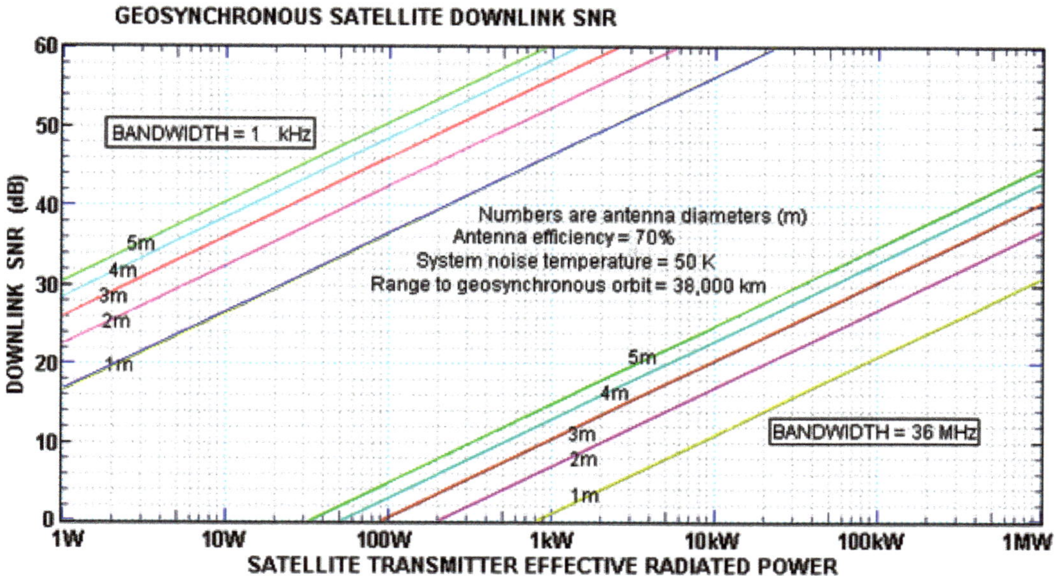

Figure 12.26 Geosat receive signal SNR.

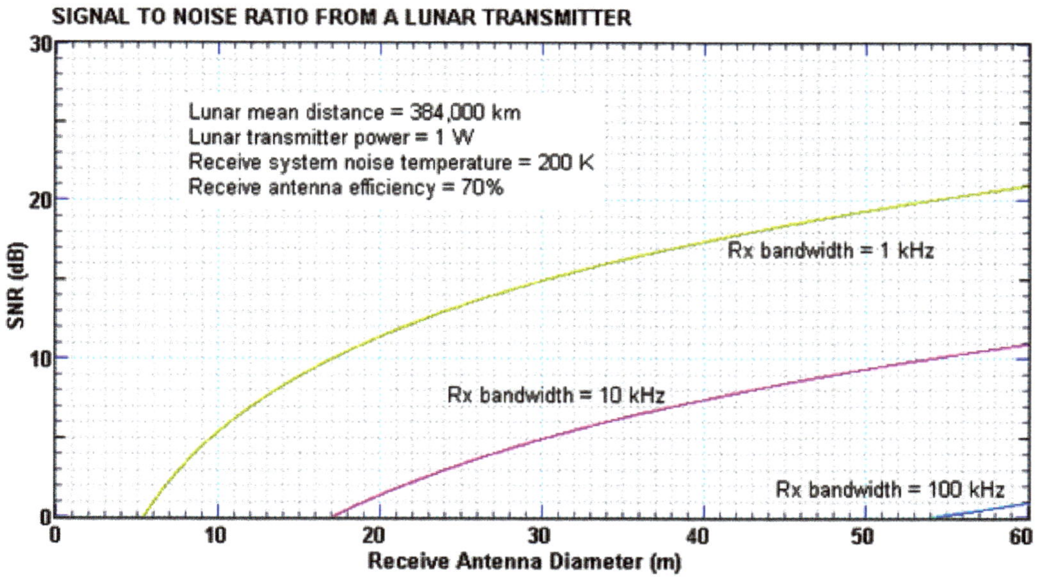

Figure 12.27 Lunar communication downlink parameters.

12.5 Reentry Communication Blackout

12.5.1 Introduction

When a reusable spacecraft or capsule leaves orbit and reenters the atmosphere as it travels to a landing site, there is a critical period of time when all communications between the spacecraft and ground are lost. This phenomenon is due to the tremendous heating experienced by the craft during reentry and is termed *reentry blackout* (Figure 12.28).

In low Earth orbit a spacecraft is travelling at almost 8 km/s. To land safely on the ground this speed must be reduced to zero by making use of atmospheric drag. What NASA calls the *entry interface* commences when a spacecraft descends to around 80-km altitude. It does this by deorbit thruster firing. This rocket burn is performed opposite to the direction of travel. The result of this is a reduction of altitude rather than a reduction in speed. The speed changes little due to an exchange of gravitational potential energy for kinetic energy.

At around 120-km altitude, the atmospheric drag increases significantly and the resultant heating increases as the shuttle kinetic energy is exchanged for thermal energy.

A shockwave forms just in front of the nose and underside of the spacecraft. Between this shock and the vehicle itself, temperatures may reach 10,000K to 12,000K. (The heat resistant surfaces of the shuttle only reach a maximum of 1600K themselves.) This very high temperature ionizes the gas close to the shuttle forming a plasma cloud or miniature ionosphere around the spacecraft. The plasma frequency (that frequency below which radio communications is not possible) may rise to many gigahertz around the lower parts of the vehicle. This gives rise to a communication blackout for direct communications between the Shuttle and ground control. This typically lasts from 25 to 12 minutes prior touchdown, a total outage of 12 to 13 minutes.

Maximum heating of the orbiter occurs during this time frame, at an altitude of 70 km and about 20 minutes prior to touchdown. Unfortunately this is also the most critical time of reentry, and if any problems occur during this phase of flight, the communications blackout prevents any diagnostic telemetry from reaching the ground. Such was the case with the catastrophic breach of the Space Shuttle *Colombia's* hull during reentry on February 1, 2003.

Figure 12.29 illustrates a typical reentry profile, showing the variation of altitude and velocity.

12.5.2 Blackout Parameters

The Saha equation may be used to calculate the ionization of the gas surrounding a deorbiting vehicle as a function of temperature:

Figure 12.28 Impression of a spacecraft reentry. (*Source:* NASA.)

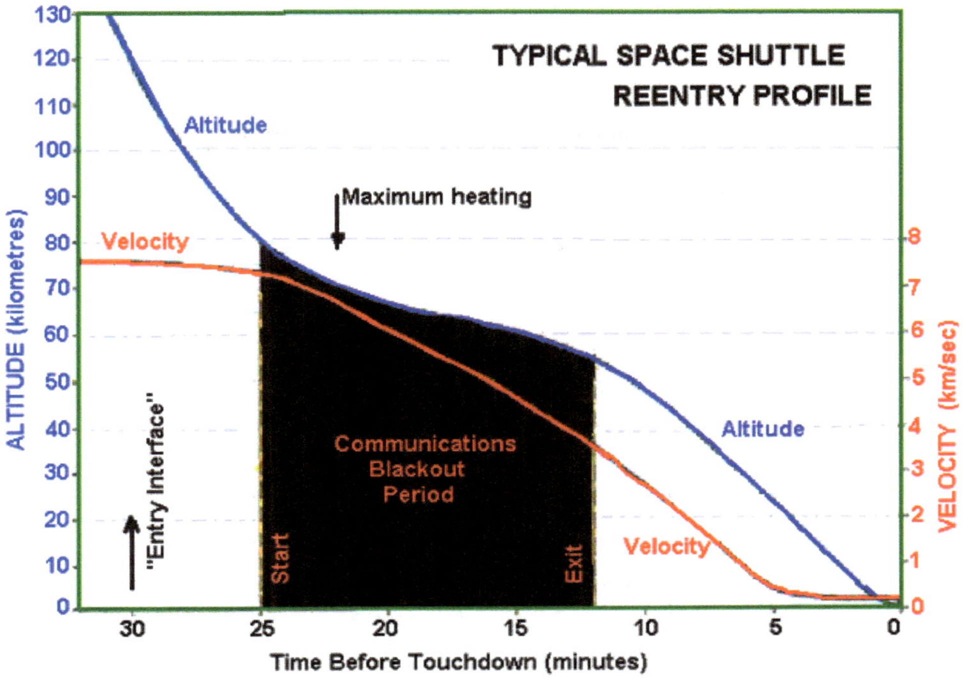

Figure 12.29 Reentry profile for a manned spacecraft.

$$N_e = 5 \times 10^{10} N_a^{0.5} T^{0.75} \exp\left(-5{,}800 \frac{V_i}{T}\right)$$

where N_e is the electron density in electrons per cubic meter, N_a is the density of unionized atoms or molecules, T is the plasma temperature, and V_i is the ionization potential of the atom/molecule. This formula is only strictly valid for less than 10% ionization.

The plasma frequency is given by f_p [Hz] = 9 $N_e^{0.5}$.

At an altitude of 70 km, where maximum spacecraft heating occurs, the atmosphere is still about 20% oxygen and 80% nitrogen. The Saha equation must be applied to these constituents separately. When this is done, the results are shown in Table 12.4 and Figure 12.30.

The Saha model becomes strictly invalid after 8000K, but the curve can be extended smoothly up to the point of complete first ionization, when the number of ions/electrons equals the total atmospheric number density (2×10^{21} per cubic meter at 70 km). Of course, higher temperatures will result in more than one electron being removed from the atoms, with a further increase in plasma frequency, but we will ignore this in the current approximation.

For a given sheath temperature, the frequencies above the curve will propagate to and from the spacecraft, whereas the frequencies below the line will not. For sheath temperatures of 10,000K to 12,000K, we see that a frequency in excess of 350 GHz would be needed for direct spacecraft communication with the ground at the time of maximum heating.

Table 12.4 Plasma Frequencies During Reentry

T (Kelvin)	N_e/m^3	f_p
1500	1.7×10^3	370 Hz
2500	2.9×10^{11}	4.8 MHz
3000	3.4×10^{13}	53 MHz
4000	1.4×10^{16}	1.1 GHz
5000	5.5×10^{17}	6.7 GHz
6000	6.7×10^{18}	23 GHz
7000	4.1×10^{19}	58 GHz
8000	1.6×10^{20}	115 GHz
9000	4.9×10^{20}	200 GHz
high	2.0×10^{21}	400 GHz

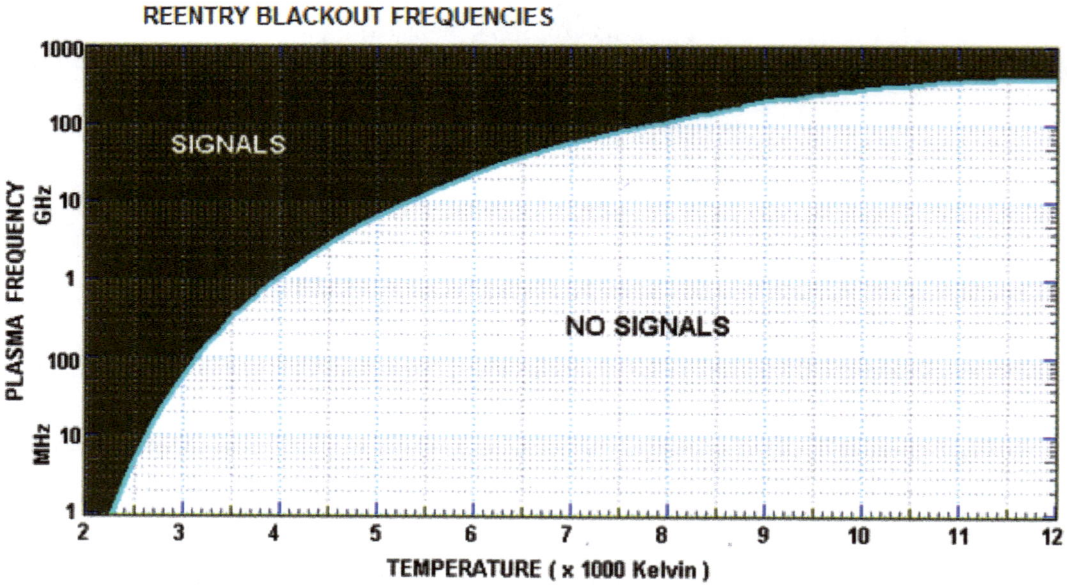

Figure 12.30 Useable frequencies during reentry blackout.

There are, however, ways in which communication is possible during this critical reentry time. The sheath temperatures on the top surface of the spacecraft are much less than those on the bottom that, at a descent angle of about 22 degrees, is where the shock is formed. Communications from the topside via a satellite relay in the gigahertz range may be possible.

A network of communication satellites well above the altitude of the reentering spacecraft could provide communications through an antenna on the upper surface of the spacecraft. A network such as Starlink would be ideal for this purpose.

Other options which have been tried are water injection into the plasma, and the use of very narrow width pulsed signals, both with some success.

12.6 Solar Radio Interference to Geosat Signals

12.6.1 Overview

No long distance communication system is free from environmental disturbances of one kind or another. High-frequency communication depends on the vagaries of the ionosphere. VHF and UHF transionospheric systems experience problems due to rapid changes in the ionospheric electron density. SHF and EHF links are subject to tropospheric attenuation, and all space-Earth links are subject to potential disruption when the Sun enters the beam of the downlink receiving antenna. This latter well-known phenomenon (sometimes referred to as a *sun outage*) is commonly associated with wideband downlinks from geosynchronous satellites, but at times of high solar activity, any electromagnetic receiving system pointed near the Sun may experience disruptive effects.

This section reviews the nature of solar radio emissions and provides appropriate conversion algorithms between solar radio astronomical terminology and useful communication parameters. Algorithms for the computation of the duration and severity of solar disturbances are provided, with graphical examples showing the effect of various parameters.

12.6.2 Electromagnetic Emissions from the Sun

12.6.2.1 General

The Sun, by nature of its high temperature, is a wideband radio emitter. Its size also makes it a source to be reckoned with when low noise electromagnetic receiving systems are pointed toward it. Furthermore, the variability of its output makes it difficult to predict its exact effect on a given system.

It is convenient to divide the solar radio emission into three separate components. These are referred to as the background component, the slowly varying component, and the burst component. Each of these can be identified with separate sources on the Sun.

Figure 12.31 is a photograph of the Sun taken in the light of the hydrogen-alpha spectral line. This shows the solar chromosphere, the *inner atmosphere* of the Sun. It is from this region, together with the lower part of the corona (the Sun's outer atmosphere), that radio emissions in the 1 to 20 GHz range originate. The featureless areas of the solar disc (or the regions above them) may be regarded as the source of the background component. The light-white areas, termed *plage*, are several thousand degrees hotter than the background areas, and these are the primary source of the slowly varying component. Plage areas and sunspot groups occur in the same regions of the Sun, both phenomena being due to the emergence of large magnetic fields from lower depths. The intensely bright area is a solar flare, an explosive process whereby magnetic energy is converted into large quantities (up to 10^{25} joule) of electromagnetic energy. It is these events that are responsible for the burst component.

When measuring the whole disc radio emission of the Sun, radio astronomers typically use the solar flux unit (SFU), which equals 10^{-22} Wm^{-2} Hz^{-1}. Figure 12.32 shows the typical range of flux densities appropriate to each of the three components.

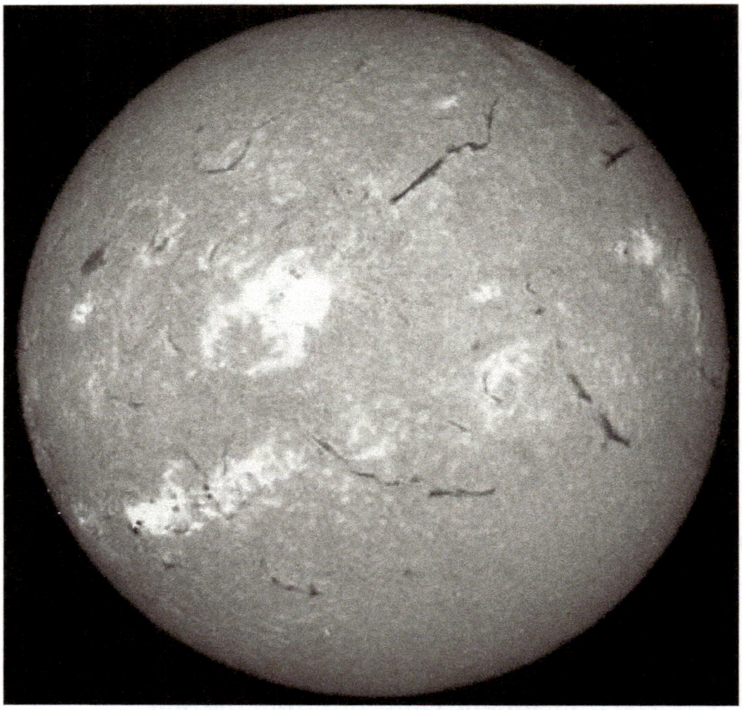

Figure 12.31 The solar chromosphere, the source of solar microwave emission. (*Source:* Learmonth Solar Observatory.)

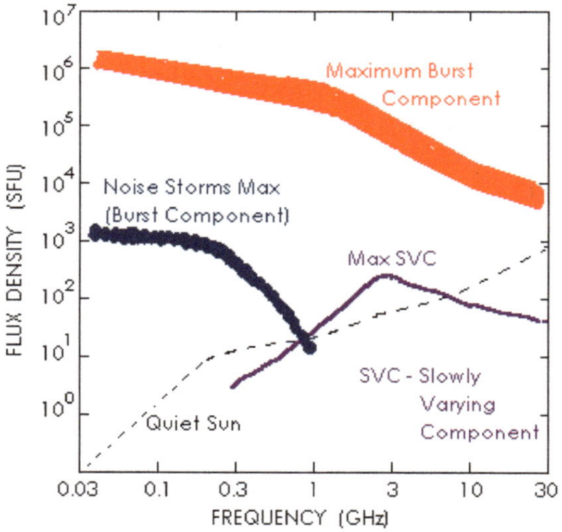

Figure 12.32 Solar radio spectra.

12.6.2.2 The Background Component

The background, or quiet component, of the solar emission is by definition the component that remains when all variable components have been eliminated. It is thus a constant at a given frequency and it is estimated to an accuracy of 10% within the range 1 to 20 GHz by the following second-order polynomial:

$$S_q = 26.4 + 12.4f + 1.11f^2 (1 < f(\text{GHz}) < 20)$$

At the higher frequencies within this range the emission is uniformly distributed across an area only a few percent larger than the visible disc of the Sun. However, at the lower end of the range, the radio Sun is rather larger than the optical Sun, and a limb brightening effect is noticeable (Figure 12.33). This is in contrast to limb darkening that occurs at visible wavelengths. The variation in radio output across the solar disc is only significant when the antenna beamwidth is less than the angle subtended by the Sun (approximately 0.5 degrees).

12.6.2.3 The Slowly Varying Component

The slowly varying component shows a variation on two quite different timescales. The first, of approximately 27 to 28 days, is the rotation period of the Sun at midlatitudes (the equatorial rotation rate is substantially greater than the polar rate) where plage is most predominant. This component is due to the uneven longitudinal distribution. As an area of intense and extensive plage rotates onto the visible disc (as seen from Earth) this will increase. A small plage area may grow and decay within a day. A large area may persist for several solar rotations. This process will also produce random fluctuations with component periods less than a rotation. The second component varies in accompaniment with the sunspot cycle in an approximately 11-year cycle (Figure 12.34).

The slowly varying component is essentially produced by specific regions on the Sun, and thus shows a significant spatial variation over the disc. Examples that are several days apart are shown in Figure 12.35. Again, this spatial variation is

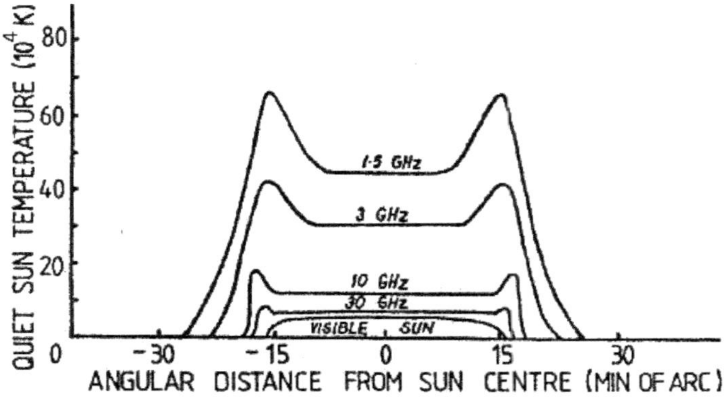

Figure 12.33 Spatial variation of solar microwave emission. (*From:* Kundu [5].)

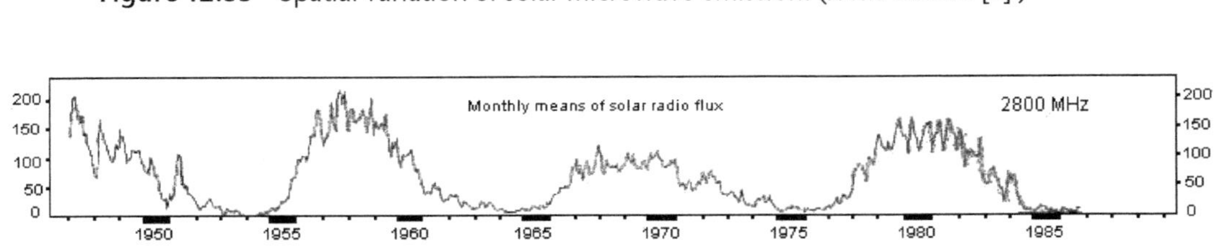

Figure 12.34 The smoothed variation of the slowly varying component with time. Individual values within a month can exceed 300 SFU. (*From:* Algonquin Radio Observatory, SGD [6].)

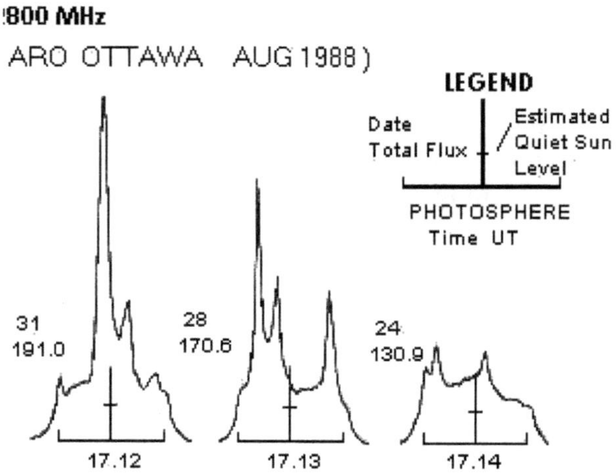

Figure 12.35 Spatial variation of slowly varying component [6].

only important when the antenna beamwidth is a fraction of the angle subtended by the Sun.

The slowly varying component has a maximum value at a frequency of around 3 GHz, and falls off at higher and lower frequencies. At 1 Ghz and 30 Ghz the value has fallen to about 25% of the value at the peak frequency. The flux density at any frequency f(GHz) can be estimated to within 10% by the following expression:

$$S_v\,(\text{SFU}) = \frac{\left[0.64(F_{10} - 70)f^{0.4}\right]}{\left[1 + 1.56\left(\ln(f/2.9)\right)^2\right]}$$

where F_{10} is the current value of the solar radio emission at a wavelength of 10.7 cm (2,800 MHz). This value is measured and made available from the Penticton Radio Observatory in Canada. It is also available on a daily basis from various space environmental services, and is broadcast on standard frequency and time services (e.g., WWV, WWVH). The function ln( ) is the natural logarithm. Over one sunspot cycle S may range from 0 to above 250 SFU. Note that F_{10} is in fact the value of $S_q + S_v$ at 2.8 GHz.

12.6.2.4 The Burst Component

This component is a transient emission associated with explosive energy releases in the solar chromosphere and corona. At meter wavelengths it may increase the total solar output by six orders of magnitude, the increase lasting anywhere from seconds to hours. The frequency at which these bursts occur varies throughout the sunspot cycle, with very low probabilities when F_{10} is at a minimum, to several times a day at times of activity at maximum. The intensity of the bursts also shows an ill-defined average correlation with burst frequency, although large isolated bursts have been known to occur irrespective of other general solar activity. In the frequency range of 1 to 20 GHz, the average maximum flux density for the largest bursts typically varies from 100,000 to 10,000 SFU, respectively. An exception to this was noted in December 2006 when two L-band bursts to around one million SFU were recorded.

No empirical expression is possible for the burst component Sb. Some space weather agencies may attempt probability forecasts for interested parties. A moderate duration high-intensity burst is shown in Figure 12.36.

12.6.2.5 Summary

The total solar radio emission is specified by $S = S_q + S_v + S_b$. However, in most applications S_b is taken as zero. A nonzero burst component is usually readily identifiable by the presence of a large and fluctuating noise level.

The nature of the solar signal is identical to the thermal (Gaussian) noise generated in receiver circuitry. The power collected by a receiving antenna is given by:

$$P = SA_e \left(\text{W Hz}^{-1}\right)$$

where A_e is the effective collecting area of the antenna.

This expression is appropriate where the total solar emission is being collected, that is, an antenna beamwidth greater than the solar diameter with the Sun totally within the beam. For large antennas on the higher frequencies, the beamwidth may be less than the solar disc (Figure 12.37). In these cases the antenna essentially assumes the temperature of that part of the Sun to which it points. Under the assumption of a uniform brightness, this temperature is given in terms of the integrated flux density S by:

$$T_{\text{sun}} = \frac{S\lambda^2}{2k\Omega_s}$$

where λ is the wavelength, $k = 1.38 \times 10^{-23}$, J K^{-1} is Boltzmann's constant, and $\Omega_s = 6.8 \times 10^{-5}$ steradians is the mean solid angle subtended by the Sun. At sunspot

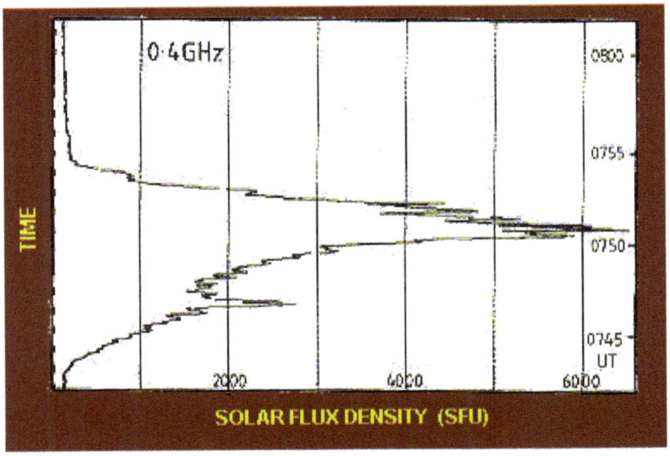

Figure 12.36 Moderate intensity solar radio burst. (*Source:* Learmonth Solar Observatory.)

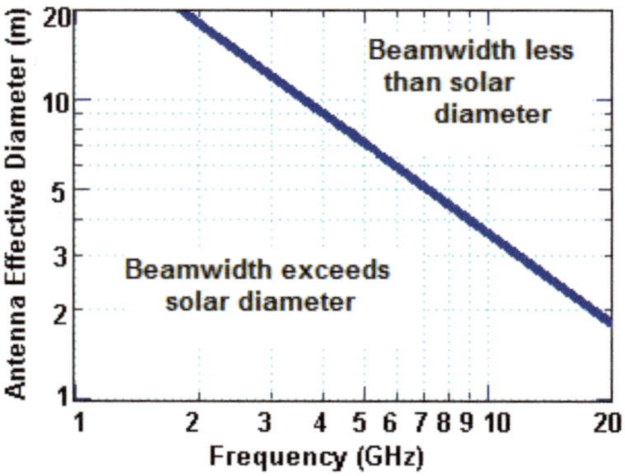

Figure 12.37 Antenna beamwidth equals mean solar diameter.

minimum, T_{sun} is approximately 10,000K at 20 GHz. The assumption of uniform surface temperature is valid at sunspot minimum, but becomes increasingly non-valid, especially around 3 GHz and lower frequencies, as F_{10} rises above 100 SFU. It is more valid at frequencies of 10 GHz and above.

12.6.3 Satellite and Solar Positions

12.6.3.1 The Nature of the Problem

The potential for solar interference with a satellite downlink circuit will occur whenever the Sun passes into the beam of the ground-station receiving antenna. For a geosynchronous orbit this will occur around the times of the equinoxes. What we are interested in are the times and durations of these occurrences, together with the magnitude of the receiving system temperature increase that will occur. From this, the degradation in the system performance is readily determined.

The first problem then is one of determining the position of both the satellite and the Sun. The former, being relative stationary, is readily accomplished. The latter, due to the changing length of our year, orbital variations, and our awkward calendar, is a little more difficult. Once these positions are obtained we then need to check for relative coincidences between the two as a function of time.

The apparent motion of the Sun across the sky renders the normal elevation-azimuth coordinate system unsatisfactory for use in solving our problem. Much better is the equatorial system of declination and hour angle. The Sun in essence then moves in an hour angle during a single day, with a much smaller annual motion in declination. The equatorial system is based about an axis projected through the poles to infinity. Hour angle is measured from local meridian (with east negative) about this axis (in essence celestial longitude), and declination is measured in a plane containing this axis. Zero declination is defined at the projection of the equator, and angles north of this are positive (Figure 12.38).

At the equinoxes the solar motion is 0.4 degrees per day in declination, and thus even with an infinitely small antenna beam, the Sun will still be in the beam for at least one day per equinox (Figure 12.39).

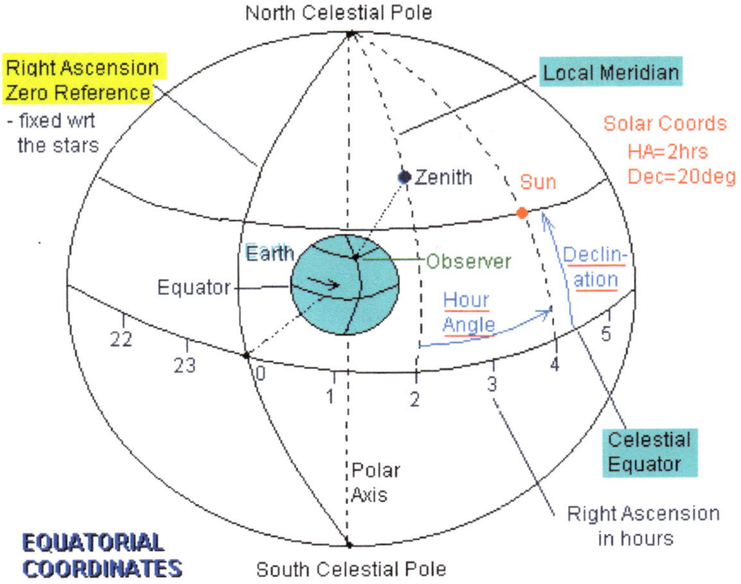

Figure 12.38 Equatorial coordinates.

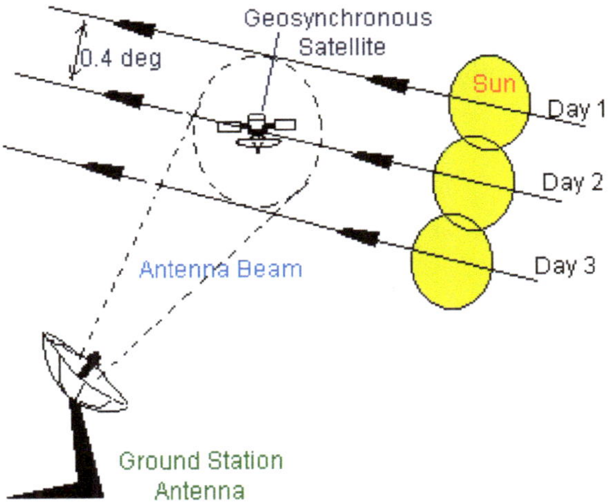

Figure 12.39 Transits of the Sun through an antenna beam on successive days around equinox.

12.6.3.2 Locational Algorithms

Calculations for geosat locations are given in Sections 5.2 and 5.6 for solar location as a function of date and time.

Calculations with an oblate Earth model show that the errors incurred from the spherical approximation amount to around 0.1 degree in azimuth and elevation, and 10 km in slant range (at midlatitudes). This angular error is one-fifth of the solar diameter and less than many station-keeping regimes. In military satellites, which are often given a considerable nonzero inclination, the errors involved will be unacceptable, but for most commercial purposes, the approximation is valid.

12.6.3.3 Range of Coincidence

To find the days on which an effect may occur, the equations in Sections 5.2 and 5.6 may be cycled until the solar declination is within a specified value of the satellite declination. The specified value is generally taken as half the sum of the antenna beamwidth ($\theta = 1.2\ \lambda/d_e$, where d_e is the effective linear aperture) plus the solar diameter (for the higher microwave frequencies, which are slightly larger (Figure 12.40) at 1 GHz). On each of these affected days the equation of time at noon may be used to find the range of affected times. One iteration to find a corrected equation of time (EoT) will be more than accurate enough for any application. The greatest inaccuracy lies in selecting the appropriate antenna contour (e.g., 3-dB beamwidth) for the computation. This can only be accurately done after computation of the maximum solar effect on the given system.

12.6.4 Downlink Calculations: CNR Changes

12.6.4.1 The Central Beam: Depth

The maximum effect will occur if and when the Sun falls centrally within the antenna response pattern. The appropriate downlink equation can then be written (neglecting tropospheric attenuation, etc):

$$\frac{C}{N} = \frac{P_r}{P_n + P_{\text{sun}}} = \frac{P_t A_e}{4\pi kT\left(1 + SA_e/kT\right)\Delta f r^2}$$

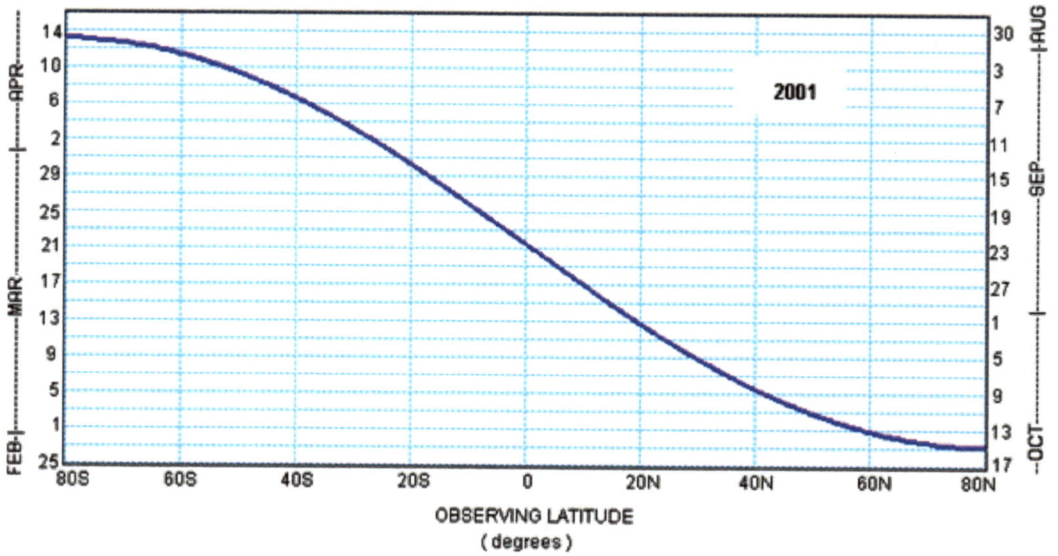

Figure 12.40 Dates of maximum solar interference to satellite signals from geostationary orbit.

where T is the system noise temperature. It suffices to note that the net effect is to raise the effective noise temperature by an amount:

$$\Delta T = 1 + \frac{SA_e}{kT}$$

and thus the decrease in the carrier-to-noise ratio will be given by:

$$\Delta CNR \text{ (dB)} = 10\log\left(1 + \frac{SA_e}{kT}\right)$$

Note that this is independent of operating frequency when expressed as shown (i.e., as a function of the system collecting area to temperature). The more commonly used G/T ratio can be computed from:

$$\frac{G}{T} = \frac{A}{T}\left(\frac{4\pi f^2}{c^2}\right)$$

Figure 12.41 shows the dB decrease in C/N for typical G/T ratios for selected frequencies and a range of solar fluxes.

The above formula only holds when the antenna beamwidth exceeds the solar angular diameter. For a narrow beam system, the appropriate relationship is:

$$\Delta CNR \text{ (dB)} = 10\log\left(1 + \frac{T_{\text{sun}}}{T}\right)$$

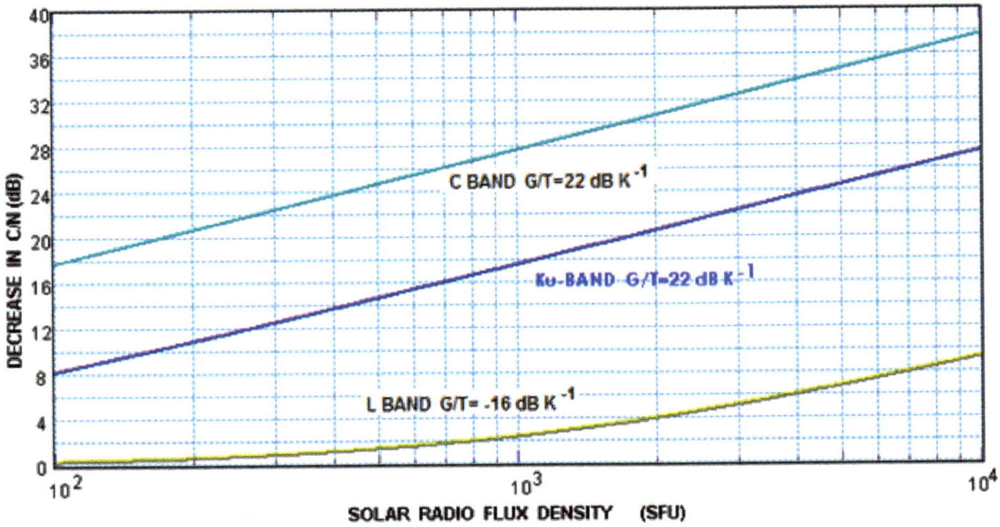

Figure 12.41 Showing C/N decrease as a function of solar noise output for three cases: (a) L-band mobile with $G/T = -16$ dB/K, (b) C-band 3.5m aperture with 60K temperature, and (c) Ku band 2.2m aperture with 250K temperature.

where T_{sun} is given in Section 12.6.2.5. The effect of going from a broad beam to a narrow beam system is essentially a lowering of the magnitude of the solar increase. Note that it is preferable to increase A rather decrease T if a minimal solar effect is desired.

12.6.4.2 Antenna Polarization

Solar flux values are reported for the total solar radio output at a particular frequency. This emission is mostly randomly polarized, although both left- and right-hand circular polarization burst components are seen around active solar regions with large magnetic fields. However, the nonburst components are generally random.

If a receiving antenna has circular polarization, either left or right hand, it will receive the full solar flux. If it uses linear polarization, however, it will only collect half the reported flux value, and thus $S/2$ should be used in all equations instead of the full flux density S.

12.6.4.3 The Peripheral Beam: Duration

The first-order calculation of the duration of the solar effect might use the antenna 3-dB beamwidth. For a central traverse of the Sun through the antenna, the duration of the effect is then equal to the sum of the beamwidth (θ) and the solar diameter (Π_s), times the rate of movement of the Sun in hour angle (4 minutes per degree). Table 12.5 shows this duration. For an off-center traverse, the time is given by:

$$\Delta t \text{ (mins)} = 4\cos^{-1}\left[\frac{\cos(\theta/2 + \Pi_s/2)}{\cos(\Delta\delta)}\right]$$

where the output of the inverse cosine function must be expressed in degrees, and where $\Delta\delta$ is the declination difference between the Sun and the antenna beam.

Table 12.5 Sun Transit Time Through Antenna Beam (Minutes)

Antenna diam	Receiving Frequency (GHz)									
(m)	*2*	*4*	*6*	*8*	*10*	*12*	*14*	*16*	*18*	*20*
0.5	86.2	44.3	30.3	23.3	19.1	16.3	14.3	12.8	11.7	10.7
1.5	30.3	16.3	11.7	9.3	7.9	7.0	6.3	5.8	5.4	5.1
2.5	19.1	10.7	7.9	6.5	5.7	5.1	4.7	4.4	4.2	4.0
3.5	14.3	8.3	6.3	5.3	4.7	4.3	4.1	3.8	3.7	3.5
4.5	11.7	7.0	5.4	4.7	4.2	3.9	3.7	3.5	3.4	3.3
5.5	10.0	6.2	4.9	4.2	3.9	3.6	3.4	3.3	3.2	3.1
6.5	8.8	5.6	4.5	4.0	3.6	3.4	3.3	3.1	3.1	3.0
7.5	7.9	5.1	4.2	3.7	3.5	3.3	3.1	3.0	3.0	2.9
8.5	7.3	4.8	4.0	3.6	3.3	3.2	3.0	3.0	2.9	2.8
9.5	6.8	4.5	3.8	3.4	3.2	3.1	3.0	2.9	2.8	2.8
10.5	6.3	4.3	3.7	3.3	3.1	3.0	2.9	2.8	2.8	2.7

If the solar noise increase is significantly larger than the downlink C/N margin, then the 3-dB beamwidth may not be appropriate. An acceptable contour must then be chosen with regard to the fraction of the solar disc covered by this contour at each moment of the traverse. Inaccuracies in antenna pointing and errors in the knowledge of the exact solar output add to the complexity of this type of calculation.

12.6.4.4 Summary

For a given collecting aperture and noise temperature, the effect of the Sun on a system increases with frequency (because of the increasing solar output). Small aperture systems are relatively unaffected by normal solar noise, although at times of high solar activity, all satellite downlinks are susceptible, and then small apertures will be affected for a longer duration. At Ka-band a relatively small aperture (e.g., 2m) has a beamwidth less than the solar diameter and the total solar radio energy collected then starts to decrease. In most cases, the link margin for tropospheric attenuation (precipitation) will cope with the solar traverse. In summary, it may be concluded that moderate apertures at midmicrowave experience the most significant reduction in carrier-to-noise ratio (with respect to the duration-depth product).

The time extent of Sun outages and their predictability makes it relatively easy to reduce the limited impact that they cause. Alternate satellite paths, if available, obviate the problem in most cases. The one exception may prove to be mobile L-band downlinks. Very limited aperture and low system noise together with limited link margins could make these systems quite susceptible to the frequent solar bursts that occur at times of high solar activity.

12.7 The Doppler Effect

The Doppler effect, with regard to electromagnetic energy, is the change in apparent frequency of a radiating source that depends on the relative velocity of the source with respect to an observer. In Figure 12.42, if the source approaches the observer, the received frequency will be higher than if the source were stationary. If the observer is located behind the source the received frequency will be lower than if the source were stationary.

Figure 12.43 gives a flat Earth approximation to the Doppler shift experienced by a ground station when a satellite makes a pass. The frequency at any time is given by the formula:

$$f_r = f_o\left(1 + \frac{v\cos\theta}{c}\right)$$

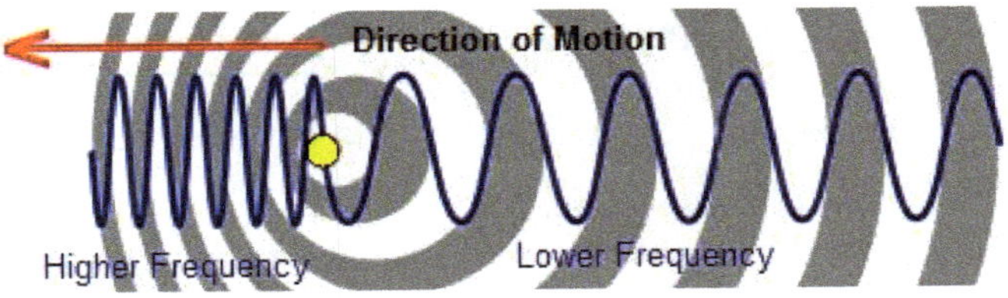

Figure 12.42 The Doppler effect.

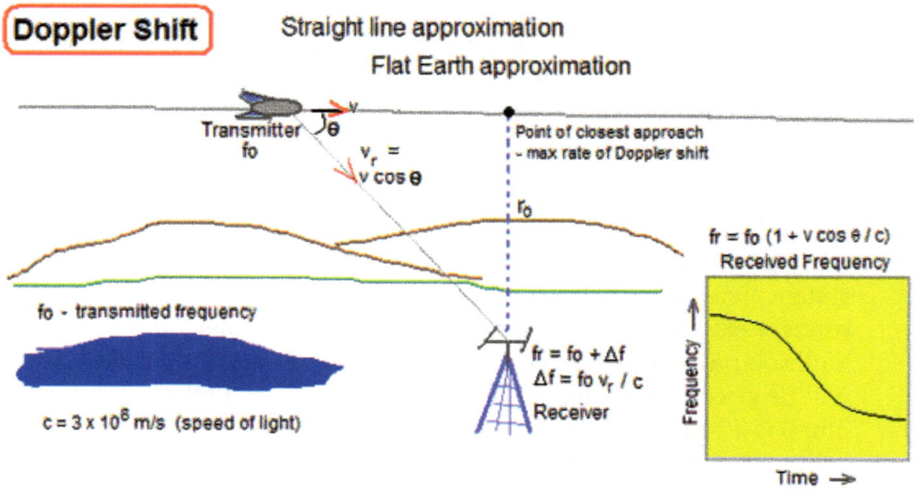

Figure 12.43 Satellite Doppler shift using a flat earth approximation.

For the extreme angles $\theta = 180°$ and $\theta = 0°$ we have a frequency difference of

$$\Delta f_{\max} = \frac{2f_o v}{c}$$

which enables us to calculate the orbital velocity

$$v = \frac{c\Delta f_{\max}}{2f_o}$$

For a LEO satellite, we might assume a circular orbit as a first approximation, which then enables us to calculate the orbital height h and orbital period T. Using the central part of the Doppler curve where the rate of frequency change $(df_r/dt|_{\max})$ is a maximum at which we can calculate the distance of closest approach

$$r_o = \sqrt{\left[\frac{f_o v^2}{c\,df_r/dt|_{\max}}\right]}$$

If this determination is made for a second pass, the orbital inclination I, and other orbital parameters, can be determined to completely specify the circular orbit.

Of course, the use of a flat Earth approximation provides first-guess estimates of these parameters. In particular, it is not always easy to determine the extreme frequency range because of horizon obstructions.

A more accurate approximation assumes a spherical Earth, and the geometry is shown in Figure 12.44.

The Doppler shift equations for this circular orbit and spherical stationary Earth are:

$$\Delta f_{\max} = 2f_o v\sqrt{\frac{\cos^2\alpha - \cos^2\beta}{c\tan\beta}}$$

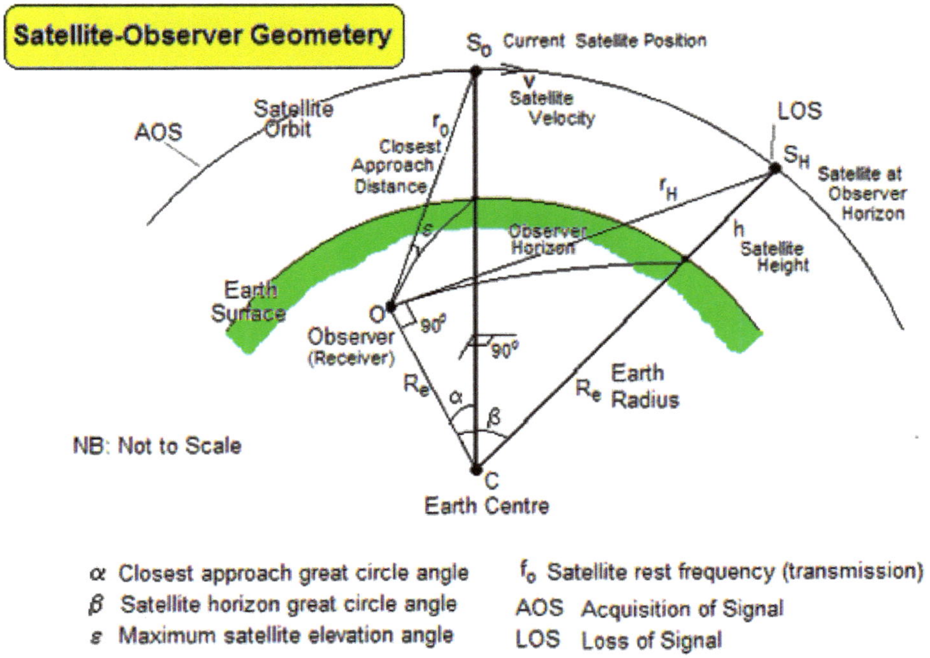

Figure 12.44 Spherical earth geometry.

and

$$\left.\frac{df}{dt}\right|_{\max} = \frac{-f_0 v R_e \cos\alpha}{c r_o (R_e + h)}$$

Digitally tuned radios are now available that make the acquisition of a Doppler frequency curve from a transmitting satellite simple. The 2m amateur band (around 146 MHz) provides an ideal source of LEO satellites for exploration into Doppler determination of orbital parameters. The receiver must be capable of CW or SSB reception. A digital frequency resolution of 100 Hz is usually adequate. Even if the satellite radiates an FM modulated signal, there is normally no difficulty in achieving the zero beat condition with an identifiable component of the signal.

Prior to the pass, the receiver is tuned to the expected frequency and the observer waits until a signal is heard. The tuning is then adjusted a little below the satellite frequency. As the signal drifts lower in frequency the heterodyne tone will pass through the zero beat frequency (minimum tone) condition. Precisely at this point the frequency and time are recorded, frequency to the nearest 100 Hz and time to the nearest second. The receiver is then tuned to a slightly lower frequency and the procedure repeated. The result should be a table of frequency versus time. This can be plotted to give a Doppler curve (Figure 12.45).

The most important parameter is obtained from the derivative of this curve (Doppler rate) (Figure 12.46). This gives the time of closest approach and the Doppler rate at this time. The first plot provides the total frequency shift (a low-accuracy parameter, particularly if horizon obstructions exist) and the total time above the horizon. The curve symmetry can help to restore missing initial values. A computer

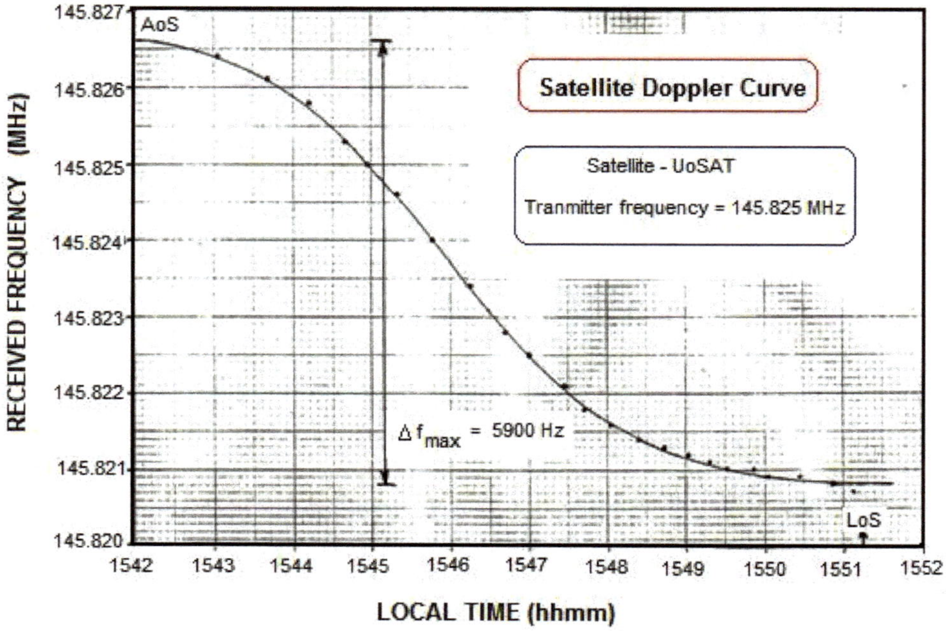

Figure 12.45 Satellite Doppler curve for University of Surrey satellite (UoSAT).

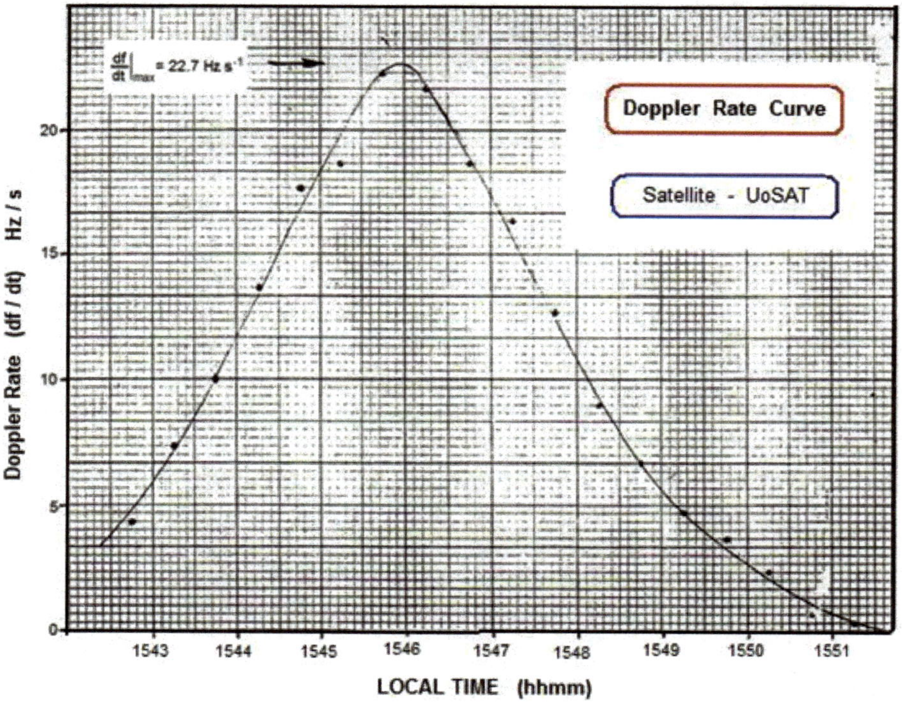

Figure 12.46 Satellite Doppler rate curve from Figure 12.45.

program based around the two equations for a circular orbit helps to analyze the results. The maximum Doppler rate and the satellite view time can be used to solve for satellite altitude, velocity, period, closest approach distance, and maximum elevation. Two consecutive passes determine the orbital inclination.

A computer program can be used to iterate a Doppler pass with different *miss* angles until an output is found that matches the observed parameters.

A typical solution to the Doppler and Doppler rate curves might be:

```
Miss angle (deg) = 8.9

Height  Velocity  Period  MaxTime  ClostApp  MaxDoppl  RateMax
MaxEl
 (km)    (km/s)    (min)   (min)     (km)      (Hz)     (Hz/s)
(deg)
  490     7.62     94.38   10.46     1138      6253      22.78
21.0
```

Satellite Doppler shift can be used to determine satellite orbital parameters, or it can be used in the reverse sense. Satellites of known orbits can be used by mobile ground stations to determine their position. Before the advent of GNSS, a Doppler satellite positional determination system called TRANSIT was in operation, mainly to allow ships to measure their position, it was too slow for land or air vehicles. This operated with several satellites transmitting at the dual frequencies of around 150 and 400 MHz. The dual frequencies were used to allow ionospheric effects to be removed.

12.8 Space Radar

A space radar uses radio-reflected energy to determine various parameters of an orbital space object. There are many types of space radar, and these may be classified in a number of different ways.

A space radar may use a wide beam or a narrow beam. The wide beam radar will direct and collect radio energy reflected from any object that passes through its beam. This does not require foreknowledge of the object or its position, and may thus be used to start a space object catalog. The received signal can be timestamped, and will usually provide amplitude information, which is roughly proportional to the size of the transiting object. The radar may also be able to measure the Doppler shift of the reflected signal and this will provide considerably more information about the orbit.

A narrow beam radar requires at least an approximate object position estimate to allow pointing of the radar. It is then able to provide a more accurate position and velocity estimate as well as a radar cross-section measurement. The radial distance estimate (i.e., the object range) can be measured more accurately than the cross-radial position. This is because the line-of-sight range is given by the time delay to return the radar signal. This is limited by the pulse width of the radar pulse (for a pulsed radar), or the correlation time for a CW radar. The position perpendicular to the line of sight is dependent on the antenna beamwidth, which depends upon the radar receiving antenna beamwidth. For a single-dish antenna, this is a direct function of the receive antenna beamwidth.

For a parabolic antenna of diameter d, the physical collecting area is $A = \pi d^2/4$. The effective collecting area $A_e = eA$ where e is the antenna efficiency. The beamwidth $\theta \approx \lambda/d$ where θ is in radians (1 radian ≈ 57.3 degrees). In more practical units:

$$\theta \approx 206{,}000\frac{\lambda}{d} \text{ (arcseconds)}$$

where the wavelength λ and the dish diameter are in the same units (e.g., meters or feet). When an antenna is used for transmitting its gain is the important parameter, and this is given by:

$$G = \frac{A_e}{4\pi\lambda^2}$$

The gain G is a unitless parameter. The antenna efficiency is a unitless number, usually between 0.5 and 0.9, depending on the nature of the reflective surface and the feed details. This is discussed more fully in Section 12.4, which also includes antenna performance graphs.

A space radar may also be classified as monostatic, bistatic, or multistatic. A monostatic radar has a collocated transmitter and receiver, and often a shared antenna. A bistatic radar has the transmitter and receiver separated by a substantial distance, and thus requires separate transmit and receive antennae. A multistatic radar has more than one receive antenna.

It should be noted that the radar cross section of a target will in general be different for a back-scattered signal (monostatic radar) than for a forward-scattered signal (bistatic radar).

The signal details for a monostatic radar are shown in Figure 12.47.

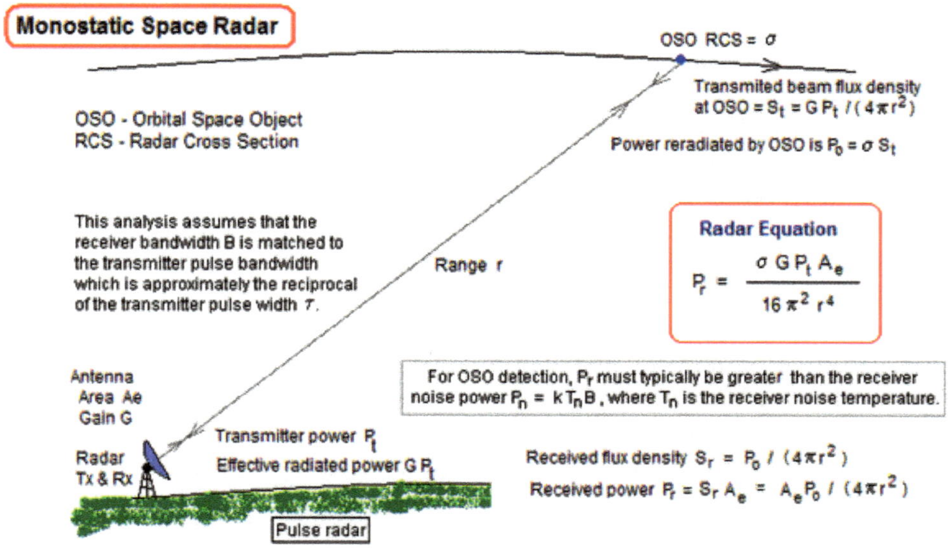

Figure 12.47 Signal details for a monostatic radar.

The monostatic radar equation shows that the received power is directly proportional to the orbital space object radar cross section, the transmitter power, the antenna gain, and effective cross-sectional area, and indirectly proportional to the fourth power of the object range.

This puts a radar at a disadvantage to optical space object detection because the latter decreases only as the square of the observing distance.

The signal details for a bistatic radar are shown in Figure 12.48.

The signal power P_r received by the bistatic radar, is the signal flux density at the receiver times the receive antenna effective collecting area A_e. That is;

$$P_t = S_r A_e = \frac{\sigma_e G_t P_t A_e}{16\pi^2 r_1^2 r_2^2 B}$$

The bistatic radar, like the monostatic radar, also shows an essentially range to the fourth-power dependence for the returned signal. Because of this, most radars are limited to low Earth orbit. They do, however, have the advantage of being able to work both in daylight and in adverse weather conditions, although the latter will affect radar performance for frequencies above S-band (~2 GHz).

Lidar is an optical version of radar, and can provide more accurate range and cross-range measurements because of the higher frequency (smaller pulse width possible) and narrower beam (a laser beamwidth is diffraction limited).

The combined use of radar with optical detection improves the overall accuracy of the measurement with radar providing an accurate range, and optical telescopes providing a more accurate plane of the sky measurement (Figure 12.49).

So far, nothing has been mentioned about the radar frequency of operation. However, the frequency is very important in that it determines the limiting size of objects that can be detected. This is not because of a frequency dependence of the propagation parameters but because the radar cross section of an object changes

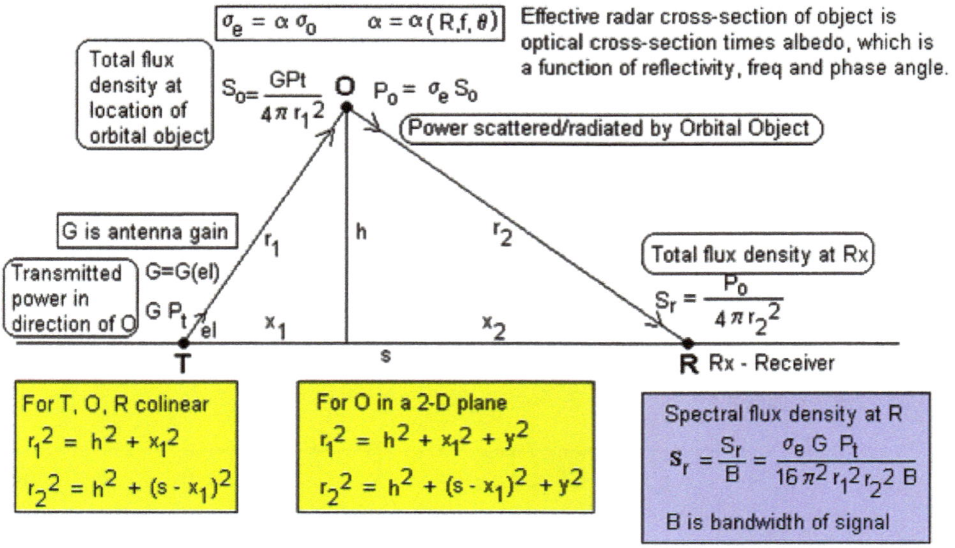

Figure 12.48 Distance details and signal flux density for a bistatic radar.

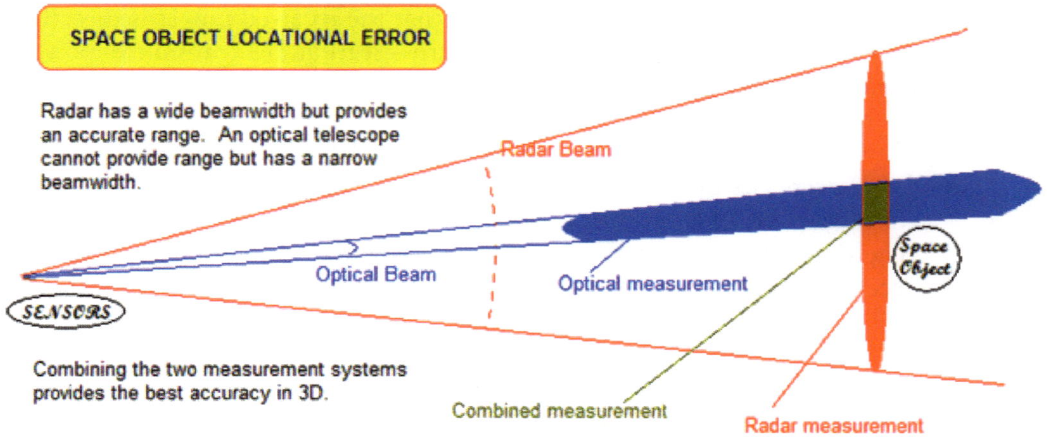

Figure 12.49 Using both radar and optical detection results in greater accuracy.

with wavelength. As long as the size of a radar target is substantially larger than the radar wavelength ($\lambda = c/f$) the radar cross section is related to the size of the object. But as the wavelength approaches the object size, there is a wild oscillatory variation in radar cross section, and when the object is smaller than the wavelength the radar cross section becomes inversely proportional to the fourth power of the wavelength in a region known as Rayleigh scattering (Figure 12.50). Thus,

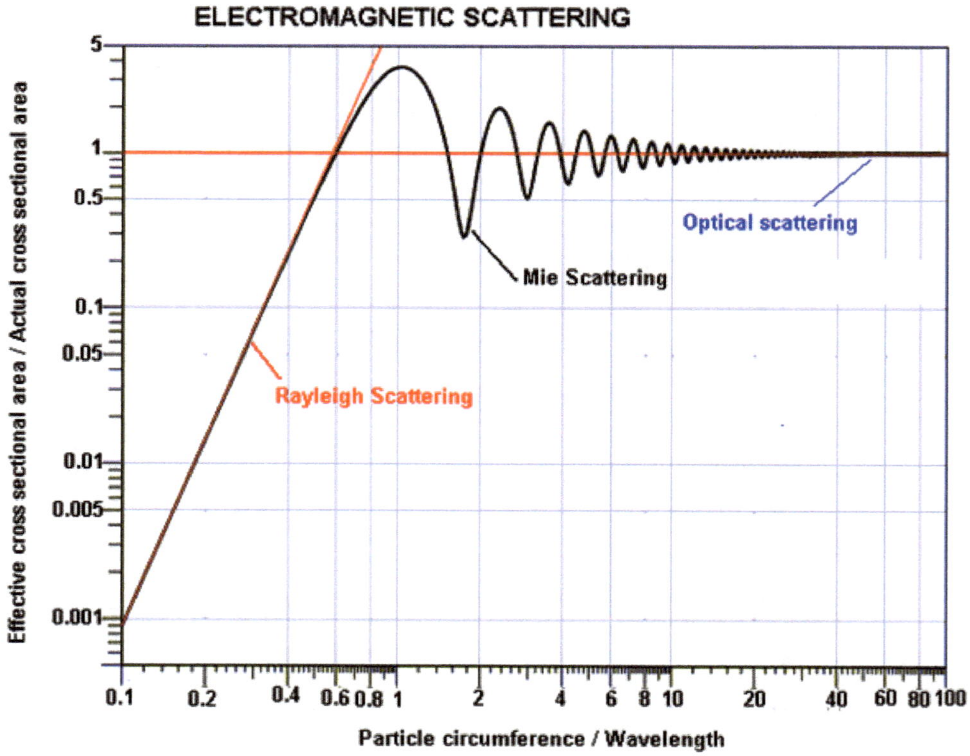

Figure 12.50 The three electromagnetic scattering regimes.

a 10-cm object will have an effective cross-sectional area of a 0.01-mm object at a wavelength of one meter.

The original NAVSPASUR radar (the space fence) operated at a frequency of 217 MHz or a wavelength of 1.4 meters. This is primarily because solid state electronics at the time could not generate high power at higher frequencies. However, to reliably detect 10-cm objects in orbit it had to use massive power (~7 GW effective radiated power) to overcome the inverse fourth power wavelength dependence in the Rayleigh scattering regime. The new S-band space fence (f = 2.8 GHz, λ = 10 cm) should have no trouble equaling the original space fence performance with moderate powers, a smaller antenna (300m array versus 3-km array), and will probably be able to detect the lower objects down to 1 cm in size.

References

[1] Cohen, A. E., "25 Years of Satellite Beacon Studies," *The Australian Physicist*, Vol. 21, 1984, pp. 224–226.

[2] Fun Cube Dongle, funcubedongle.com.

[3] ITU frequency allocations, www.itu.int.

[4] Australian Communication and Media Authority, www.acma.gov.au.

[5] Kundu, M. R., *Solar Radio Astronomy*, New York: Wiley, 1965.

[6] US National Geophysics Data Center, ngdc.noaa.gov.

Selected Bibliography

Calcutt, D., and L. Tetley, *Satellite Communications: Principles and Applications*, Edward Arnold, 1994.

Inglis, A. F., and A. C. Luther, *Satellite Technology: An Introduction*, Focal Press, 1997.

Logsdon, T., *The NAVSTAR Global Positioning System*, Van Nostrand Reinhold, 1992.

Martin, J., *Communications Satellite Systems*, Prentice Hall, 1978.

Roddy, D., *Satellite Communications*, McGraw-Hill, 2006.

Ryan, M. J., *Principles of Satellite Communications*, Argos Press, 2004.

CHAPTER 13

Rules of Thumb and Data For Space Debris Studies

13.1 Space Debris Velocities

Average orbital velocity in LEO is 7 km/s.

The average collisional velocity in LEO is 10 km/s.

13.2 Debris Impact Craters

The diameter of an impact crater is typically from 10 to 20 times the diameter of the impactor.

Most hypervelocity impacts produce circular craters, irrespective of the angle at which the impactor strikes the target.

If an impactor penetrates a target, the egress hole will be irregular and considerably larger than the ingress hole.

13.3 Space Debris Density

Low Earth orbit, spanning an altitude range of about 200 to 1500 km, has a total volume of around one trillion (10^{12}) cubic kilometers.

As of 2024, the current orbital space object density (>10 cm in size) in LEO was approximately 10^{-72024} per cubic kilometer.

13.4 Space Debris Collisions

Any impact with debris greater than 1 cm in size has serious consequences.

The average debris impact produces approximately 100 pieces with sizes greater than 10 cm.

Catastrophic impact generally occurs when the mass of the impactor is greater than 1/1000 the mass of the target.

The average collision frequency with a piece of space debris may be calculated from the product of the target area, the debris density, and the collisional velocity:

- $\text{freq}_{\text{collision}}\ (s^{-1}) = A_{\text{target}} \cdot \rho_{\text{debris}} \cdot v_{\text{collision}}$;
- The cross-sectional target area is in square meters;

- The debris number density is number per cubic meter;
- The average collisional velocity is in meters per second.

For the International Space Station in the year 2023, this works out at one collision with a 10-cm piece of debris roughly every year. And indeed, ISS maneuvers are now made every month or so to avoid such collisions. The difference between the above computation and reality is due to the former being an average over all of LEO space. The actual value at a particular orbital altitude may be more or less than this value.

13.5 Reentry Hazard

An orbiting spacecraft or piece of orbital space debris in general needs to have an initial mass greater than 1 ton to at least partially survive atmospheric reentry and deposit pieces on the Earth's surface.

In general, bodies above 10 ton will definitely partially survive reentry whereas objects less than 1 ton will not.

However, some particular classes of objects of much lower mass do make it (partially) to the Earth's surface. These include objects made of high melting point materials (e.g., titanium spheres), and objects with a very high area to mass ratio (e.g., thermal blankets). They also include objects contained with other objects (e.g., in a space station) that are not exposed to the reentry environment until the outer surface is breached.

The reentry interface altitude, where significant ablation of a reentering orbital object occurs is generally around 75 km.

13.6 Debris Mass and Size Distributions

The number distribution of a debris population is given by:

- $N(>m) = Am^{-a}$ where m is the object mass, and
- $N(>s) = Bs^{-b}$ where s is the object size.
- a, A, b, B are all constants specific to the given population.

CHAPTER 14

Space Weather for Space Situational Awareness

14.1 Introduction

Space situational awareness, in its truest sense, refers to a knowledge of our near-space environment, and thus includes a knowledge of space weather (the natural component of the space environment). Space weather affects systems both in space and on the ground. In this chapter we will concentrate on the aspects of space weather that affect space-based assets and their products. We will consider both the positive and negative sides of space weather, that is, the benefits and opportunities as well as the hazards and the threats.

14.2 Overview

A narrow view of space weather considers the Sun as the sole source of disturbances in our space environment. However, particularly with respect to SSA, there are most definitely extrasolar sources that influence the environment around the Earth. And as we move assets beyond Earth's orbit we need to consider these even more.

Figure 14.1 attempts to provide an overview of those aspects of space weather that are significant for SSA. It considers four sources that both form and modify the space environment. These are (1) the Earth, (2) the Sun, (3) the solar system, and (4) the galaxy. Each of these, and their mutual interactions, will be discussed in the following paragraphs.

Joe Allen, working at the US National Geophysical Data Centre (NGDC), compiled a list of satellite anomalies [1]. In the two decades, from 1971 to 1989, just under 3,000 were recorded. It is not always easy to ascertain the cause of each anomaly, particularly as manufacturers are usually reluctant to admit any vulnerability in their products, but it is estimated that up to 25% of all spacecraft failures are due to the space environment. This is an indication of the significance of space weather on our assets.

14.3 The Earth

Although the Sun is often quoted as the prime (or even only) source of space weather, its role is essentially to modify what already exists in the near-Earth

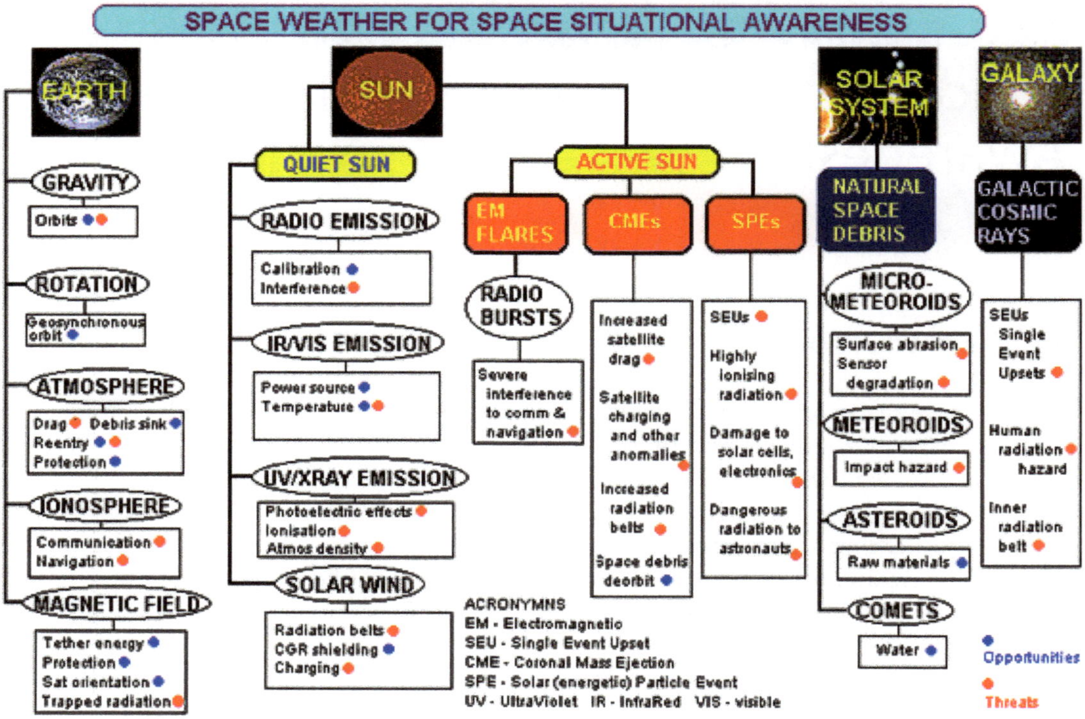

Figure 14.1 Aspects of space weather of concern to SSA.

space environment. Even without a solar influence the nature of the Earth's gravity, atmosphere, and magnetosphere influence what we can do in space.

14.3.1 Gravity

The mass distribution that we call the Earth determines the gravitational field that exists around it. On the one hand, the gravitational field makes it difficult for us to leave the surface of the Earth, but on the other, it makes it possible for satellites to orbit the planet without any expenditure of energy.

For many purposes we can regard the Earth as a sphere, but the difference between a true sphere and the actual mass distribution of the Earth changes the elements of a satellite orbit so that orbital predictions would be useless unless this difference is taken into account. Satellites have helped refine our knowledge of the terrestrial gravitational field, which in turn have allowed more accurate orbital predictions.

14.3.2 Rotation

The Earth's rotation makes possible one particular very useful orbit: the geosynchronous orbit. This orbit, where a satellite appears to be stationary (or nearly so) with respect to a point on the Earth, means that the ground station does not have to track the motion of the satellite, a considerable savings in hardware and cost.

The geosynchronous orbit also lies at an altitude where approximately one-third of the Earth's surface is visible. If the geosynchronous orbit was at a much lower altitude it would not be nearly so useful. The trade-off, of course, is the higher power required for communication from this orbit.

14.3.3 Atmosphere

The Earth maintains an atmosphere by virtue of its gravity, with a secondary role being played by its magnetic field, as will be discussed later.

The atmosphere has both positive and negative aspects with respect to space assets. On the beneficial side, it allows the return of spacecraft to the Earth's surface with minimal energy expenditure. After an initial deorbit burn, the atmosphere can be used to absorb the large kinetic energy of the returning craft, and allows aerodynamic lift for a soft landing (or the use of a parachute if the reentry vehicle does not have aerodynamic surfaces), without the expenditure of any fuel.

The atmosphere totally ablates most space debris on an uncontrolled reentry, thus protecting the inhabitants on the surface below.

Gravity and thermal pressure work to give the atmospheric density an exponential fall-off with altitude (see Chapter 7). By the time we reach a height of 20 km, 99% of the atmosphere lies below us, and at orbital altitudes we have a better vacuum than most research laboratories can achieve

This lack of atmosphere gives rise to three main problems. A pressure differential between the interior and exterior of a spacecraft (particularly if pressurized for humans), gives rise to a mechanical strain. Thermal control is more difficult in space. Heat can only be transmitted by radiation. The vacuum is not conductive and the lack of gravity in an orbiting satellite does not allow convection. A totally different mechanism of dissipating heat from electronic equipment must be employed. Heat pipes to radiative surfaces are one solution.

The vacuum also allows outgassing of materials that can return to the spacecraft and contaminate surfaces. This is why many satellites are assembled in clean rooms on Earth.

Although most of the atmosphere is gone at orbital altitude, enough still remains to create drag at low Earth orbit, slowly reducing the orbital energy and altitude of a satellite. Either fuel must be carried and used to occasionally reboost the satellite to higher orbit, or a reduced lifetime must be accepted. Figure 14.2 plots the altitude of the International Space Station over one year, showing frequent reboosts.

The residual atmospheric atoms also collide with spacecraft surfaces at hypervelocity, which produces physical sputtering of the surface, chemical attack from atomic oxygen, and a spacecraft glow from large surfaces that can interfere with low-light experiments (Figure 14.3).

14.3.4 Ionosphere

The ionosphere is created by the ionizing effect of solar EUV and X-rays on the Earth's upper atmosphere. This occurs between the altitudes of about 100 to 500 km. The ionosphere is a weak plasma—only about one percent of the neutral atmosphere is ionized, but this is sufficient to affect radio signals passing through it.

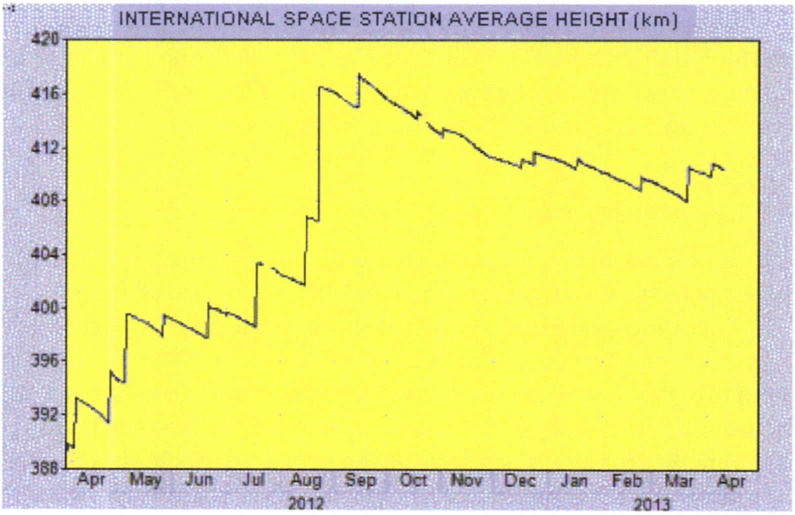

Figure 14.2 Variation in ISS altitude showing the many reboosts necessary to keep the Space Station in orbit.

A stable ionosphere enables high-frequency communications between stations on the Earth's surface, but for space assets the ionosphere, which shows a strong diurnal and other variations, displays only undesirable effects.

Transionospheric communication and navigation signals suffer delays and distortion when passing through the ionosphere. Refractive effects displace objects from their true positions for space-looking radars. Satellites that use polarization as a means of multiplying the number of channels in a given frequency band can experience polarization rotation that couples undesirable power into the orthogonal channel.

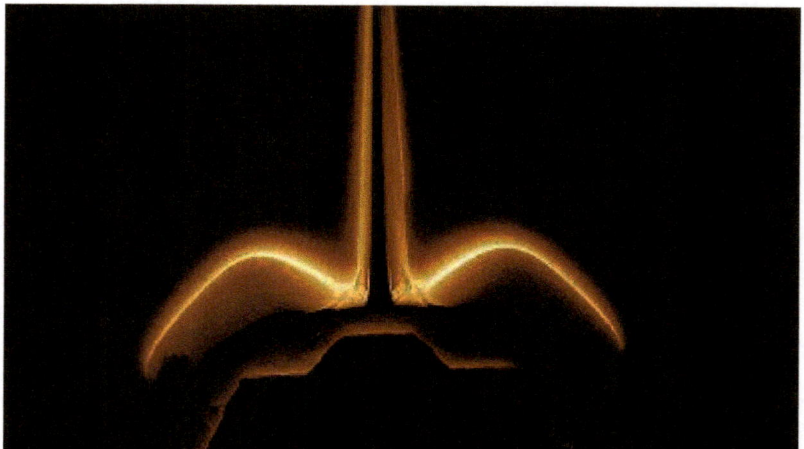

Figure 14.3 Glow on Space Shuttle surfaces facing the direction of motion. (*Source:* NASA.)

The ionosphere is a dispersive medium and the magnitude of the distortions varies with frequency. Higher frequencies are affected less than lower frequencies. C-band frequencies are only infrequently affected, and higher bands are usually immune to ionospheric effects.

14.3.5 Geomagnetic Field

The Earth's magnetic field has two components. The internal component is generated within the Earth's core by movements of molten conductive material (mostly liquid iron). The external component is generated by particles circulating in orbit. The internal component shows a slow variation with time, on the order of years. The external component is subject to large variation as the solar wind and coronal mass ejections make rapid changes in the particles injected into the Earth's magnetosphere.

The internal component of the geomagnetic field is several orders of magnitude larger than the external component. Over midlatitude Australia, the internal component is around 50,000 nanotesla (nT), whereas the largest external component rarely exceeds 500 nT.

Some space assets use the Earth's magnetic field for attitude stabilization, and/or for gross location determination. Tethers (Figure 14.4) can be used to generate power from the field and may be useful to deorbit space assets at the end of their life, reducing the proliferation of orbital debris.

14.3.6 Radiation Belts

The Earth's magnetic field traps charged particles from the Sun and those of galactic cosmic rays, forming radiation belts around the Earth. These belts were discovered in 1958 by a radiation detector on the first US satellite. This was due to the space

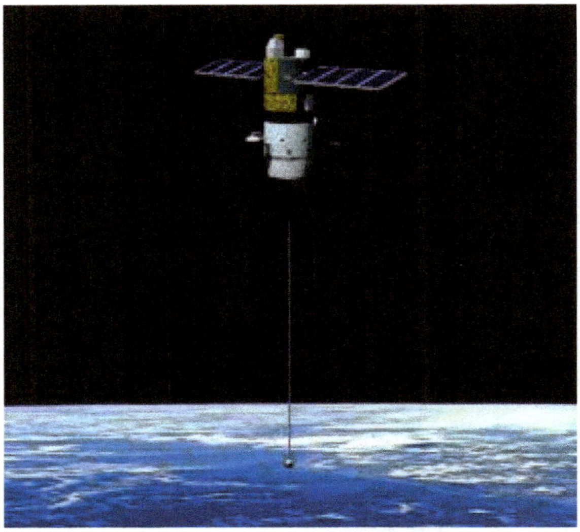

Figure 14.4 Artistic rendering of space tether. (*Source:* NASA.)

physicist James Van Allen, and the belts are often known as the Van Allen radiation belts.

There are two main belts (Figure 14.5). The outer belt is mainly populated by solar-charged particles. The inner belt is mainly populated by protons arising from the decay of albedo neutrons. These neutrons are produced by interaction of galactic cosmic rays with the Earth's atmosphere and then reflected back to space. They decay to positively charged protons with a half-life of about 13 minutes.

The trapped radiation belts define the upper limit of low Earth orbit at around 2,000 km. The outer belt ends (fortunately) just below geosynchronous orbit at 36,000-km altitude. The region between these two limits is one of high-radiation levels that causes rapid deterioration of unhardened electronics and must be traversed rapidly by manned missions to the Moon or beyond to limit the radiation dose received. Global navigation satellites (such as GPS) that utilize this MEO regime must use radiation-hardened electronics to survive in this environment. Gallium arsenide semiconductors replace silicon devices as they can withstand much higher radiation doses.

Because the Earth's magnetic field is centered about an axis that is offset, and rotated with respect to the rotational axis, one part of the inner belt comes closer to the Earth than the rest. This occurs in a region to the southeast of South America and is termed the South Atlantic Anomaly (SAA). LEO satellites passing through this region experience higher radiation levels, and astronauts report seeing flashes when their eyes are closed. This occurs when ionizing particles activate cells in the retina.

14.4 The Sun

The Sun continually gives off both electromagnetic and particulate emissions. It is convenient to divide these into quiet and active components. However, it is important to realize that the quiet component is not constant. It simply varies at a slower rate than the active component.

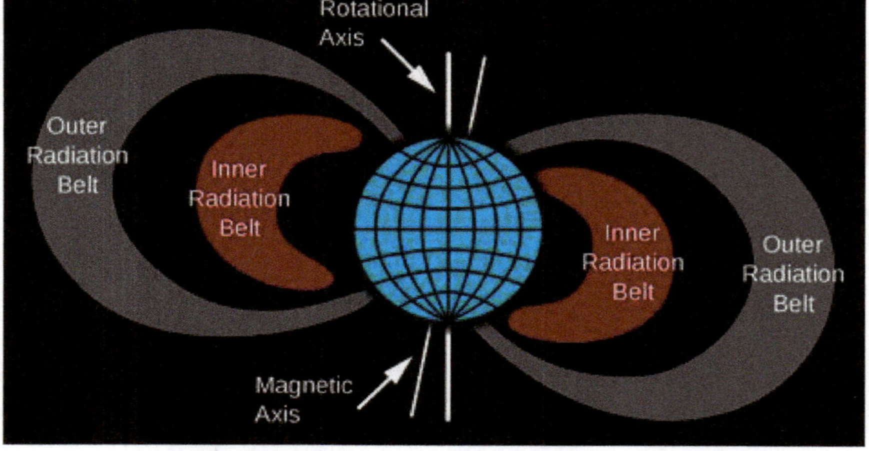

Figure 14.5 The Van Allen radiation belts.

14.5 The Quiet Sun

Emission from the quiet Sun spans the full electromagnetic spectrum. The particulate emission is usually referred to as the solar wind. Variations in both types of emission occur with periods of 11 years (the sunspot cycle period) and shorter term periods of around one month (the solar rotation period of about 28 days), and of several months (due to the growth and decay of solar magnetic features such as large sunspot groups and coronal holes). The period due to solar rotation is of course an observational artefact at the Earth's position as it orbits around the Sun.

14.5.1 Solar Radio Emission

Solar radio emission as detected at the Earth's orbit is measured in SFUs, where in SI units 1 SFU = 10^{-22} W m^2 Hz^{-1}. Solar radio flux increases with frequency, being about 10 SFU at 200 MHz and 500 SFU at 15 GHz. This variation is shown in Figure 14.6.

Figure 14.6 indicates solar radio emission at the time of sunspot minima with no spots visible on the solar disc. Added to this is a slowly varying component, which peaks around 3 GHz and can increase the flux at this frequency by 400% during times of large sunspot maxima. This emission does not come from the sunspots themselves, but from chromospheric plage associated with high-magnetic fields.

The quiet radio emission from the Sun provides both an opportunity and a hazard. It is used as a calibration source for radars and microwave communication

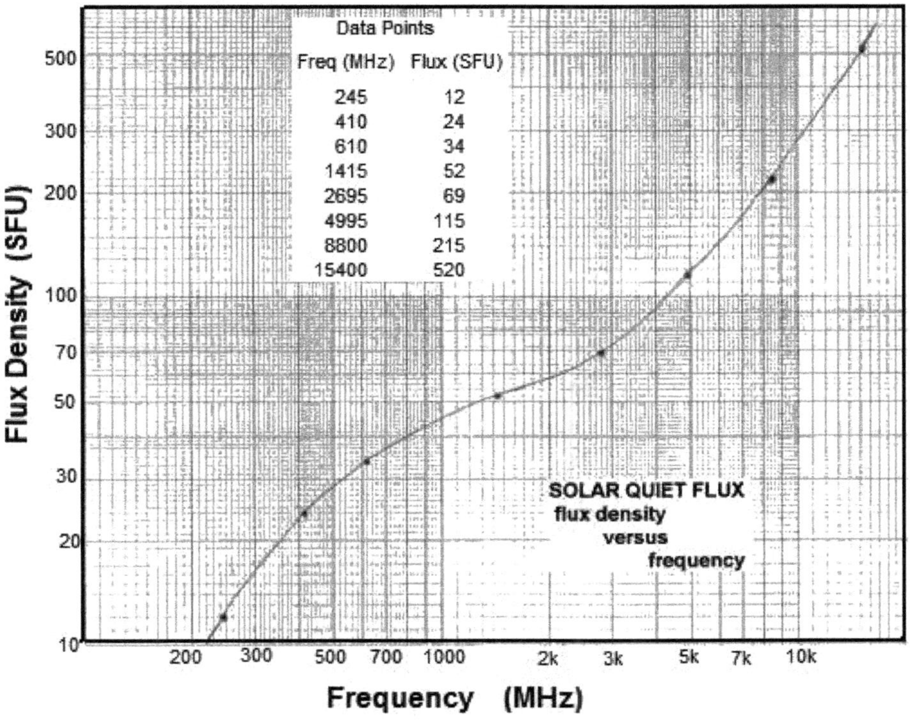

Figure 14.6 The quiet Sun radio output in solar flux units versus frequency.

systems. This requires ground-based calibrated solar radio telescopes that constantly monitor and report the solar radio fluxes.

It is also a source of interference to any electromagnetic system where the Sun appears in the beam. An example is the Sun outage phenomenon where geosynchronous TV downlinks are subject to this interference around the equinoxes, when the Sun appears behind a geosat for several tens of minutes over a period of a few days. The solar radio signal is typically 20 dB or so greater than the satellite signal, and no TV reception is possible during this time (Figure 14.7).

14.5.2 IR/Visible Radiation

Most of the solar electromagnetic output lies in the visible and infrared spectral bands. This is constant to within about 0.1%, and at the Earth's orbit has a value of 1370 W m^{-2}. It is thus a good source of power (Figure 14.8) for space systems, and most satellites and spacecraft use solar photovoltaic cells to provide all their system power.

Infrared radiation from the Sun is also a factor in the temperature stabilization of a satellite.

Since Peter Glaser's seminal paper in 1968 [2], solar power satellites (SPS) have been intensively investigated, whereby energy from the Sun is collected by huge arrays in space and converted to microwave energy for downlink to Earth (Figure 14.9). Although there are many technical and political hurdles to overcome before SPS can become a reality, the Japanese government has made it an integral part of their space policy.

14.5.3 UV/X-Ray Emission

The extreme ultraviolet and soft X-ray emission from the Sun ionizes the Earth's upper atmosphere creating a weak plasma known as the ionosphere (Figure 14.10). Different wavelengths of this emission interact with different atomic and molecular species in the atmosphere to produce the different ionospheric layers (D, E, and F).

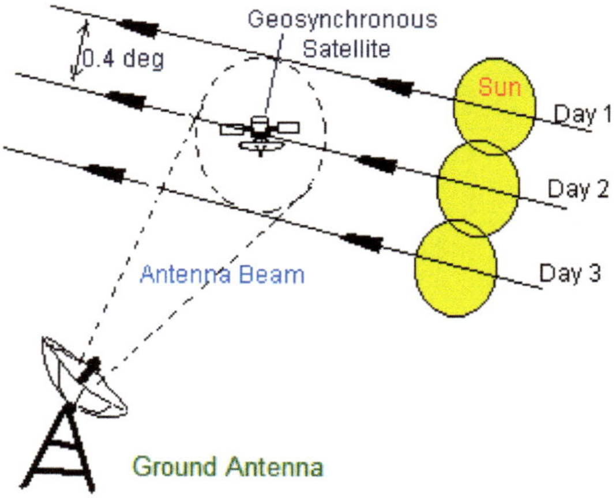

Figure 14.7 The Sun can interfere with satellite communication when it lies within the antenna beam.

Figure 14.8 The International Space Station uses many solar panels to produce the power required for its operation. (*Source:* NASA.)

Ultraviolet radiation photons also have sufficient energy to release electrons from surface metals on satellites leaving them charged. If a differential charge of sufficient magnitude builds up between two different surfaces, a sudden discharge may occur that can damage components in and around the discharge.

14.5.4 The Solar Wind

The Sun's outer atmosphere, the corona, has a kinetic temperature of 2 million kelvin. The individual particle energies at this temperature can easily escape the large gravitational field of the Sun and boil off into the interplanetary medium in a stream we call the solar wind.

In fact, it is not the Sun's gravity but its coronal magnetic fields that control this outflow.

Figure 14.9 Concept of a solar power satellite. (*Source:* Wikipedia.)

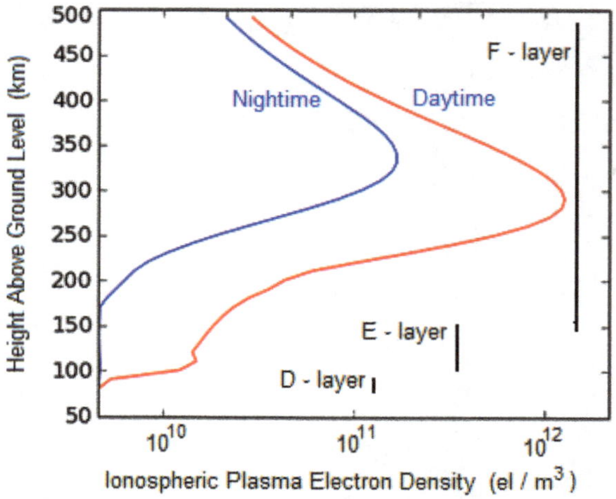

Figure 14.10 Ionospheric layers.

At the Earth's orbit, the solar wind consists of around 7 particles per cubic centimeter (7 million per cubic meter) traveling at an average speed of 400 km/s. However, these values can vary by over a factor of two either way.

The solar wind is usually monitored by a satellite at the Earth-Sun central Lagrange point (Figure 14.11). The US Space Weather Prediction Center (SWPC) displays solar wind parameters in a dial-pointer display.

The pressure produced by the solar wind modifies the shape of the Earth's magnetic field. Most of the time the magnetopause (the boundary between the solar

Figure 14.11 The Atmospheric Composition Explorer monitors the solar wind at the Earth-Sun Lagrange point. (*Source:* NASA).

wind and the Earth's magnetosphere) lies well outside geosynchronous orbit (Figure 14.12). However, during extreme conditions in the solar wind (usually produced by a transient coronal mass ejection) the magnetopause may be pushed inside geosynchronous orbit (Figure 14.13). This exposes geosats to a different particle and

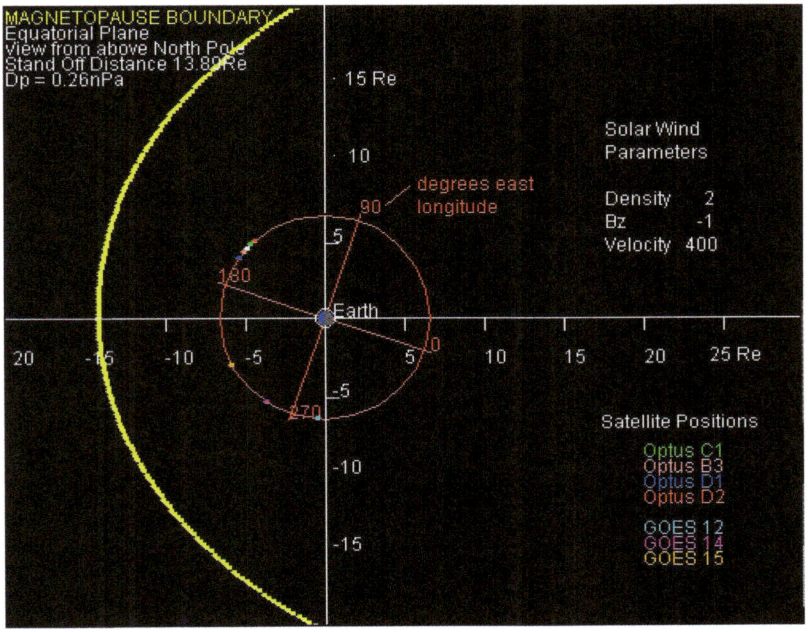

Figure 14.12 Normal magnetopause position.

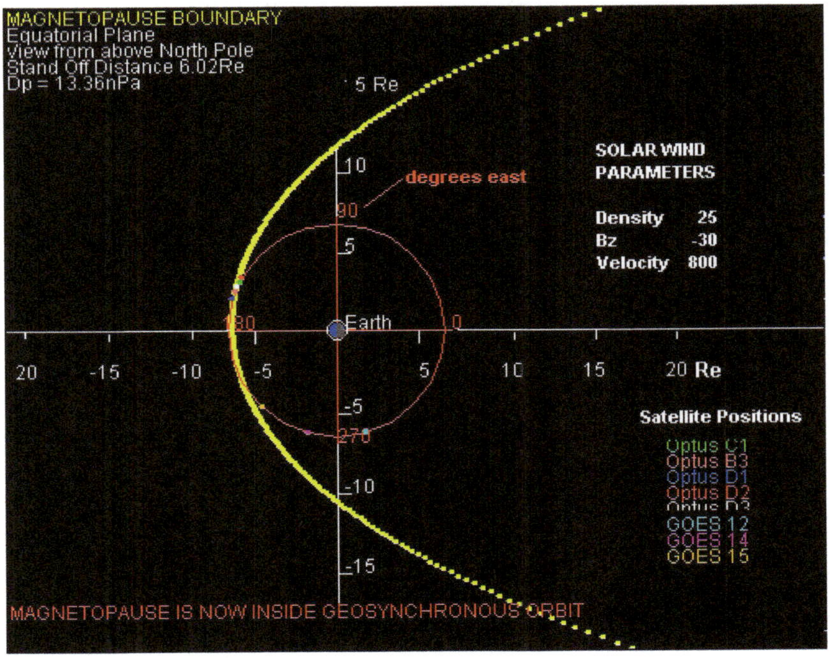

Figure 14.13 Compressed magnetopause position.

radiation environment and can cause problems to any satellite that relies on magnetic stabilization to any extent. It has yet to be seen how such conditions will affect the new electric thrusters that will be used for geosat station keeping.

The solar wind extends out and defines the far limits of the solar system or heliosphere. The heliopause is the boundary where the pressure of the solar wind is just countered by the pressure of the interstellar medium. Very high energy but very low frequency (~1 kHz) emissions are produced around this interaction. We cannot receive this emission at Earth because it is shielded by the plasma in the interplanetary medium. Distant spacecraft, such as Pioneer and Voyager, detect this energy.

14.6 The Active Sun

The total energy emitted by the Sun varies by less than 0.1%, and most of this energy is emitted in the infrared and visible parts of the electromagnetic spectrum.

However, this stability is not evident in either end of the spectrum. Both radio and X-ray emissions may vary in a transient fashion by up to six-orders of magnitude. This emission comes in bursts or flares and may last from minutes to hours.

14.6.1 Radio Bursts

The highest amplitude radio bursts occur at the lower frequencies. At 100 MHz, a very large radio burst may reach one million SFU, whereas at 20 GHz the largest bursts do not exceed 20,000 SFU due to a process known as synchrotron self-absorption.

Any space communication or navigation system that has the Sun in its beam (or even near the beam) during one of these large bursts will invariably be disrupted to some extent. In most cases, the link will fail completely. Fortunately, such events are rare and do not last long, hours at the most.

Before 2006, it used to be thought that the GPS global positioning system was immune to interference from even the largest solar radio bursts. GPS operates at L-band (1 to 2 GHz), and the largest solar bursts ever recorded in this band were around 100,000 SFU. It just so happens that calculations showed that this was about the threshold at which we might expect GPS degradation. And so it was thought that GPS was immune to solar electromagnetic interference (although it was well-known that ionospheric irregularities in equatorial regions could cause signal dropout). GPS is more immune than most space EM systems because it uses spread-spectrum techniques that compress to a very narrowband of about 100 Hz in the receiving processor.

In December of 2006, the Sun surprised everyone by producing a series of extremely large bursts (Figure 14.14) on two separate days. These bursts were estimated to have flux densities of 1 million SFU at L-band. Many GPS receivers in the sunlit sector of the Earth were affected by these bursts.

Although X-ray flares from the Sun cause additional ionization in the lowest layer of the ionosphere (the D-region), this mostly affects subionospheric communication, as the amount of additional ionization is considerably less than that in the normal upper F-layer, which is the layer that usually affects transionospheric communication.

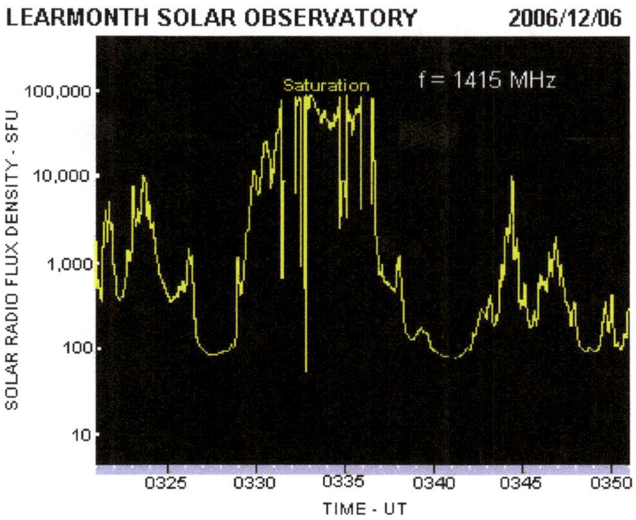

Figure 14.14 Large L-band solar radio burst of December 2006 that affected some GPS systems. (*Source:* Learmonth Solar Observatory.)

Apart from a slightly higher absorption (less than 1 dB) on space-ground circuits, and the blinding of X-ray sensors on astronomical satellites, even the largest X-ray flares do not have the same deleterious effect as a very large solar radio burst.

14.6.2 Coronal Mass Ejections

The solar wind carries away from the Sun a mass of about 1 million tons (10^9 kg) per second. However, on occasion it can also spew out a large cloud of plasma with a total mass on the order of 5 billion tons. This can be seen using a white-light coronagraph onboard a spacecraft, and is called a coronal mass ejection or CME. Occasionally a CME may be seen with a ground-based coronagraph on top of a high mountain, above most of the Earth's atmosphere, which scatters sunlight to an extent that CMEs are normally invisible from the ground.

CMEs carry away solar magnetic fields that are trapped through the plasma and help keep it together (Figure 14.15).

These CMEs are millions of kilometers in extent and travel out from the Sun with velocities from about 200 to 2000 km/s. If they are traveling toward the Earth their travel time may range from 1 to 5 days, although 2 to 4 is more common.

What happens when they arrive at the Earth depends on the direction of the magnetic field they carry along with them. If this field has a southward component, the CME will couple into the Earth's magnetosphere and large quantities of plasma particles will be dumped into the Earth's upper atmosphere. This creates a geomagnetic storm. This is seen by ground-based magnetometers.

During the storm the Earth's magnetic field will fluctuate wildly as particles move around the Earth with some precipitating to lower layers. A large storm may last for several days creating intense aurora, ionospheric storms and irregularities, neutral atmospheric density changes, spacecraft charging, and other effects on space assets.

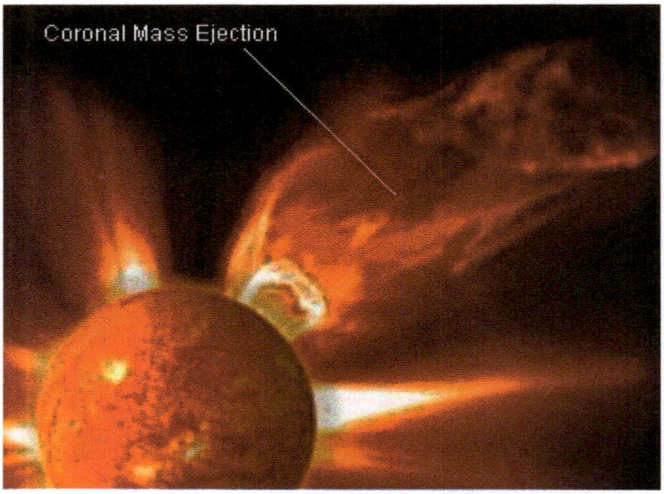

Figure 14.15 Artist's impression of a solar coronal mass ejection. (*Source:* NOAA/SWPC.)

In March 1989, when a large CME caused one of the largest geomagnetic storms seen in 50 years, the US Air Force Space Surveillance Center in Colorado lost track of 30% of the space object population in low Earth orbit. What happened was that although the sensors still observed these objects, the geomagnetic storm had so increased the atmospheric density at LEO that the objects were not in their expected orbital positions. And so they became decorrelated. That is, the known space objects (active satellites and debris) could not be matched, or correlated, with the objects observed. It took the SSC three weeks to recorrelate the objects they had previously been tracking.

CMEs have the beneficial effect of shielding the inner solar system from the lower energy galactic cosmic rays (Figure 14.16).

14.6.3 Solar Energetic Particle Events

Every so often a large number of solar protons (hydrogen nuclei), together with a smaller percentage of heavier nuclei (mainly helium) are accelerated to energies of a few million electron volts (MeV) and even up to around 1 GeV (although this is rare). This acceleration seems to occur by at least three different mechanisms. It can be a direct acceleration near the Sun, associated with CME shock waves near the Sun and sometimes with CMEs out in the interplanetary medium.

Although SPEs rarely make it to the Earth's surface, occasionally a very high energy event (with protons to 100s of MeV) will produce a ground level event (GLE).

A lot more SPEs do affect space assets in Earth orbit. They can cause permanent degradation in solar cells, sensors, and electronic equipment, and they pose a severe hazard to astronauts. Spacecraft in high-inclination orbits are most affected as they receive little protection from the Earth's magnetic field in polar regions where the field lines are open.

One large SPE event in 1990 reduced the efficiency of solar cells in several communication geosats by up to 30%. This can represent $100 million per satellite in lost revenue.

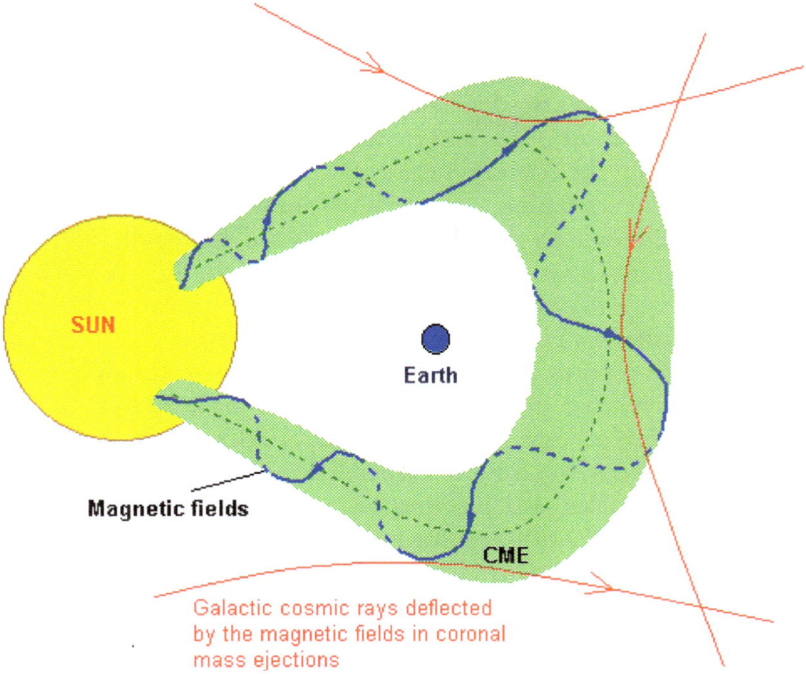

Figure 14.16 Coronal mass ejections shield the inner solar system from low energy cosmic rays.

On interplanetary trips, the highest energy SPEs represent a significant threat to humans, although because of their lower energies they are easier to shield than galactic cosmic radiation.

14.7 The Solar System

The main bodies of the solar system are the planets, which orbit around the Sun, and whose massive bulk provides the primary gravitational field source controlling motion in the system.

However, there are numerous smaller bodies in the solar system. These include asteroids, comets, and innumerable small pieces of rock and dust. Some of these pose hazards to heliospheric space operations, but others may offer opportunities in the way of material resources for future space operations.

Some of these smaller particulate bodies are also discussed in Chapter 8 ("Space Debris Reentry Hazards"), Chapter 9 ("A Debris Reentry Model"), and Chapter 16 ("Near-Earth Objects and Planetary Defense") but are revisited here in more detail for completeness and in the context of space weather.

14.7.1 Micrometeorites and Interplanetary Dust

The smaller bodies in the solar system usually come from pieces evaporated from comets or asteroidal collision fragmentation.

The smallest particles are micron-size pieces that are termed micrometeorites (or probably more correctly micrometeoroids), and collectively form a population known as interplanetary dust. The population of interplanetary dust particles (IPD) can be seen from Earth due to the sunlight they scatter. This is known as zodiacal light. It is seen most often, and more clearly from, near equatorial (tropical) sites. It appears during the evening (in the west) or morning (in the east) twilight hours as a cone of light stretching from the horizon, where the Sun has just set, up to sometimes 45 degrees in elevation. This is scattered light from IPD, which lies mainly in the plane of the ecliptic between the Earth and the Sun. A much rarer phenomenon is the gegenschein, which appears around the zenith at midnight. This is from IPD in orbit further away from the Sun than the Earth.

Micrometeorites are encountered as they pass through low Earth orbit, but the particle size is so small that impact damage is confined to very limited surface abrasion. Although a single particle thus does negligible damage, continued erosion of sensitive surfaces over time can reduce the efficiency of a sensor such as a telescope.

These particles also account for, along with meteoroid atmospheric ablation products, about 10,000 tons of matter that floats down to the surface each year. Some of this undoubtedly accounts for the nickel found on the ocean floor.

14.7.2 Meteoroids

Meteoroids are bodies less than a millimeter to say 10 meters in diameter that orbit the solar system and occasionally are captured by the Earth's gravity. When they meet the atmosphere they burn up, producing the visual meteor phenomenon. A one-gram meteoroid produces a meteor of around zeroth magnitude (the apparent brightness of our nearest star, Alpha-Centauri). At the beginning of the space age it was thought that meteoroids might pose a serious threat to space operations. This has not proved the case, although a Space Shuttle windscreen was cratered by such an object. At the size of 1 centimeter diameter, where serious damage will result if a spacecraft suffers an impact, artificial space debris is now more prominent and a greater danger than natural meteoroids. Figure 14.17 shows the cumulative impacts of small debris on a panel of the NASA Long Duration Exposure Facility (LDEF) over a few years. This included both artificial and natural debris.

14.7.3 Asteroids

The dividing size between a meteoroid and an asteroid is not well defined, but could be considered to be about 10 meters. Asteroids are pieces of rock that vary from this size up to hundreds of kilometers in diameter. The smaller asteroids are much more prolific, with only a few thousand asteroids known or thought to exceed 1 km in size.

Because of their small numbers, asteroids are unlikely to pose any significant impact hazard to spacecraft, even when they pass by the Earth below geosynchronous altitude. Asteroids are more likely to be useful as a source of raw materials in future space operations, and private companies have recently been established to explore this potential (Figure 14.18).

Figure 14.17 Small meteoroid and debris impacts on a $1m^2$ panel from the *Long Duration Exposure Facility* spacecraft, which was in low Earth orbit for 6 years. (*Source:* NASA ODPO.)

Figure 14.18 Deep Space Industries [3] was a private company that planned to mine asteroids for the resources they contain. (*Source:* DSI.)

14.7.4 Comets

Comets are low-density objects that, in contrast to asteroids, have highly elliptical orbits. It is thought that the source of comets is the Oort cloud, located right on the fringes of the solar system. It is estimated that this cloud contains about one billion comets. Occasionally, gravitational perturbations send one of these comets into the inner solar system. Some of these comets are captured by planetary gravitational fields (usually by Jupiter) and remain within the planetary system.

The US astronomer Fred Whipple in the early 1950s described a comet as being a "dirty snowball," indicating the presence of a large quantity of ices, including water ice. It has been suggested that comets may have delivered water to the surfaces of planets in the past and that they could be used in the future as sources of water for large-scale space operations.

Comets (Figure 14.19) are rarer than asteroids and their capture, due to their highly elliptical orbits, will be much more difficult than asteroidal capture.

14.8 The Galaxy

To most people concerned with the practicalities of space operations and space assets in the near-Earth space environment, the Milky Way Galaxy seems too distant to have any relevance to space situational awareness. This, however, is not so. The galaxy provides a vast playground of immense size where even small magnetic fields and gas clouds can accelerate masses of subatomic particles (mostly protons) to immense energies such that they provide a radiation hazard to space operations around the Earth, and in the solar system. In fact, galactic cosmic radiation (GCR) is probably the single most important limiting factor to human space travel beyond Earth orbit.

14.8.1 Galactic Cosmic Rays

On the Earth's surface we are exposed to a continuous low level of ionizing radiation. Some of this comes from materials that compose the terrestrial regolith, but

Figure 14.19 Comets may be useful as water resources in the distant future. (*Source:* John A. Kennewell.)

some of it comes from space. This was proven by the Austrian physicist Victor Hess in 1911, when he took a very simple radiation detector up in a balloon. He found that for a few hundred meters the radiation level decreased, but that as he continued ascending the dose rate increased substantially. In fact, at the levels flown by commercial airliners, the radiation levels are 20 to 50 times that experienced on the surface. Although not a problem for passengers, aircrew who frequently fly certain routes need to have their radiation dose monitored. Some countries have put laws in place that limit flying hours based on radiation dose. In space, the dose received by astronauts is often well above the maximum levels accepted in industry.

Space radiation comes from three sources: (1) radiation trapped in belts or zones by a planetary magnetic field (e.g., the Van Allen belts), (2) infrequent transient bursts of solar energetic particles, and (3) galactic cosmic radiation.

Galactic cosmic rays are a relatively constant source of radiation, no matter where you are in the solar system, and have the highest energies of the three.

Galactic cosmic rays are mainly protons with energies of several hundred MeV to over 10^{21} electron volts (Figure 14.20). Not even the largest particle accelerators on Earth (e.g., the Large Hadron Collider) can reach these energies. Particles are accelerated to these energies by galactic magnetic fields over 100,000 light-year distances, and by interactions with interstellar shock waves associated with large plasma clouds. A small percentage of GCR are alpha particles and nuclei from atoms of higher atomic mass.

Galactic cosmic rays are not present in large enough numbers to be a bulk hazard to spacecraft materials and electronics, but an individual particle has enough energy to cause a single event upset (SEU), usually in semiconductor material. Because

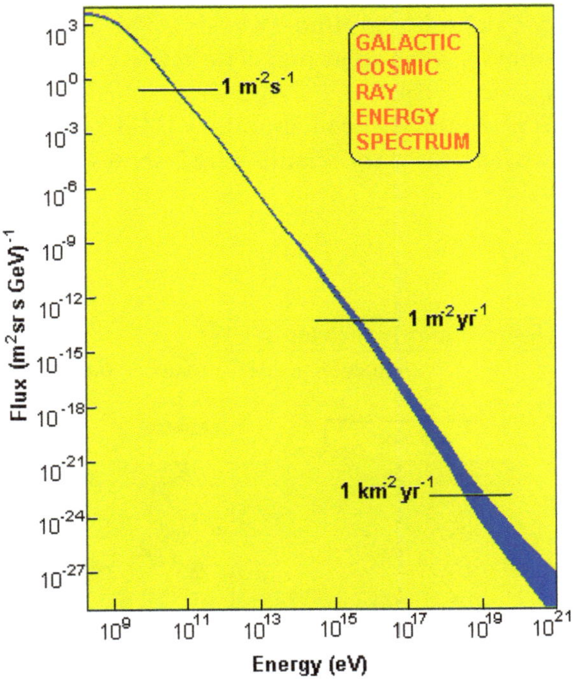

Figure 14.20 The flux of galactic cosmic rays.

humans are more sensitive to radiation than inorganic materials, GCR does pose a considerable threat and even a limitation to human presence in space.

An SEU happens when a GCR particle, usually a proton in space, passes into a piece of semiconductor material, a transistor or integrated circuit, which contains many active circuit elements (Figure 14.21).

The particle may pass through the material, creating a limited amount of ionization in its path, or it may deposit all its energy near the transistor elements. When this occurs, a large number of charge carriers will be created and an undesirable transient current may flow. This will create an unexpected signal in the processor, memory, or other circuitry in which the transistor resides. In memory elements, what is known as a bit-flip may occur, wherein a one may be changed to a zero or vice versa. These upsets are usually transitory, but if they occur at the wrong time, they may be interpreted as a rogue command to the spacecraft. In the worst cases, they may cause a signal latch-up that requires the power be removed from the system to correct the problem. This may or may not be possible. Several cases of total loss of a spacecraft (including a Russian Phobos mission to Mars) have been attributed to this mechanism.

On the surface of the Earth, humans are mostly shielded from GCR by the atmosphere and the geomagnetic field. In low Earth orbit, the dose is much larger, but the magnetosphere stills provides a measure of protection. Outside this protection the GCR dose is higher still.

The interplanetary magnetic field does prevent the lower energy GRC particles from making it to Earth's orbit. The amount of shielding provided depends on the intensity of the magnetic fields carried out into the solar system by the solar wind and coronal mass ejections. The lowest energy of GCR, which makes it to Earth's orbit, and thus the total GCR intensity shows a variation from about 300 MeV to 1 GeV with a period of around 11 years., This is an inverse variation to the sunspot cycle; that is, when solar magnetic activity is at a minimum, GCR intensity is greatest, and vice versa.

The NASA Mars Science Laboratory (MSL) carried a radiation detector on its six-month voyage to Mars. Figure 14.22 shows the radiation levels encountered during this time.

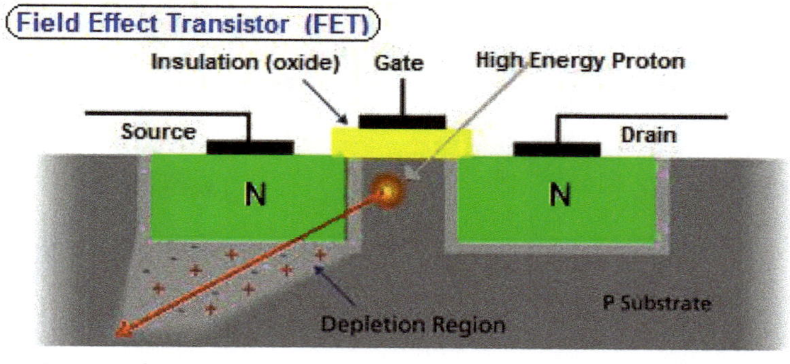

Figure 14.21 Semiconductor damage causes current pulse due to incident cosmic rays.

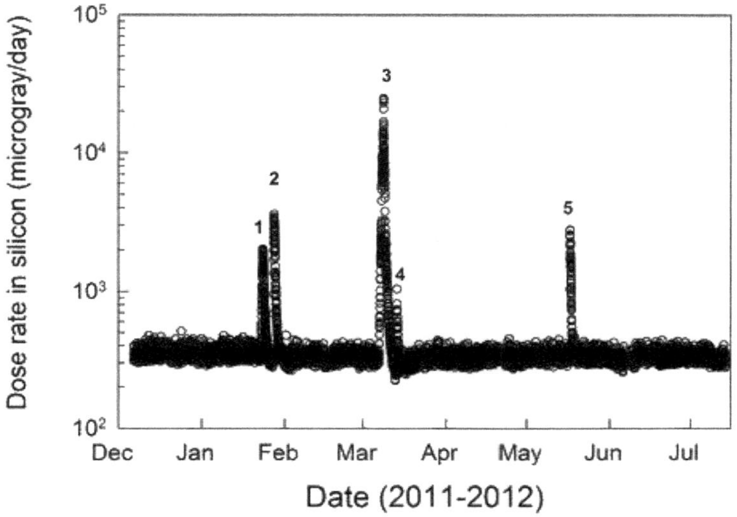

Figure 14.22 Mars Science Laboratory radiation levels. (*Source:* NASA/JPL/MSL.)

The constant background of about 350 microgray per day is due to GCR. The five peaks are due to solar energetic particle emission. Most SPE radiation is transitory and does not add significantly to the total dose (although large SPE events do). NASA/Jet Propulsion Lab (JPL) estimate that the human dose equivalent on this six-month journey was 300 millisievert (mSv). Allowing for a return trip and maybe a year on the Martian surface, the total dose to an astronaut would be around 1.2 Sv. This is 25% of the acute lethal dose that will kill 50% of a human population in 30 days (the $LD_{50\text{-}30}$). GCR is probably the single most limiting factor to human space travel outside the Earth near-space environment.

14.9 Summary of Natural Space Weather

There are many aspects of the natural space environment that must be considered for safe and reliable operation of space assets. Some aspects of the environment may be utilized to advantage and some must be mitigated. The most dangerous dimension of space weather is the radiation from high-energy particles. Figure 14.23 illustrates the various sources of space radiation.

High-energy solar particle events pose the greatest hazard because of the high fluxes possible. The danger is to hardware and to humans. Low-energy particles from coronal mass ejections are probably the next most hazardous space weather condition, and these occur significantly more frequently than SPEs. Galactic cosmic rays can cause SEUs and are definitely a limiting factor in long duration human presence in space.

At the current time, space weather poses a greater economic hazard to space operations than does the artificial space debris population in Earth orbit. This may change in the future as that population increases.

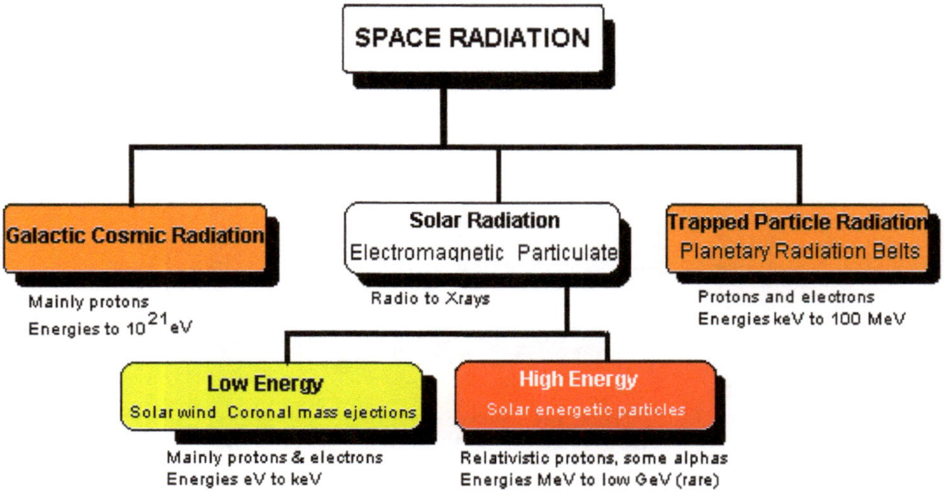

Figure 14.23 Sources of radiation in the space environment.

14.10 Anthropogenic Space Weather

Anthropogenic space weather [4] refers to changes in the space environment that are caused by humans. Examples include:

1. Artificial space debris;
2. Upper atmosphere chemical releases;
3. Ionospheric modification–from the ground;
4. Ionospheric modification–from space;
5. High-altitude nuclear explosions.

Item (1) has already been discussed extensively. Item (2) generally relates to scientific research of upper atmosphere and magnetosphere conditions and changes. Sounding rockets are sent to an appropriate altitude and release a colorful chemical tracers that spread out indicating the presence, direction, and speed of the neutral atmospheric winds (for atomic tracers), or magnetic fields for ionized tracers.

Experiments in ionospheric modification have been undertaken. These may use high-power ground-based radio transmitters as ionospheric heaters. The most famous of these is the HAARP (High-frequency Active Auroral Research Program) facility in Alaska. In space, the US Space Shuttle used its rocket exhaust to modify the ionosphere, with ground stations in Tasmania and elsewhere to monitor the hole created in the ionosphere.

The most extensive modification of the upper atmosphere and near-space environment, however, is cause by a nuclear explosion at high altitudes [5]. Both ground and space effects of a high-altitude nuclear explosion (HANE) can be extensive, and will be the concern of this chapter.

It has been suggested that anthropogenic changes in tropospheric weather or climate might produce changes in space weather. However, the latter is mostly affected by changes in the upper atmosphere, be they solar induced or from human modification of the same.

14.11 High-Altitude Nuclear Explosions

A high-altitude nuclear explosion is usually defined as a nuclear explosion above 100 km but below 1000-km altitude. Such detonations pose no direct danger to life on the Earth's surface but can severely damage technical systems on the ground and in space, through electromagnetic and particulate radiation. They also give a rogue state the opportunity for asymmetrical warfare.

The effects of a HANE include:

- Electromagnetic pulse (EMP);
- Ionospheric disruption;
- Global geomagnetic field disturbances;
- Aurora;
- Immediate nearby LEO satellite damage;
- Increased radiation belts;
- Delayed LEO satellite deterioration;
- Little effect on GEO satellites;
- Synchrotron radio emission.

Electromagnetic pulse is a high intensity transient or pulse of energy in the radio part of the electromagnetic spectrum that occurs within the first second of detonation of the device. It has frequency components of 10^{-3} Hz to 100 Mhz with ground field intensities peaking at over 10,000 volts per meter. This pulse will induce large voltages in open (unshielded) conductors. Power transmissions lines are particularly susceptible to EMP. It has been surmised that vehicle electrical systems will also be affected. Although such effects were observed in the 1960s HANE tests, it is unclear how modern vehicles, with more shielded electrical systems, will fare. Certainly induced voltages in nearby LEO satellites will destroy electronics and render them useless.

The ionosphere will be disrupted to the extent that all high frequency (HF) systems will be rendered temporarily useless. This includes over the horizon radar (OTHR). It is possible that increased ionization in some ionospheric layers may increase to the extent that over the horizon propagation will occur at VHF. VLF propagation, used for submarine communication, will suffer sudden phase changes but may be usable.

Global geomagnetic field disturbances will affect geomagnetic surveys, including searches for submarines. Aurora will be prolific in places where aurora is not normally observed (i.e., at low latitudes). The main affects to LEO satellites not in the immediate vicinity of the explosion will be due to increased high-energy particulate radiation. This will damage electronics and solar cells. Any astronauts in space at the time will be particularly affected. There will be little affect at geosynchronous orbit. New and increased radiation belts will result when the magnetosphere captures the lower energy particles. Synchrotron radiation from particles spiraling around the magnetic field lines will create interference to radio telescopes for periods of months to years.

Figure 14.24 shows the main types of radiation released in a HANE. These include both prompt and delayed emission.

Only nine of the atmospheric nuclear test explosions (Figure 14.25) were true HANE events. Seven of these were due to the United States [6]. The largest of these

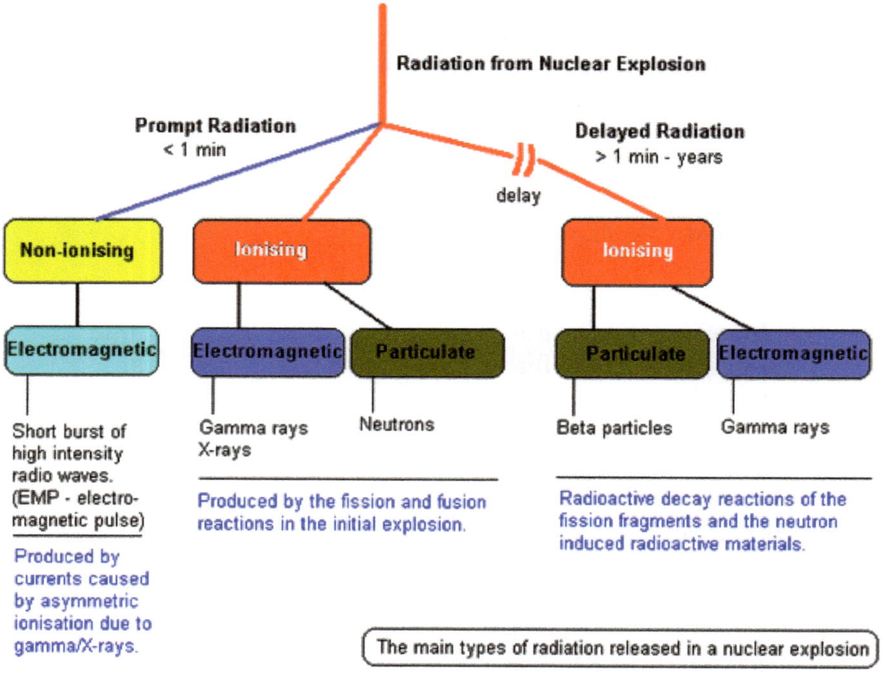

Figure 14.24 The main types of radiation in both prompt and delayed emission from a HANE.

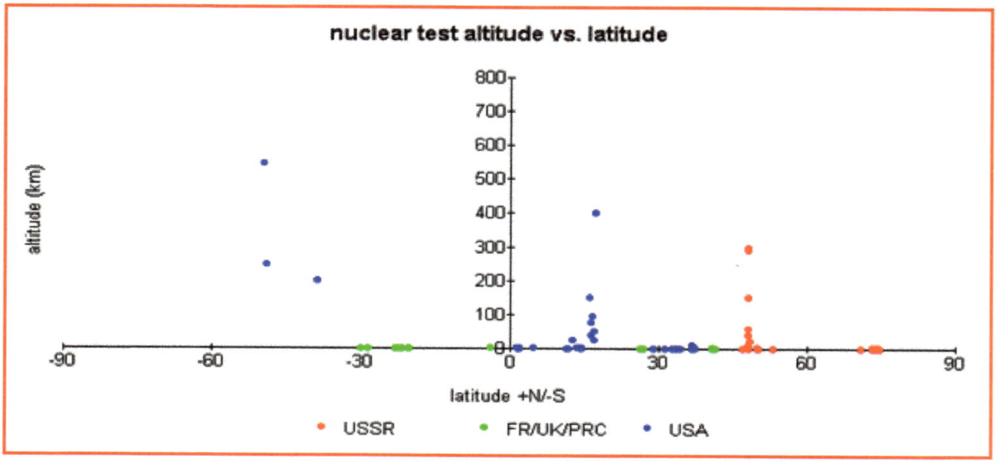

Figure 14.25 Atmospheric nuclear test explosions that have been conducted above ground.

was a 1.4 Mt explosion on July 9, 1962, at 09.00.09 UT over Johnston Island in the Pacific Ocean near Hawaii, at an altitude of 400 km. The event was referred to as Starfish Prime and was part of Operation Dominic.

The large EMP produced induced large voltages in long conductors in Hawaii that caused power failures, deactivated street lighting, activated burglar alarms, and damaged telephone circuits.

The explosion produced a large number of high-energy electrons and protons that became trapped in the Earth's magnetic field, creating an intense artificial radiation belt that lasted until the early 1970s. Seven satellites were destroyed within 7 months. In particular:

Transit 4B	Solar cell degradation
Traac	Solar cell degradation
Ariel	Solar cell degradation
Telstar	Command decoder failure

A graphical summary of US and USSR effects is shown in Figure 14.26.

The time and frequency spectrum of a typical nuclear EMP is shown in Figure 14.27.

The later occurring and lower frequency EMP penetrates the ground and can cause problems in underground cables.

14.11.1 Ionospheric Disturbances

The D-region of the ionosphere suffers an immediate increase in ionization due to prompt gamma rays and X-rays. The gamma rays penetrate to lower altitudes.

This increase in ionization causes a *short wave fade* to frequencies up to at least 30 MHz.

At the same time, there is a general increase in ionization in the E- and F-layers, and OTH communication becomes possible at low VHF.

Electrons will cause later ionization at D-region heights and below. This recurs at periods of an hour or so as the radiation particles move around the Earth.

Systems affected will include HF comms and low VHF comms, OTHR, VLF comms, and position errors in satellite navigation systems.

A monitoring station set up near Hobart by Len Evans (Figure 14.28), and chart record (Figure 14.29) showing immediate cessation of WWVH high-frequency signals due to increased D-region absorption from the Starfish Prime explosion.

A NASA monitoring station located at Carnarvon, Western Australia, was preparing for the Apollo missions. Included in the space monitoring equipment at the SPAN (Solar Particle Alert Network) space weather site was a riometer. A riometer is a relative ionospheric opacity meter to monitor ionospheric absorption. It is used to monitor particle precipitation into the upper atmosphere. It was included as part of SPAN to monitor radiation due to atmospheric nuclear explosions—man-made, not solar radiation. Peter Davies, one of the Australian physicists monitoring the equipment, wrote this in the log for the August 24, 1968: "We saw absorption on one of our riometer receivers from the French H-bomb test between 1907 and 1909 Z."

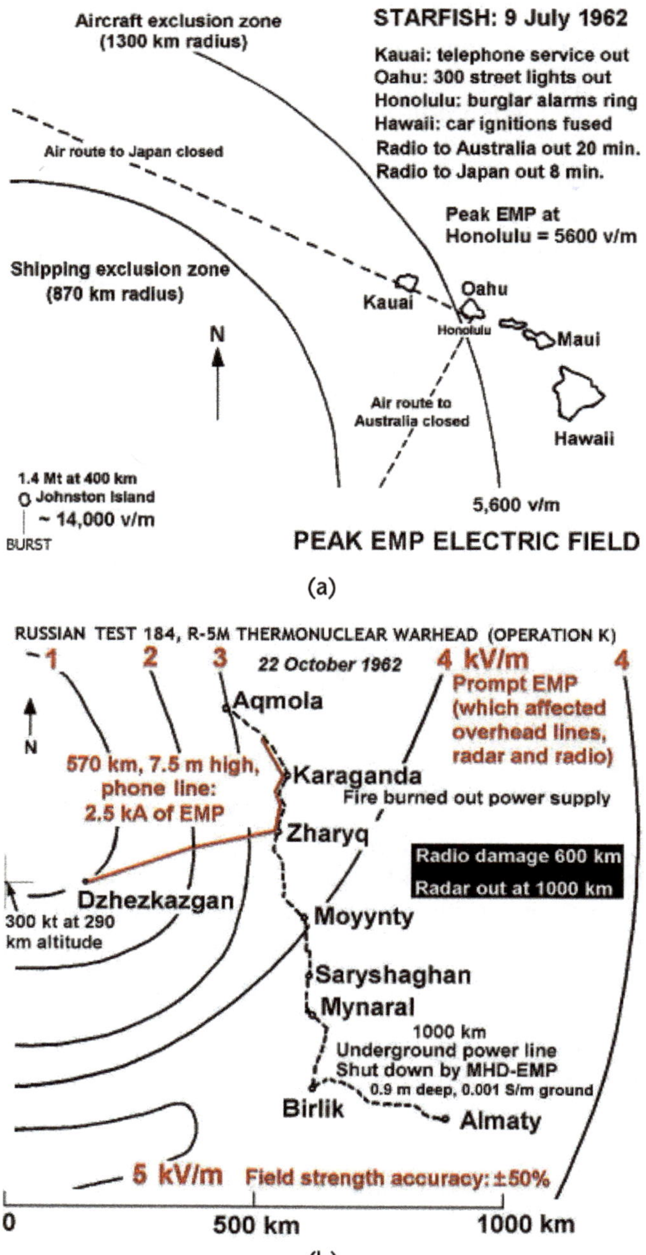

Figure 14.26 Overview of HANE by the United States and USSR.

14.11.2 Geomagnetic Disturbances

Global geomagnetic disturbances were recorded from the Starfish Prime explosion. A large volume around the explosion excludes the geomagnetic field (Figure 14.30). The most intense field changes occur at the footprints of the magnetic field lines associated with the explosion (Figure 14.31), but significant changes are recorded at antipodal points.

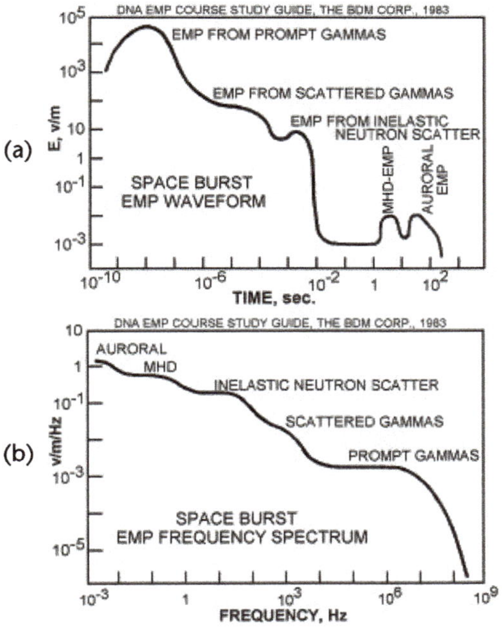

Figure 14.27 (a) Time, and (b) frequency spectrum of a nuclear explosion in the upper atmosphere.

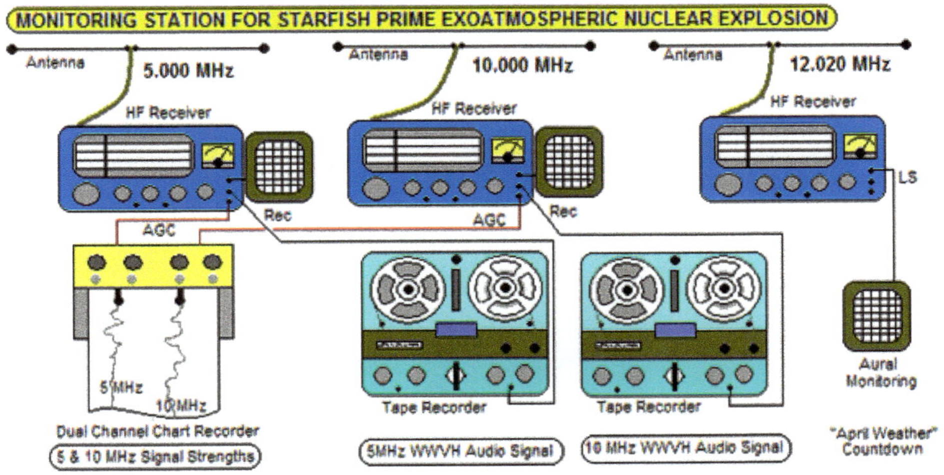

Figure 14.28 Starfish Prime monitoring station.

14.11.3 Radiation Belt Effects

Increased radiation mostly comes from decay of neutrons and fission fragments producing both high-energy electrons and protons (Figure 14.32). Buildup continues for several days and poses a hazard to satellite solar panels and electronics for months.

The US Defense Threat Reduction Agency (DTRA) estimates that a 10–20 kT HANE would incapacitate 90% of all LEO satellites within a few months (Figure 14.33).

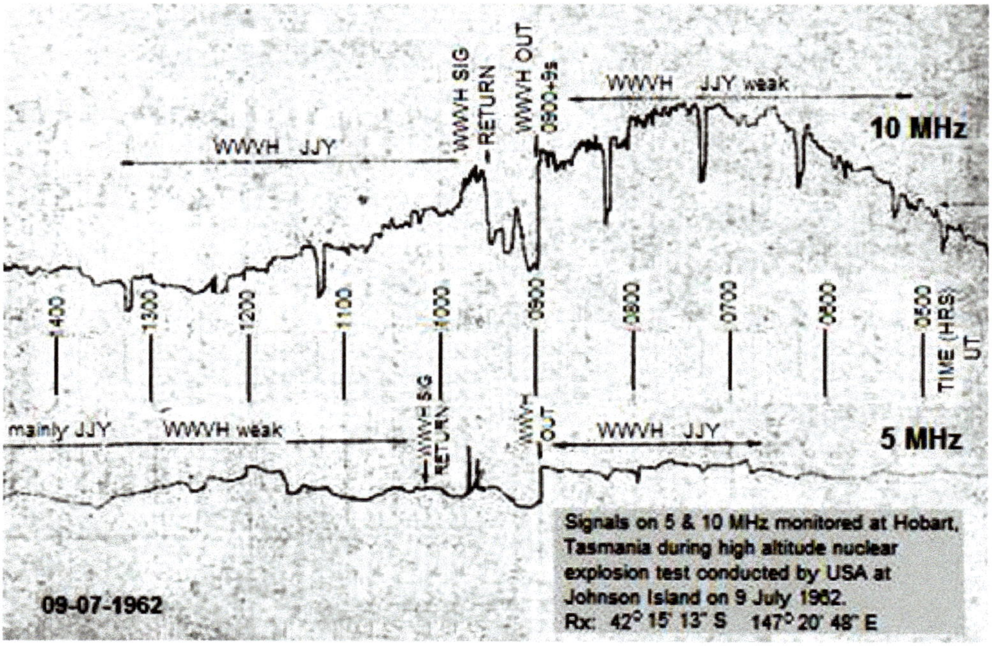

Figure 14.29 Graph of station WWVH during Starfish Prime HANE. (*Source:* Len Evans, Amateur Radio.)

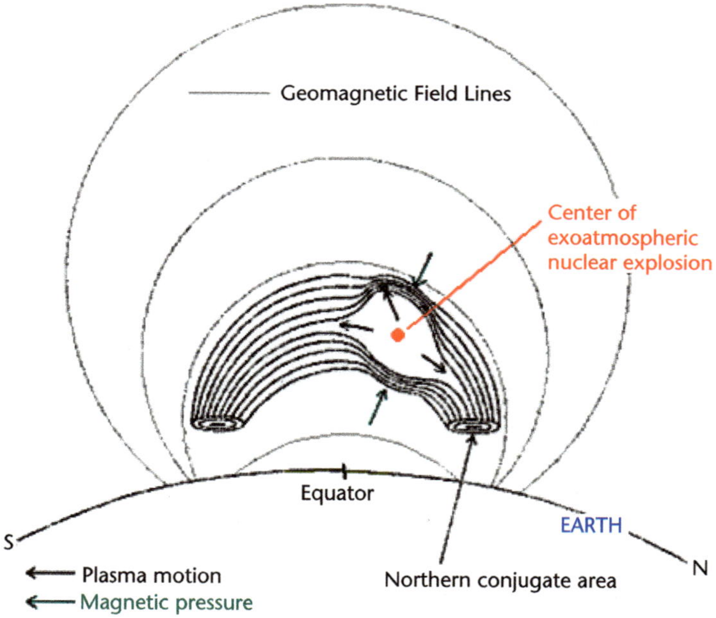

Figure 14.30 Magnetic effect of exoatmospheric nuclear explosion.

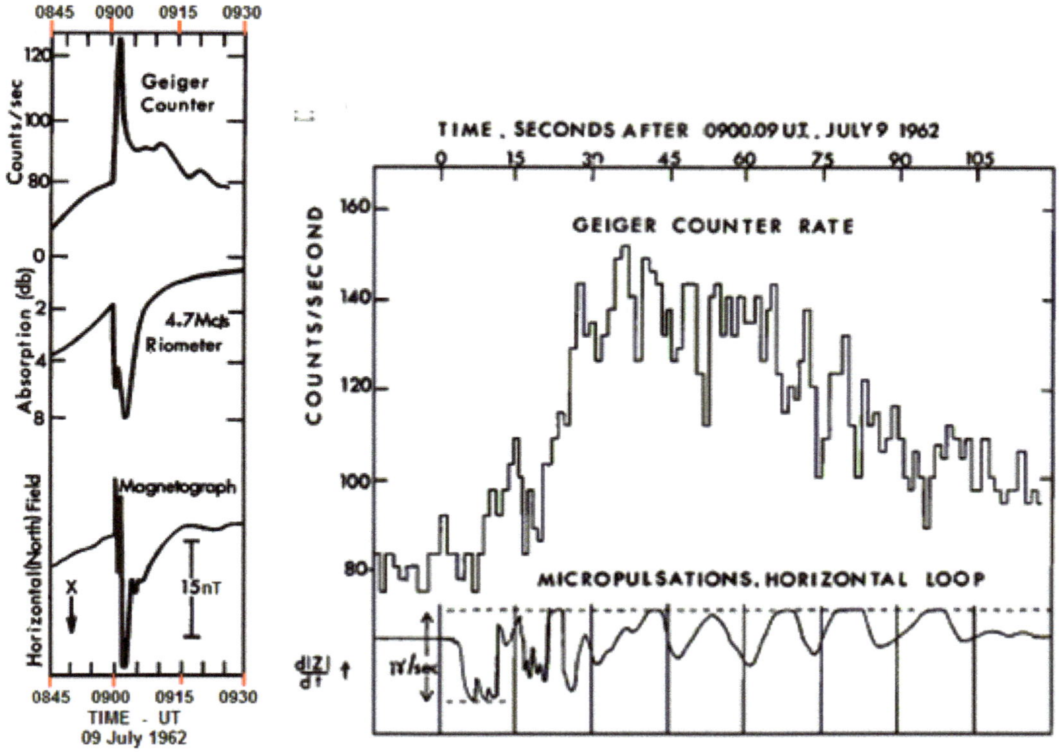

(a) (b)

Figure 14.31 Geophysical parameters measured at Hobart following the Starfish Prime explosion. A mini geomagnetic storm is apparent in this fluxgate magnetometer record together with micropulsations (loop). (*Source:* University of Tasmania.)

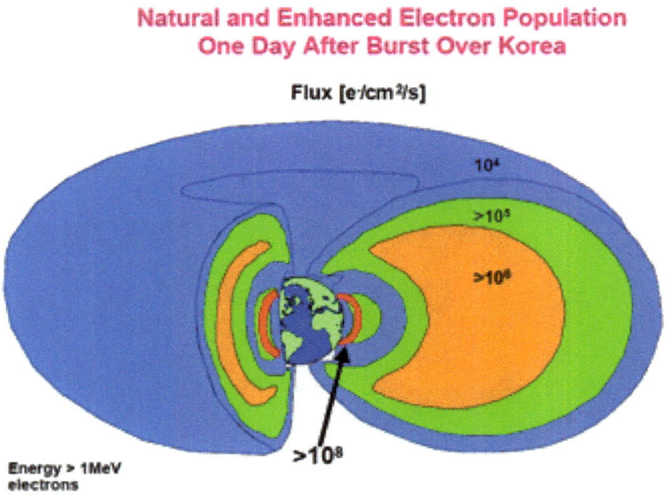

Figure 14.32 Radiation belt enhancement following a HANE above Korea.

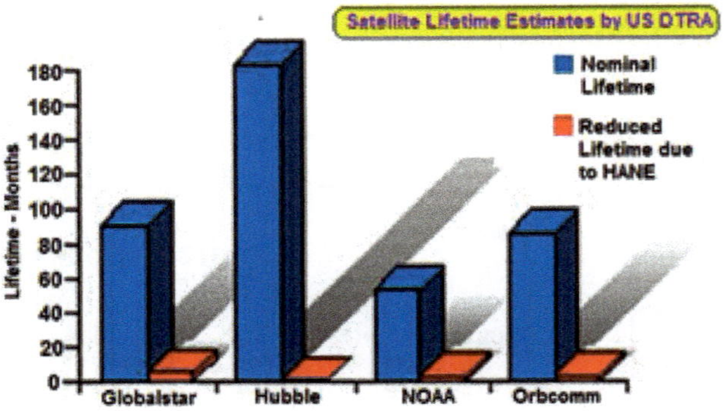

Figure 14.33 Reduction in LEO satellite lifetimes following nuclear explosion in space. (*Source:* US Defense Threat Reduction Agency.)

14.12 Hane Preparation and Mitigation

- Reinstall midlatitude riometers and possibly shortwave fade monitors.
- Ensure space weather forecasters know the different effects of natural and anthropogenic space weather.
- EMP protection is very similar to protection against close lightning strikes.
- Frequency agility is important (HF comms and OTHR).
- Do not rely solely on GNSS.
- Radiation belt remediation (see following section).
- Near immediate evacuation capability for any humans in orbit.

14.13 Radiation Belt Remediation

The trapped particles are mostly protons and electrons. These spiral around the Earth's magnetic field lines, bounce back and forth along field lines, and drift around the Earth (Figure 14.34). As the particles approach the Earth the magnetic

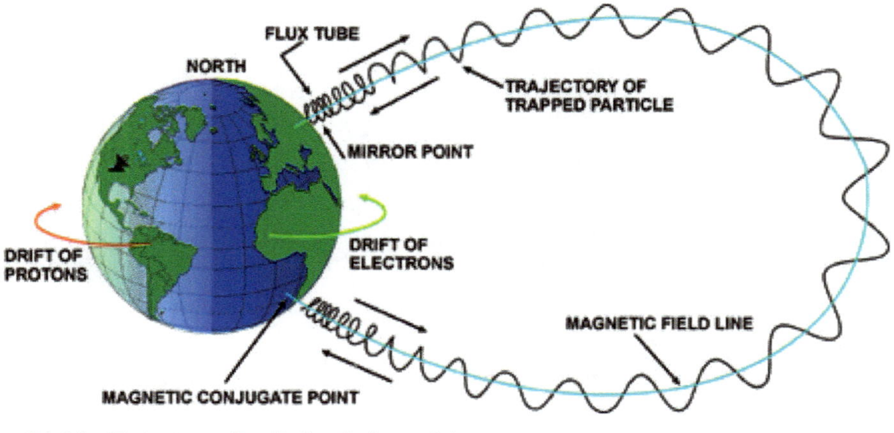

Figure 14.34 Trajectory of radiation belt particles.

field increases. This decreases their spiral radius, and eventually turns them around (like a magnetic mirror), and sends them back out into space. The altitude at which this occurs depends on their *pitch* angle at the equator.

The motion of particles in radiation belts can be affected by VLF radio waves. The principal effect is to the pitch angle of the particles and thus their reflection altitude. If this can be lowered below 100 km, the atmosphere will absorb the particles.

The right frequency and modulation causes the particles to absorb energy and move lower into the Earth's atmosphere as they bounce between the poles. At low enough altitudes they will collide with molecules in the upper atmosphere and be removed harmlessly.

Figure 14.35 shows the precipitation of radiation belt particles due to a high-power transmitter call sign NWC located at North West Cape in Western Australia. Note that there is also a similar but more diffuse precipitation at the magnetic conjugate location in northern China.

The precipitation interaction with the atmosphere occurs in the upper atmosphere. There is no significant radiation enhancement at ground level.

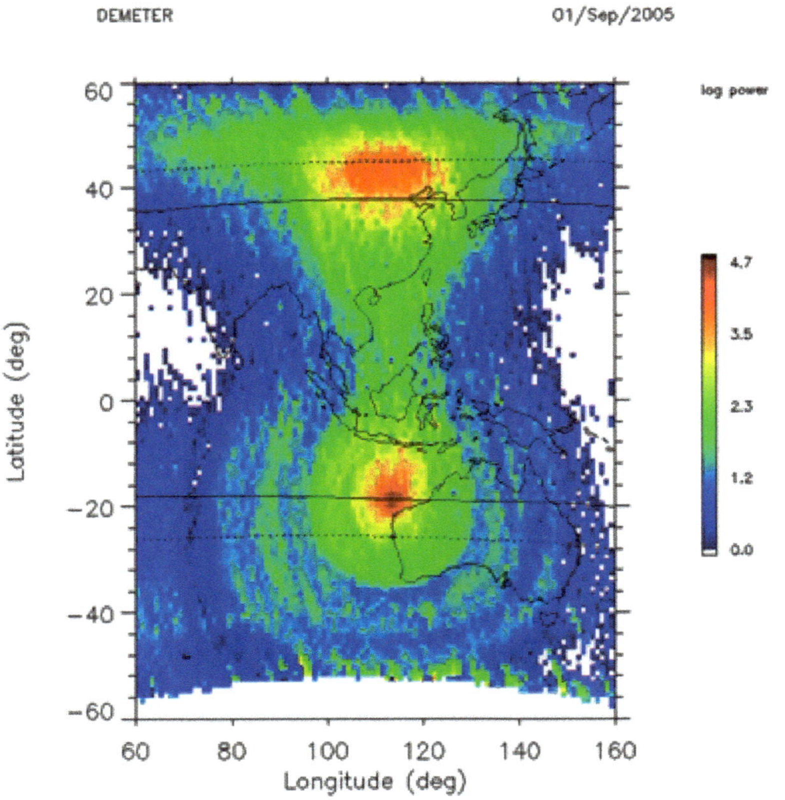

Figure 14.35 Precipitation of radiation belt particles due to a high power (1 MW) VLF transmitter at North West Cape in Western Australia, and at the conjugate magnetic location in China.

References

[1] Satellite Anomaly Database, http://www.ngdc.noaa.gov/stp/satellite/anomaly/satellite anomaly.html.

[2] Glaser, P., "Power from the Sun: Its Future," *Science*, Vol. 162, 1968, pp. 857–61.

[3] deepspaceindustries.com.

[4] Gombosi, T., et al., "Anthropogenic Space Weather," *Space Science Reviews*, Vol. 212, No. 3, 2017, pp. 985–1039, arXiv: 1611.03390v3.

[5] Dupont, D. G., "Nuclear Explosions in Orbit," *Scientific American*, June 2004.

[6] Hoerlin, H., "United States High Altitude Test Experiences," LASL Monograph, LA-6405, 1976.

Selected Bibliography

Books

Aldrin, B., and J. Barnes, *The Return,* New York: Forge Books, 2000.

Carlowicz, M. J., and R. E. Lopez, *Storms from the Sun*, Joseph Henry Press, 2002.

Freeman, J. W., *Storms in Space*, Cambridge University Press, 2001.

Garrett, H. B., and A. Whittlesey, *Guide to Mitigating Spacecraft Charging Effects*, Jet Propulsion Laboratory, 2011, descanso.jpl.nasa.gov.

Hastings, D., and H. Garrett, *Spacecraft-Environment Interactions*, Cambridge University Press, 1996.

Jursa, A. A. (ed.), *Handbook of Geophysics and the Space Environment*, Air Force Geophysics Laboratory, National Technical Information Service Accession number ADA 167000, 1985.

Lang, K. R., *The Sun from Space*, Springer, 2000.

Poppe, B. B., and K. P. Jorden, *Sentinels of the Sun*, Johnson Books, 2006.

Tribble, A. C., *A Tribble's Guide to Space*, Princeton University Press, 2000.

Tribble, A. C., *The Space Environment: Implications for Spacecraft Design*, Princeton University Press, 1995.

Wilson, P. R., *Solar and Stellar Activity Cycles*, Cambridge University Press, 1994.

Websites

Space situational awareness space weather straight from the Space Pole in Belgium, http://swe.ssa.esa.int/.

Space weather for satellite operations from the Australian Space Weather Service, http://www.sws.bom.gov.au/Satellite.

US source of space weather data with many space sensor feeds, http://www.swpc.noaa.gov/.

CHAPTER 15

Rules of Thumb in Space Weather Forecasting

From early times man has observed the weather and formulated rules to predict it, "Clear skies mean fine, gray clouds may rain," or the more poetic: "Red skies at night, sailor's delight; red skies in the morning, sailors take warning."

Although we have been trying to predict space weather for only around 70 years, forecasters have developed a substantial lore of the space environment to help them predict various aspects of their trade.

Some of these rules are based on sound physics, and some have been verified by later research. Others have grown up from forecasters' gut-feelings and may or may not have stood the test of time.

Rules of thumb may be divided into various categories: solar, magnetospheric, ionospheric, along with others, and may be applicable to short, intermediate, or long-term forecasts. Some rules are encapsulated in a phrase, some are expressed graphically, and some are given by a simple formula. Here is a selection.

15.1 Early Rules

Probably the earliest rules are the five formulated by Max Waldmeier in his 1935 doctoral thesis concerning sunspot cycles:

1. The higher the maximum, the shorter the rise time;
2. The higher the maximum, the longer the fall time;
3. The higher the maximum, the stronger the activity 5 years post max;
4. The higher the maximum, the higher the descent area;
5. The ascent area is almost independent of the intensity of the maximum.

The average period of the sunspot cycle is close to 11 years (Figure 15.1).

15.2 Solar Flares

Before the satellite age, flares were arguably the most spectacular events observable on the solar disc. They were also suspected of being associated with geoactivity since the giant flare of 1859. Flare predictions were usually made by reference to sunspot features and related features in white light and H-alpha images (Figure 15.2), and later from photospheric magnetograms [1, 2].

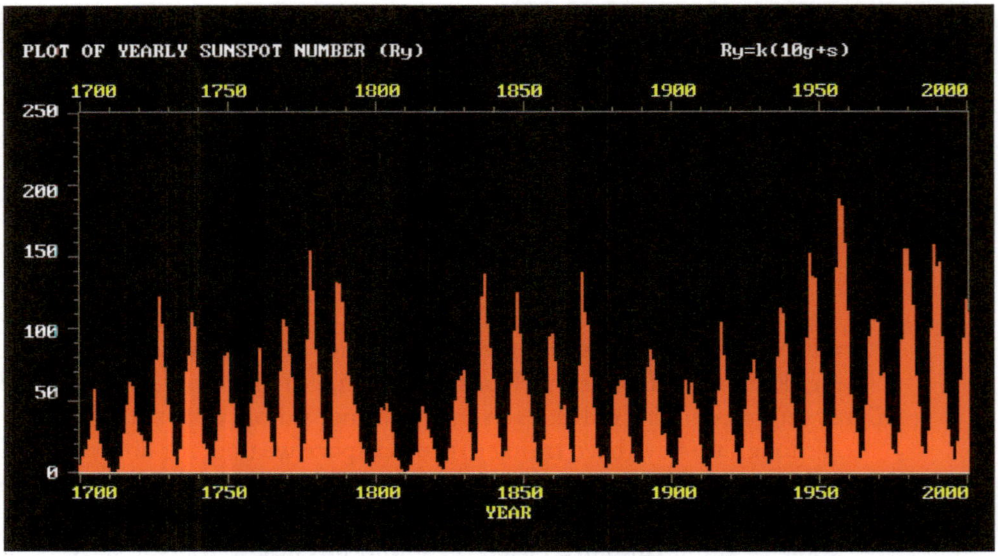

Figure 15.1 Three centuries of sunspot cycles.

The flare potential of a region has variously been related to:

- Complexity of a sunspot group (white light and magnetogram);
- The area of a sunspot group (white light);
- Magnetic field strength (magnetogram);
- Magnetic field gradients (magnetogram);
- Large number of kinks in the inversion/neutral line (H = alpha);
- Delta configuration (opposite polarity umbra in same penumbra);
- Proper motions of the sunspot group (white light);
- Rapid growth or decay of the sunspot group (white light area);
- Bright plage (H-alpha);
- Plage brightness fluctuations (H-alpha, X-ray);
- Neutral/inversion line orientation (H-alpha);
- Rising X-ray flux.

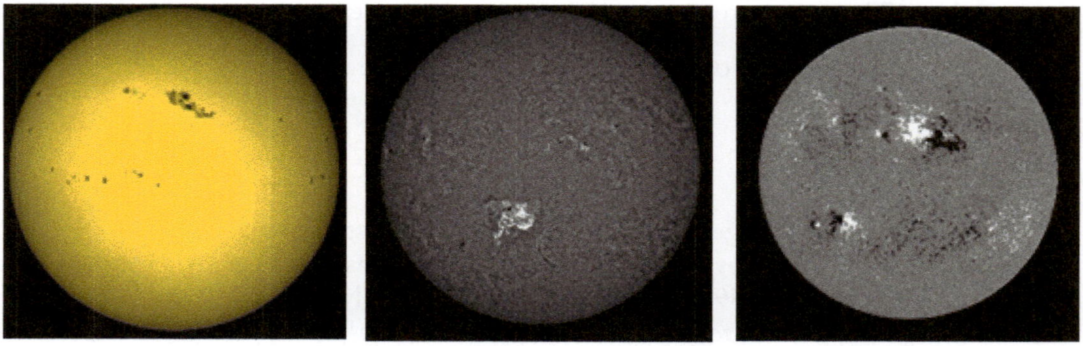

(a) (b) (c)

Figure 15.2 The Sun in (a) white light (*Source:* NASA), (b) H-alpha light (*Source:* US National Solar Observatory), and (c) magnetic fields (*Source:* US National Solar Observatory).

One of the most important rules of flare forecasting is *persistence*. If a region is relatively stable (i.e., it maintains approximately the same configuration) the likelihood it will flare tomorrow is strongly related to its flare activity today; and the maximum class of flare will be similar to those previously experienced. Wheatland has confirmed that persistence is one of the strongest flare forecasting tools with a Bayesian analysis of a large number of active regions.

15.3 Coronal Mass Ejections

Satellite observations have revealed a CME as the mechanism by which ejected solar plasma is carried to the Earth, resulting in geomagnetic storms and related phenomena (Figure 15.3).

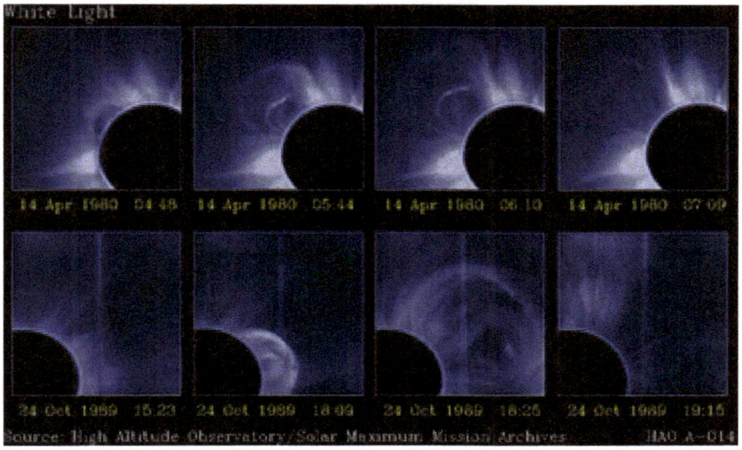

(a)

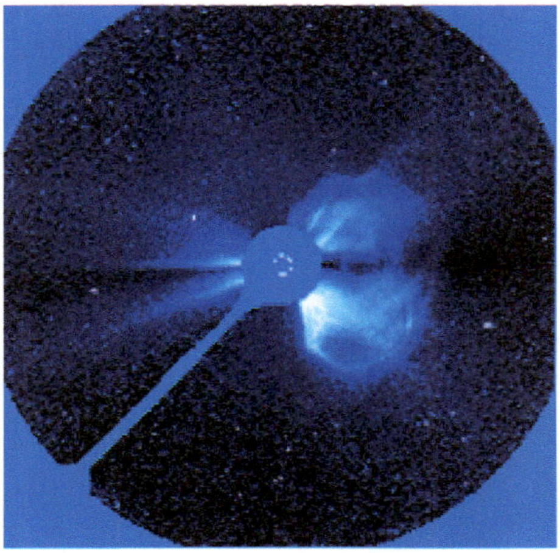

(b)

Figure 15.3 Coronal mass ejections observed by SMM. (*Source:* NASA/SMM, NASA/ESA/SOHO/LASCO.)

CMEs are observed with coronagraphs, where the bright light of the solar photosphere is removed by an occulting disc. In most cases the coronagraph must be located outside the Earth's atmosphere. Because the dataset of CMEs is much smaller than that for flares, and really only started in the late 1970s, CME prediction rules are relatively scarce. However, because there is a very large correlation between flares and CMEs, particularly the larger variety where the association is almost 100%, forecast rules for flares are also applicable to CMEs.

One notable rule promoted by Canfield specifically for CMEs is that a sigmoid (S-shaped) feature observed in solar X-ray images has high predictive value for a CME occurrence (Figure 15.4).

15.4 Geomagnetic Storms

During times of high solar activity, most geomagnetic storms are the result of CMEs intersecting the Earth in its orbit. Given the occurrence of a mass ejection from the solar corona, prediction rules are then related to how geoeffective this CME might be, and to the time delay between the solar event and the subsequent storm. A selection of these rules:

- Geomagnetic storms only occur when the interplanetary field is southward allowing direct coupling to the Earth's magnetosphere;
- The average delay between a CME and a geostorm is pi (= 3.14) days;
- Geomagnetic storms are more frequent around the equinoxes;
- Type II / IV radio bursts are a predictor of geomagnetic storms;
- Halo CMEs generally produce geomagnetic storms;
- Forbush decreases of cosmic rays are good indicators of magnetic storms.

During times of low solar activity, when CME production is also very low, most geomagnetic storms (Figure 15.5 and 15.6) are related to solar coronal holes, regions where the magnetic field is open to space and which have a lower density of solar plasma. These are visible in X-ray images as dark areas. Coronal holes are some

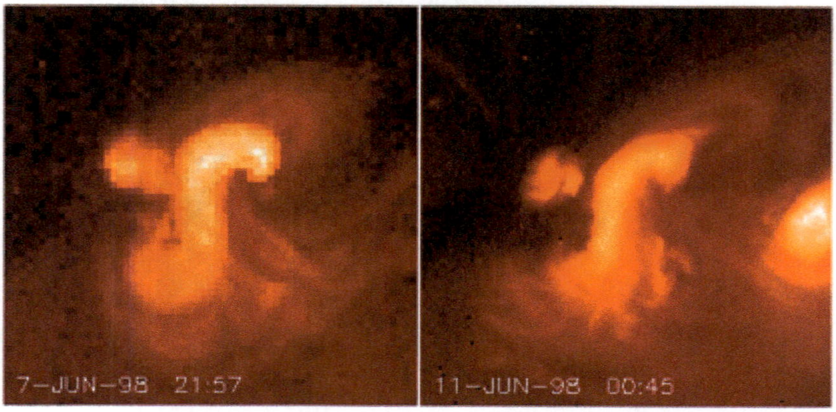

(a) (b)

Figure 15.4 Coronal mass ejections observed by (a) NASA/SMM and (b) NASA/ESA/SOHO/LASCO.

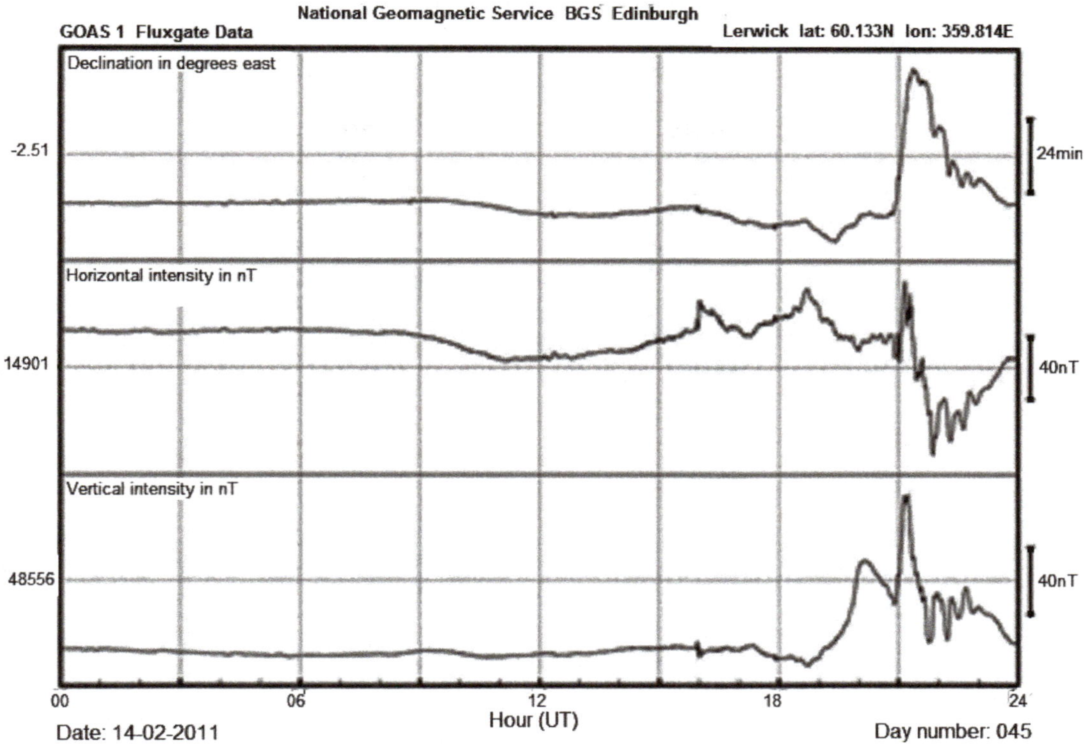

Figure 15.5 Edinburgh magnetic record showing start of a magnetic storm.

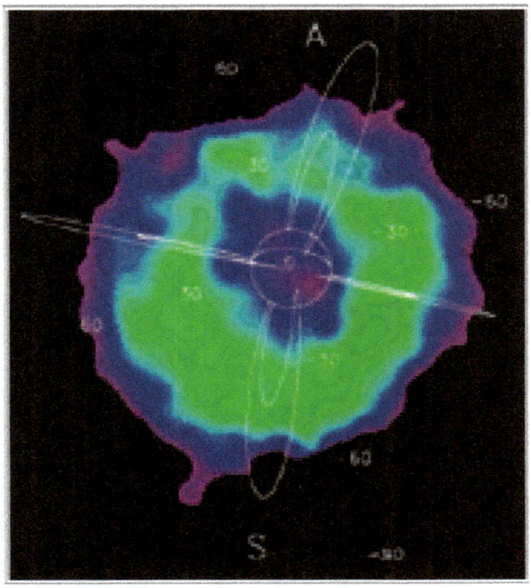

Figure 15.6 The Earth's magnetosphere under storm conditions.

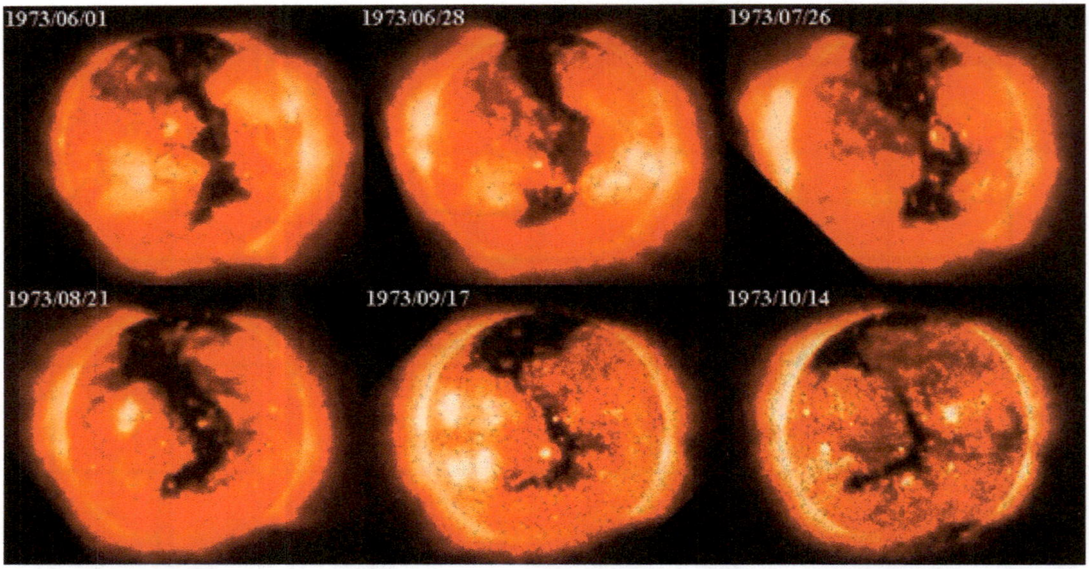

Figure 15.7 *Skylab* X-ray images showing the stability of a large coronal hole over a 4-month period. (*From:* NASA/Skylab.)

of the most stable features on the Sun and may last for several solar rotations (i.e., many months (Figure 15.7)). Because a geomagnetic storm will occur whenever the interplanetary extension of a coronal hole sweeps past the Earth, recurrence with a 27-day period is the primary rule for forecasting these types of geomagnetic disturbances. The formation and disappearance of these holes as seen in solar X-rays is used to indicate the start and end of these recurrence periods.

15.5 Geomagnetic Effects

Geomagnetic effects cause many phenomena including the production of aurora (Figures 15.8 and 15.9). Geomagnetic storms can also induce large currents in long

Figure 15.8 Aurora as seen from the ISS. (*Source:* NASA.)

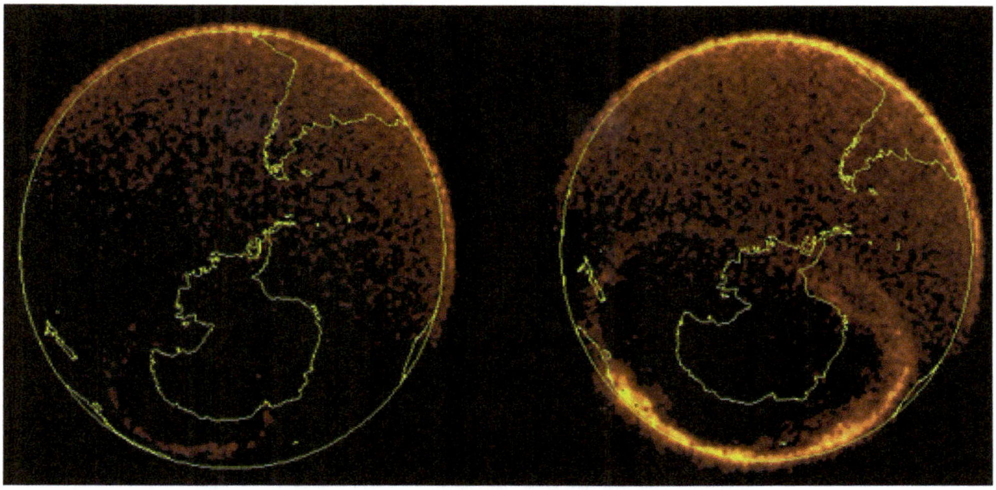

Figure 15.9 Aurora under quiet (left), and active (right) conditions.

pipelines and extensive electrical grids (Figure 15.10). The latter can damage power stations and associated substation equipment (Figure 15.11).

15.6 Ionospheric Effects

The ionosphere is affected both by the solar X-ray output and by the solar particle output (CMEs). X-ray flares on the Sun increase the absorption of radio waves traveling through the lower part of the ionosphere (the D-region), producing short

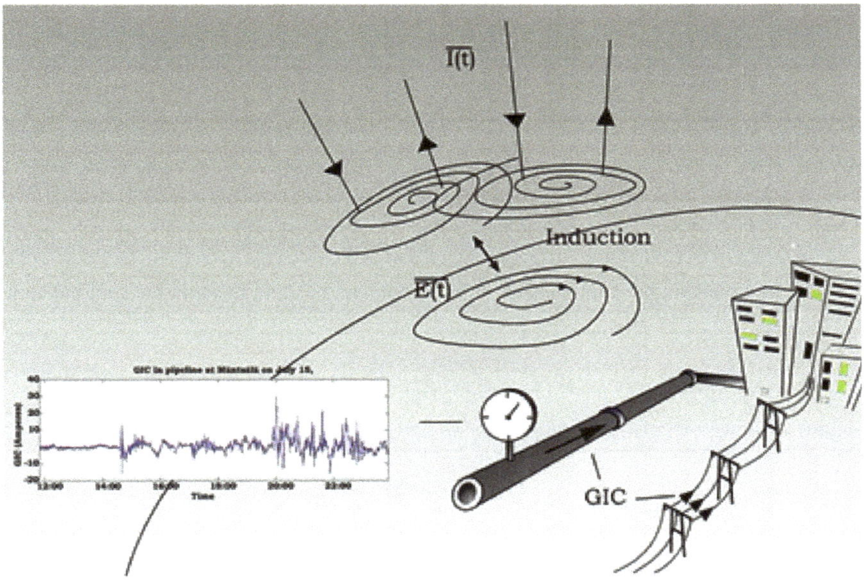

Figure 15.10 GIC in theory with an experimental record. (*Source:* Finnish Meteorological Institute.)

Figure 15.11 GIC damaged large substation transformer, March 1989. (*From:* US Space Environment Center.)

wave fadeouts. The rule we present for this effect is a graph (Figure 15.12) developed by the US Air Force.

The ionosphere is also subject to storms, mostly driven by geomagnetic activity in the magnetosphere above the ionosphere. Some rules for forecasting ionospheric storms:

- Ionospheric storms typically follow geomagnetic storms;
- Positive storms in winter, negative storms in summer (midlatitudes);

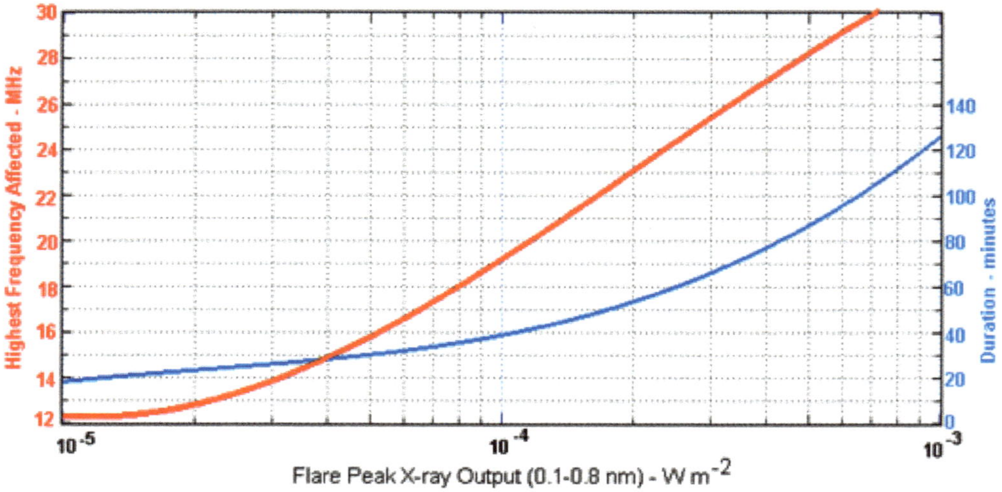

Figure 15.12 Solar X-ray flare statistics.

- Larger magnetic storms extend ionospheric storm effects equatorward;
- Ionospheric disturbance is delayed with respect to geomagnetic storm main phase onset time (T_m) by:

$$t\ (\text{hr}) = 0.2T_m + 22 \qquad \text{for times between 08-18LT and}$$
$$t\ (\text{hr}) = T_m + 4.8 \qquad \text{for times between 19-07LT.}$$

15.7 Solar Particle Events

Relativistic particles from the Sun, mostly protons, contribute the greatest hazard of any solar activity (Figure 15.13), particularly to space operations. Spacecraft systems such as solar cells can suffer permanent and significant degradation from a large SPE, and the radiation dose to manned spaceflight outside of low Earth orbit can be fatal.

Many rules have been devised for the prediction of SPEs. The event probability for most is not very high (some may be necessary but not sufficient) as is the false alarm rate. We list a few of these rules:

- A large X-band solar microwave burst indicates an SPE;
- A Castelli-U radio spectral burst strongly indicates an SPE (Figure 15.14);
- A type II/IV radio spectral burst indicates an SPE;
- A sunspot group with an east-west inversion/neutral will produce an SPE;
- A large CME produces an SPE;
- prediction_success = a * [Goal] + b * [Funding]^2 + c * $\sqrt{\text{[Enthusiasm]}}$ + d * [Luck].

The last rule was formulated by veteran SPE researchers Shea and Smart to indicate the low detection rate and high false alarm of previous rules!

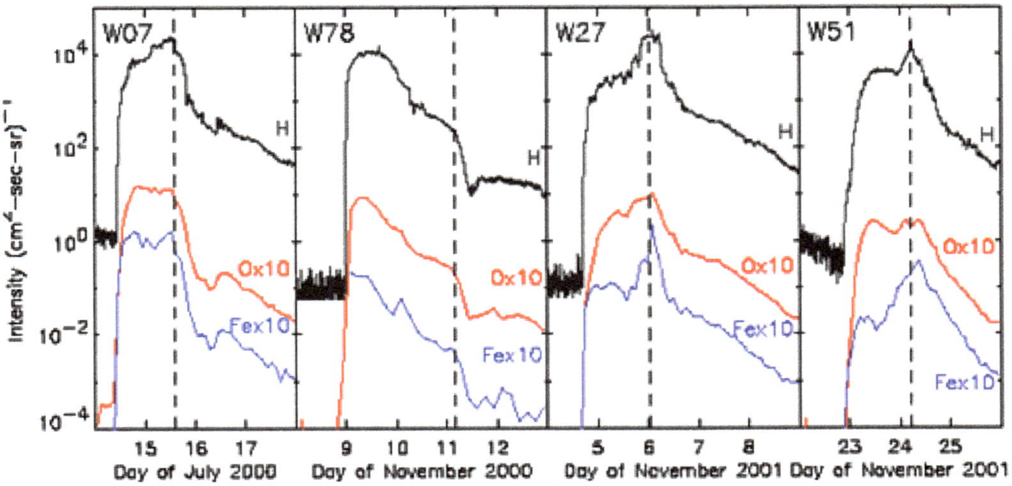

Figure 15.13 A selection of four solar proton events.

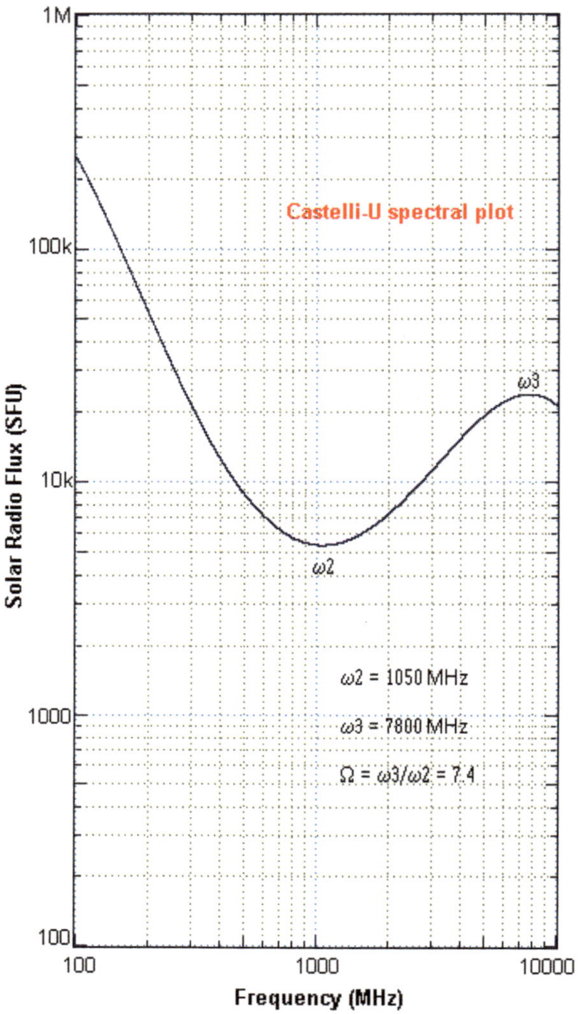

Figure 15.14 A Castelli-U radio burst, sometimes thought to have a high association with a solar proton event.

Our current understanding is that there are several different mechanisms that accelerate particles to high energy. One of these is related to the initial magnetic reconfiguration event on the Sun (flare/CME production), whereas another involves Fermi acceleration around the CME leading shock front. This latter mechanism may occur well away from the Sun.

A recently developed rule claims a high association between type 3L radio bursts and SPEs.

15.8 Spacecraft Effects

Satellites are affected by several aspects of the space environment. The neutral atmosphere produces drag on low Earth orbiting objects. Both the soft X-ray and particle output from the Sun heat the base of the thermosphere increasing the atmospheric density at LEO satellite heights. When these factors increase, satellite lifetime is reduced.

Relativistic particles from SPEs have been known to reduce the power output from solar cells by 30%. These particles also affect solar imaging sensors, producing a snowstorm effect (Figure 15.15).

High energy electrons carried in the solar wind can cause satellite charging, with resultant surface and deep dielectric discharges causing significant damage to satellite electronics.

The US Air Force space weather operations developed the following rule:

Potentially dangerous conditions exist for charging and discharging events to satellites if the 2-MeV electron flux at geosynchronous orbit (numbers per square centimeter per steradian) meets either of the following conditions:

- Greater than 3×10^8 per day for three consecutive days;
- Greater than 10^9 for a single day.

It has also been noted that such conditions often occur about 2 days after the onset of a large geomagnetic storm.

Solar particle effects on satellites are shown in Figures 15.16 and 15.17.

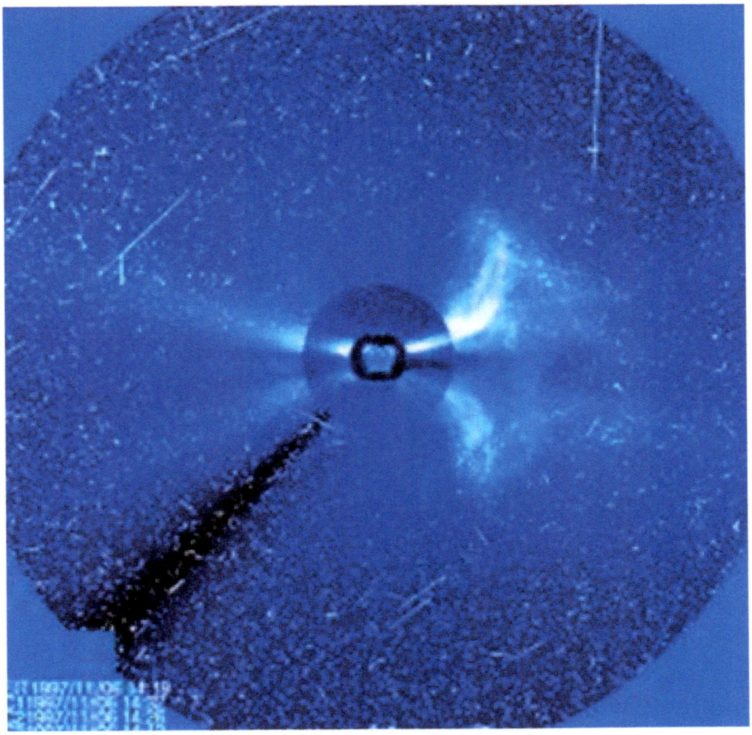

Figure 15.15 Solar proton event induced snowstorm in the LASO satellite coronograph. (*Source:* NASA/ESA/SOHO/LASCO image.)

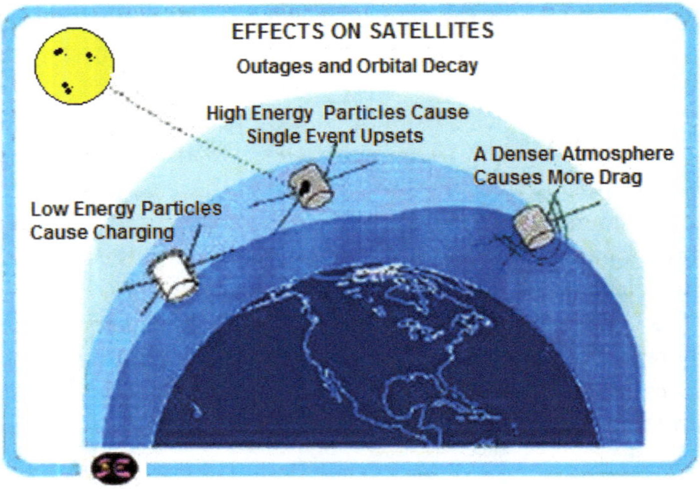

Figure 15.16 Satellite effects due to solar particle emission. (*Source:* US Space Environment Center.)

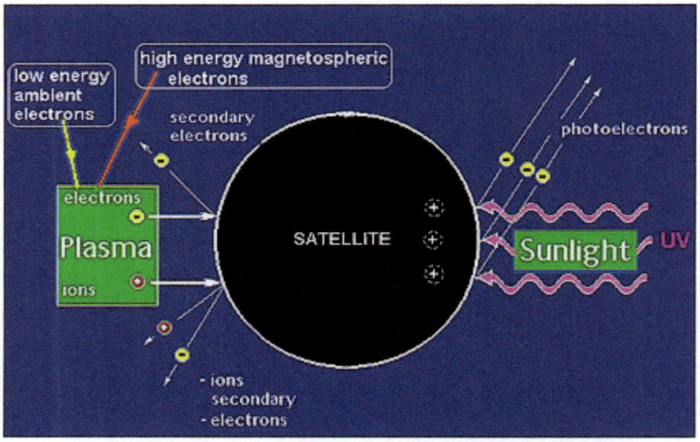

Figure 15.17 Space environment interactions with satellites.

15.9 Acknowledgments

I would like to thank the people who replied to an initial request for space weather forecasting rules of thumb with useful material. In particular, I would like to thank Joe Kunches, Garth Patterson, Richard Marshall, Shinichi Watari, Risto Perjola, and Graham Steward. I would also like to thank the many unknown people who initially developed some of these rules through long and diligent observation of the space weather phenomena. I would lastly like to thank some pioneers of space weather forecasting: Ron Giovanelli, Frank Cook, Gary Heckman, Joe Hirman, and various members of the US Air Force space weather community.

References

[1] Bartels, J., "Terrestrial magnetic activity and its relation to solar phenomena," *Journal of Geophysical Research*, Vol. 37, 1932, pp. 1–52.

[2] Sammis, I., F. Tang, and H. Zirin, "The Dependence of Large Flare Occurrence on the Magnetic Structure of Sunspots," *The Astrophysical Journal*, Vol. 540, 2000, pp. 583–587.

Selected Bibliography

Sawyer, C., J. W. Warwick, and J. T. Dennett, *Solar Flare Prediction*, Colorado Associated University Press, 1986.

CHAPTER 16

Near-Earth Objects and Planetary Defense

16.1 Introduction

A knowledge of near-Earth objects is the third part that completes the triad of space situational awareness. An awareness of both the threat and the opportunity that they present.

Near-Earth objects comprise the population of both near-Earth asteroids (NEA) and near-Earth comets (NEC):

$$NEO = NEA + NEC$$

These are moderately large objects (10m to 10+ km) that come close enough to the Earth that they could pose a collisional risk in the future. They could also provide a source of raw materials for future space operations.

Comets have been known for thousands of years due to the long tails they form when approaching the inner solar system. As a comet gets closer to the Sun, the surface is warmed and volatile material is evaporated from the comet and streams out into space, forming a tail or tails millions of kilometers in length that is directed behind the comet due to both the solar wind and light pressure from the Sun.

It is believed that comets initially come from a reservoir of billions of comets, or proto-comets, well beyond Pluto, but still loosely bound to and orbiting the Sun in what has been called the Oort cloud. The speculated distance of the Oort cloud is on the order of 25 to 50 thousand astronomical units (AU) or 4 to 8 trillion (10^{12}) km. These may have their orbits perturbed by a passing star over periods of millions of years, with some being thrown toward the Sun where they become visible to us on Earth. Some of these comets may be captured by the Sun and become what is known as periodic comets. Comet Halley is the best known of these with an orbital period on 76 years.

Comets have for a long time been known as harbingers of doom, with some speculating that this may have come from a distant collective memory when the Earth suffered a collision with a comet. Immanuel Velikovsky wrote widely on this theme over 70 years ago. Even today, comets can still inspire fear among the population at the close approach of a comet. At the 1910 appearance of comet Halley, astronomers using spectroscopes attached to their telescopes discovered that the comet tail contained small amounts of cyanide gas. A French astronomer speculated that this could impregnate the Earth's atmosphere, which led to newspaper reports of death for the Earth's population.

Even today, when it appears that we may be able to deal with an approaching asteroid before it was to hit the Earth, this is not at all certain for a doomsday comet. The reason is that we now have a good knowledge of the thousands of potentially hazardous asteroids, and will thus have sufficient time to send a mission to deflect any potential impact. This is not the case for an errant comet. This author photographed comet Hyakutake (Figure 16.1) at its closest approach to Earth, which occurred only 50 days after its initial discovery. If the comet had been on an Earth impact trajectory, instead of passing by us 15 million kilometers away, there would have been absolutely no hope of mounting a deflection mission.

16.2 Early History

Unlike comets, asteroids were unknown before the nineteenth century. Their discovery resulted from an apparent discrepancy in the distances of the planets from the Sun.

In 1772 the German astronomer Johann Bode popularized a law relating to the distances of the planets from the Sun. Using the astronomical unit (1 AU ~ 150 million km), which is the average distance of the Earth from the Sun, we can construct a series as shown below:

Start with:	4	4	4	4	4	4	4	4
Add series:	0	3	6	12	24	48	96	192
	4	7	10	16	28	52	100	196
Divide /10	0.4	0.7	1.0	1.6	2.8	5.2	10.0	19.6
Planets	0.4	0.7	1.0	1.5	–	5.2	9.5	–
	Mercury	Venus	Earth	Mars		Jupiter	Saturn	

Figure 16.1 Comet Hyakutake at its closest approach to the Earth in March 1996. (*Source:* John A. Kennewell.)

This can be expressed by the mathematical formula:

$$a = 0.4 + 0.3 \times 2^n \quad \text{(AU)}$$

where $n = -\infty, 0, 1, 2, 3, \ldots$

Note that this series was first introduced by Johann Titius in 1766, and thus is now often known as the Titius-Bode law. It is of course, not a law of nature, but simply a mathematical relation.

This series correctly predicted the orbit distance of Uranus, which was discovered by William Herschel in 1781. And thus Bode's law passed the first test of any useful scientific theory, the ability to correctly predict. This gave the law significantly more credence in the eyes of the late eighteenth century astronomers. And so, with this success, astronomers became convinced that there had to be a planet in position 5 of the table. (God would not have allowed such an anomaly to exist.) And so, as the eighteenth century drew to a close, a search began for the missing planet.

In 1800 a group of German astronomers under the leadership of Johann Schroter and sponsored by Baron Franz Xavier von Zach, took on the task of a directed search for the missing planet. They called themselves the Celestial Police. Each astronomer was assigned a particular area of sky to search. The Celestial Police started with 6 members but soon added another 18 to make two dozen, and included some of the most influential astronomers in Europe.

However, it was not the Celestial Police, but an Italian astronomer Guiseppe Piazzi who made the first discovery. On the first day of the nineteenth century (January 1, 1801), at the observatory of Palermo in Sicily, Piazzi discovered a new planet (it moved with respect to the stars) which he called *Ceres*.

Within a few months the mathematician-astronomer Gauss had computed the orbit of Ceres to lie at 2.8 AU, just as predicted! However, something was wrong. Ceres was too small—it appeared in a telescope as a point (starlike or asteroidal) not as a disc like the other planets. The Celestial Police continued to search. In March, one of them, Heinrich Olbers, found a second asteroid, which was named *Pallas*. This was even fainter than Ceres (and thus most probably even smaller). It was surmised that once there may have existed a planet in this position that had then fragmented. The search then continued to find mote fragments:

- 1804—*Juno*;
- 1807—*Vesta*;
- 1815—No more found and the Celestial Police disbanded;
- 1830—*Atsraea* (found by Prussian amateur Hencke);
- 1890—A total of 300 asteroids had been found;
- 1891—Max Wolf started a photographic survey for asteroids.

The name *asteroid* was adopted for these bodies, which is quite unfortunate for they are anything but starlike. They may appear as points, even in large telescopes on the Earth, but the name planetoid (planet-like) would have been far better. This was suggested by several people and even adopted in limited circles, but it never gained general acceptance.

Today, over one million main-belt asteroids have been discovered. That is, asteroids that cluster around a distance of 2.8 AU. However, there are significant numbers of asteroids discovered that have orbits in other regions of space, including many in the inner solar system around the orbits of the planets Mercury, Venus, the Earth, and Mars. These are classified into orbital subgroups. Data from the Minor Planet Center in August 2023 is summarized in Table 16.1.

The current asteroid population is summarized graphically in Figure 16.2. Only a fraction of the numbers can be realistically be shown here.

16.3 The First Near-Earth Asteroid

The first near Earth asteroid to be discovered was *433 Eros*. This was discovered by Carl Gustav and his assistant Felix Linke in 1898 at the Urania Observatory in Berlin. Confirmed asteroids are given both a number (in order of naming) and a name. *Eros* is an irregularly shaped stony asteroid (designated as S-type) that has main axes dimensions of 33 × 13 × 13 km. Its current orbit brings it to a closest approach distance of about 22 million kilometers from Earth.

Not only was *Eros* the first NEA to be discovered, it was also the first asteroid to be orbited by a spacecraft and the first to have a spacecraft land on it.

A NEO is any small body in the solar system that comes close to the Earth. This is usually defined as a natural heliocentric space object that has a perihelion distance of less than 1.3 AU. If the NEO actually crosses the Earth's orbit and is larger than 140 km in size, it is also classified as a potentially hazardous object (PHO/PHA/PHC). There are currently over 2,000 known PHAs.

The orbital classification of asteroids is shown in Figure 16.3.

Asteroids are also given a geological classification according to the material of which they are composed. This is related to the classification of meteorites, which are speculated to come from each asteroid type. The most common types are S (stony), C (carbonaceous), and M (metallic). These classifications come principally from spectroscopic analysis of the reflected sunlight from the asteroids.

Table 16.1 Numbers of Asteroids Known at August 2023

Inner/Mid Solar System	*Count*
Atens	2,645
Apollos	17,215
Amors	13,792
Hungarias	28,356
Mars-crossers	24,315
Mid/Outer Solar System	*Count*
Main-belt asteroids	1,196,068
Hildas	5,717
Jupiter trojans	12,697
Distant objects	5,123

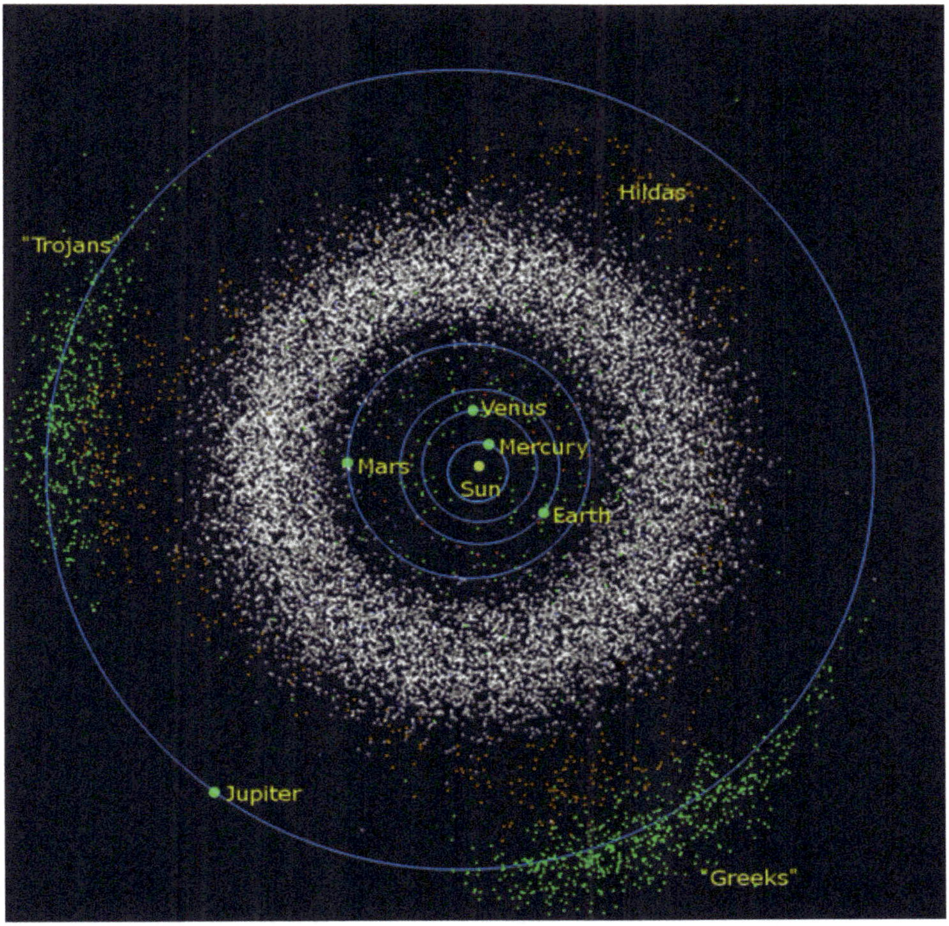

Figure 16.2 A representation of the currently known asteroid population. (*Source:* NASA/JPL/CNEOS.)

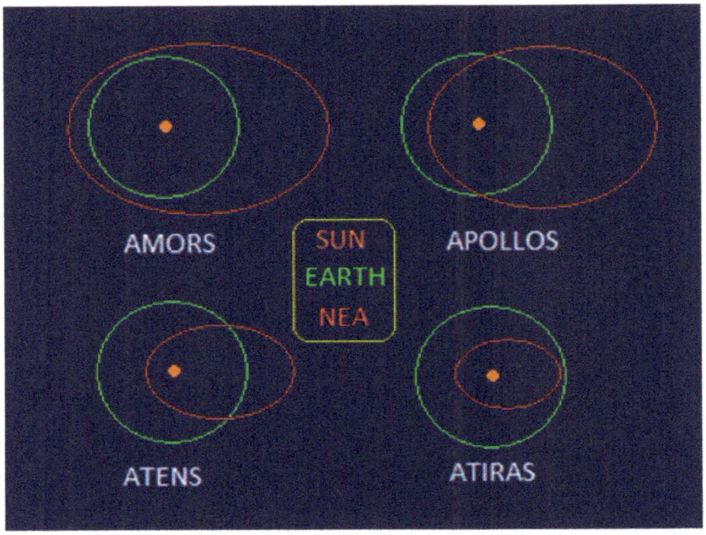

Figure 16.3 Orbital classification of near-Earth asteroids.

16.4 Continuing Discovery

The discovery of near-Earth objects has been continuous but slow for most of the twentieth century. The upturn in discoveries really started to increase with the establishment of the Skywatch Observatory on Kitt Peak, Arizona by Tom Gehrels and Robert McMillan in 1980. A magnitude increase started in 1998 with the establishment of the Linear Observatory by MIT Lincoln Laboratory team led by Grant Stokes, then the Catalina Sky Survey, and the Pan-Starrs Observatory (Figures 16.4 and 16.5).

Despite over 2,000 potentially hazardous now known, the only PHA that has a remote possibility of hitting the Earth in the near future is *99942 Apophis*. This asteroid was discovered in 2004 and is estimated to be about 340 meters in size. Continued observations and orbit refinement have now given the closest approach distance to the Earth's surface of ~32,000 km (closer than geosynchronous orbit) on 2146 UT, April 13, 2029. (See Section 11.8 for the probability of collision with a geosat.)

16.5 Asteroid Mining and Planetary Defense

NEOs offer us both opportunities and threats. The opportunities come in the form of water and mineral resources that might be extracted or mined from comets and asteroids. This includes rare Earth elements that are essential to modern

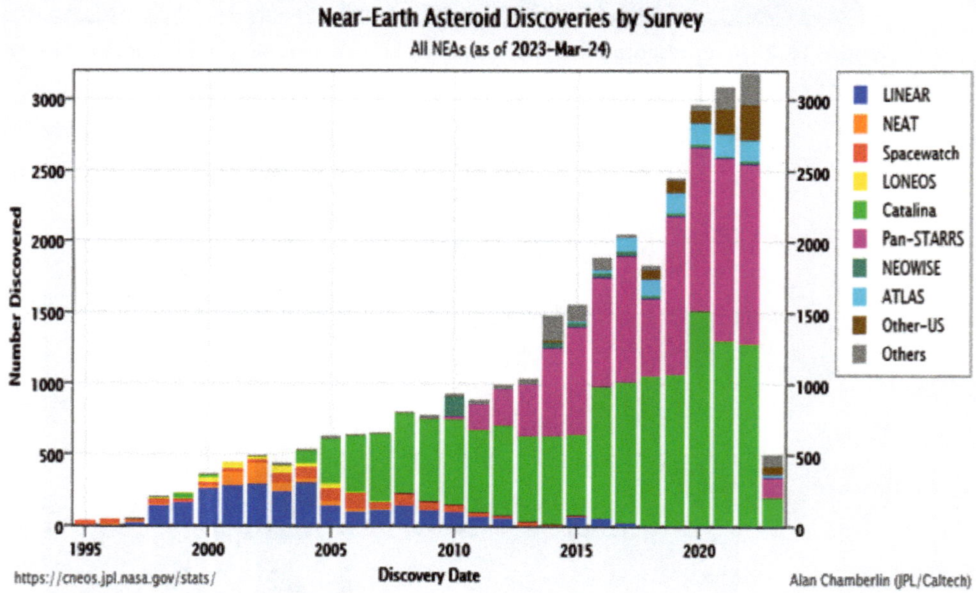

Figure 16.4 Discovery by year and observatory of near Earth asteroids. (*Source:* NASA/JPL.)

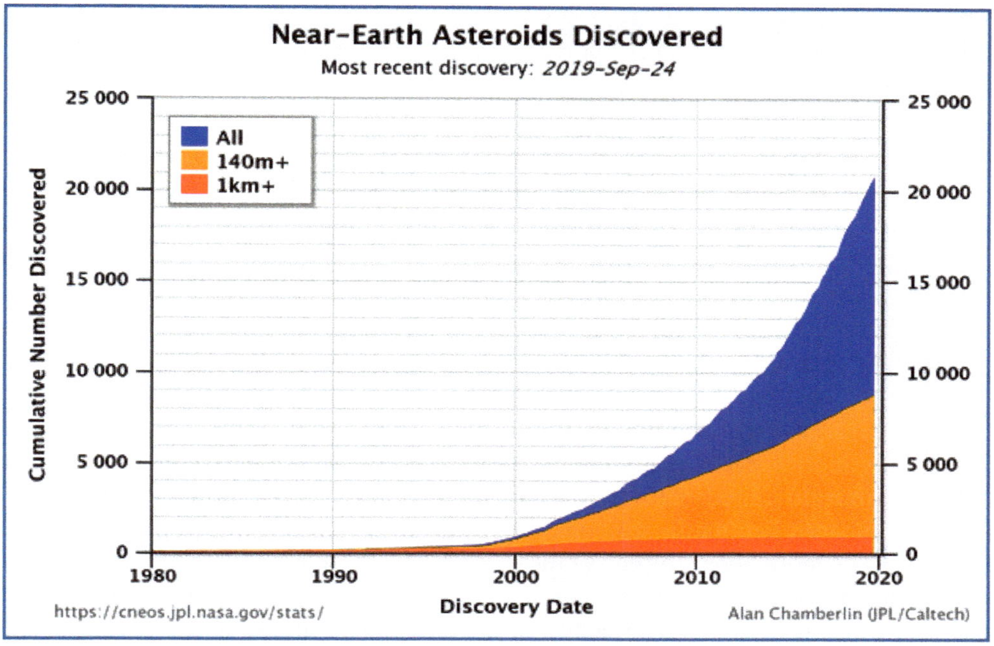

Figure 16.5 Cumulative discovery of NEAs by size. (*Source:* NASA/JPL.)

technology. The threats are from potential impacts with the Earth and the resultant damage caused.

Probably the first mention of asteroid mining was in a science fiction book written in 1892 by Garrett Serviss in which a group of Martians are mining an asteroid for its resources. There have been many science fiction books and movies since based around resource mining in various space locations.

The first proposal to exploit near Earth asteroids was explored by Dandridge Cole and Donald Cox in the 1964 book *Islands in Space: The Challenge of Planetoids*. A signature book was written by Gerard O'Neill in 1976 titled *The High Frontier*. Based on the interest in this book NASA published a 1977 study by Richard Johnson and Charles Holbrow titled *Space Settlements: A Design Study (NASA SP-413)*. John Lewis, professor of planetary sciences at the Lunar and Planetary Laboratory of the University of Arizona at Tucson, has been responsible for a large number of publications in this area (see the reference section at the end of this chapter).

Two private companies in the United States were established to promote and investigate asteroid mining. The first and largest of these was Planetary Resources. The second company was Deep Space Industries. Unfortunately, both companies were established before their time and now no longer exist.

Australia jumped on the space mining bandwagon in 2013. The Australian Centre for Space Engineering Research (ACSER) at the University of New South Wales held the world's first Off-Earth Mining Forum, repeated every two years (with a break for Covid). The last OEMF was held in 2022. These forums have generated a large amount on interest in the country and from overseas. Australian mining companies have embraced these forums and presented many robotic techniques used in remote Australian mines, which may be applicable to mining asteroids in space.

The threat of asteroid and/or comet impacts has had similar history in the world of fiction. The first nonfiction book to address the issue of the Earth's response to collision with an asteroid was probably that of the geologists Allan Kelly and Frank Dachille. In 1953 they privately published *Target Earth: The Role of Large Meteors in Earth Science* in which they analyzed global evidence suggesting that one or more giant asteroids had impacted the Earth in the past, causing a shift in the angular momentum of the Earth, shifting the rotational axis accompanied by global flooding, giant firestorms, atmospheric occlusion, and the extinction of the dinosaurs.

They too were before their time and it was not until 1980 when Luis and Walter Alvarez suggested a similar scenario for the so-called Cretaceous-Tertiary (K/T) extinction event around 65 million years ago, which led to the disappearance of the dinosaurs. Evidence has now confirmed that a 10- to 15-km asteroid hit the Mexican Yucatan Peninsula at Chicxulub around the right time releasing about 4.2×10^{23} joule, equivalent to the explosion of 100 million megatons of TNT and creating an impact crater of around 200 km in diameter (mostly underwater).

It was in the last decade of the twentieth century that a collision between the Comet Shoemaker-Levy 9 and Jupiter that really brought the reality of asteroid collisions in the solar system. One US Air Force officer, following a conference on the issue, came up with the term *planetary defense* to describe the range of issues related to an asteroid collision with Earth and the mitigation techniques to prevent such an impact. This officer was Lindley Johnson. It was also promoted by Colonel Simon (Pete) Worden of the US Air Force Space Command 50th Space Wing. Lindley Johnson is now planetary defense officer of NASA's Planetary Defense Coordination Office.

Planetary defense was a term to indicate that we are under bombardment from outer space. Not an attack from an alien race, but a sporadic bombardment by naturally occurring objects in the solar system: asteroids and comets, fragmentary space debris that occasionally are perturbed onto a collision course with Earth. The term furthermore was a recognition that our technology has now advanced to the point where not only can we search for and track such objects, but where we may also be in a position to deflect them from their intended path, and thus avert a disaster or catastrophe.

Planetary defense may be broken into three phases:

Phase 1: Awareness and acknowledgment. The need to fully realize that there is a real threat and the magnitude of that threat.

Phase 2: Surveillance. A study of the threat and the properties of the objects that pose that threat.

Phase 3: Mitigation. Evaluation of technology to assess the assets required for diversion of the threat.

These three phases are discussed in more detail in the following sections.

16.6 Awareness and Acknowledgment

In 1801, the annual *Leonid* meteor shower became a storm (something we now know happens every 33 years or so) where over 1,000 meteors were observed per hour. Some meteorites were discovered around the same time, which may or may not have been related to the storm. Such was the prejudice at the time that a comment was made, "I would rather believe that two Yankee professors had lied than that God would allow stones to fall from the heavens."

This prejudice was very much apparent in the assignation of large craters to volcanism rather than to impacts. It was a belief shared by geologists and astronomers alike (lunar craters were originally thought to be of volcanic origin). This explains why Kelly and Dachille's 1953 publication had to be privately published and was then ignored until the Alvarez's work in 1980. Even now, there are still some geologists who believe the K/T event was due to vulcanism over a lengthy period of time. Much of this thought can be traced to the religious debate over Darwinism and the hypotheses of *uniformitarianism*: all processes continue as we see them today, but with sufficient time we can achieve all things. Catastrophism was anathema to the early Darwinians as it smacked of biblical religion. We can see volcanoes, we do not see impacts. At least not until the Comet Shoemaker-Levy 9 impact with Jupiter.

The first phase of planetary defense is an awareness and an acknowledgment that the Earth has been subject to a cosmic impact hazard in the past, and will be the target of hypervelocity impacts from comets and asteroids in the future. Evidence comes from astronomy, which shows us the cratered state of other bodies in the solar system; from geology, which has found craters on the Earth and explains their limited number in active processes; from paleontology, which provides proof of species mass extinctions; and from physics, which can extrapolate orbits and calculate damage.

Throughout the solar system everywhere we look we find craters. Not only on the planets and their moons, at least the ones with solid surfaces, but also on the asteroids, bodies much too small to support volcanic processes. On Earth we can now recognize impact craters from their surroundings, which show shock metamorphism in their rocks. Bodies in the solar system that do not show extensive cratering can be shown to be either geologically active or are gas giants in which there are no permanent features. In recent history, we have the Tunguska event with convincing evidence that it was an approximately 12-megaton airburst explosion that occurred near the Podkamennaya Tunguska River in Yeniseysk Governorate, Russia, on the morning of June 30, 1908. And in very recent history, we have the Chelyabinsk event of February 2013, recorded by multiple dash cameras throughout the region.

Evidence from US defense satellites (e.g., DSP) show that there are several atmospheric impacts with an energy greater than one kiloton TNT occurring each year.

The energy released in an NEO impact is essentially the kinetic energy of the NEO.

At hypervelocity (>4 km/s) a body has an energy in excess of the explosive energy that would be released by the same mass of chemical explosive (e.g., TNT). Table 16.2 shows the asteroid energies (in megatons of TNT) assuming an NEA velocity of 20 km/s and a density of 2,500 kg/m^3.

Table 16.2 Asteroid Sizes, Energies, and Impact Effects

Diameter (meters)	*Mass (million tons)*	*Energy (Mton TNT)*	*Effect*
100	1.3	62	Destruction of a large city
1,000	1,300	62,000	Destruction of a country/region
10,000	1,300,000	62,000,000	Global extinction event

Crater sizes from hypervelocity impacts may be determined by comparison with craters produced in nuclear explosions. A 1-kT TNT explosion produces a 40m crater. Crater size scales as energy to the 0.3 power, is typically 10 to 20 times the size of the impactor that produced it. (See Table 16.3.)

Figure 16.6 shows the standard dimensions for a crater created by a 1-kT explosion or impact. This may be scaled using the $E^{0.3}$ scaling rule.

The actual effects of a large asteroid impact will depend greatly on whether the impact occurs on land or sea (Figure 16.7). The probability for the latter is higher with an ocean to land ratio of 3 to 1.

Table 16.3 Crater Size Related to Impactor Size

NEO Diameter	*Crater*
10m	Minimum size of an iron asteroid for a hypervelocity impact
50m	Minimum size of a comet for a hypervelocity impact
100m	1–2 km
500m	4–10 km
1 km	10–20 km
5 km	40–100 km
10 km	100–200 km

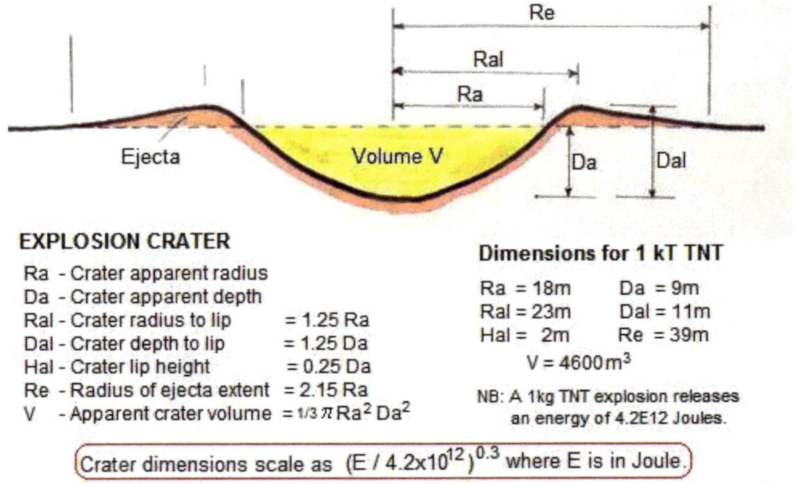

Figure 16.6 Crater size for a 1k energy release.

Figure 16.7 Impression of a large asteroidal ocean impact. The shock wave spreads out ahead of falling ejecta. (Painting by Don Davis.)

16.7 Surveillance

The second phase of planetary defense involves the acquisition of information about NEOs:

- Acquire orbital information on all potentially significant NEOs.
- Follow, track and extrapolate NEO orbits for possible Earth intersections.
- Spacecraft observation of asteroids and comets to explore physical properties that influence phase 3.
- Continuous surveillance is required to detect main belt asteroids perturbed into NEO orbits, and to watch for long period comets. (Orbital mechanics of asteroids is chaotic enough that predictions cannot be made accurately beyond a few thousand years.)

Astronomical observations are made with a telescope and camera and can be used for imaging, position determination, and brightness measurements. The latter are particularly valuable to determine the rotational periods of an asteroid. Astronomical analysis of the observations is used for orbit determination, and analyzed for possible impact. In cases of close NEA passes, radar may be used to image the asteroid and help determine rotational axes and periods (from Doppler effects) (Figure 16.8).

Figure 16.9 shows the trail left by an NEA. Typical exposure times may be from 1 to 5 minutes using a 16-inch reflecting telescope.

The position of the asteroid is determined by comparison with the accurately known positions of other stars in the field of view. Computer programs are available to automate this process. Accuracies less than one second of arc (1/3600 of a degree) are achievable with good images.

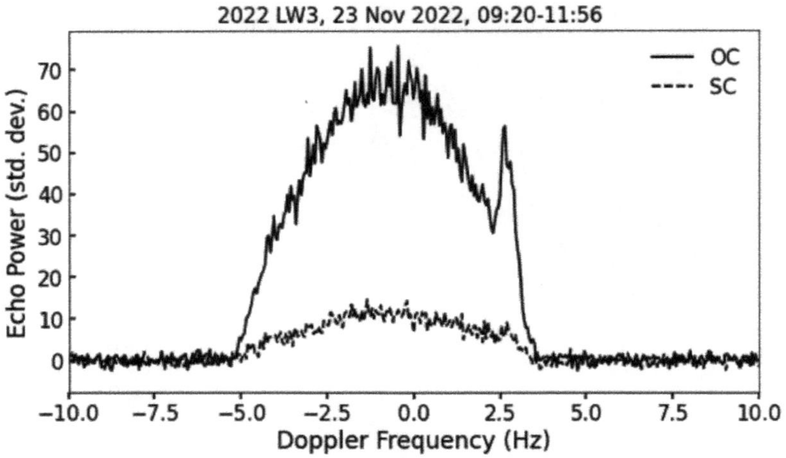

Figure 16.8 Doppler spread radar return reveals small closely orbiting companion asteroid (peak on the right of the main asteroid return) that optical observations could not resolve. (*Image:* Southern Hemisphere Australian Radar Project.)

Figure 16.9 Asteroid trail on a CCD camera image. (*Image:* Arie Verveer.)

As with astrometry, the brightness of an asteroid is determined by comparison with the known brightness of a comparison star or stars. A CCD camera is a very linear device, and the number of electrons collected is a given fraction of the number of photons incident on each pixel. Observations are made over many hours and even days to determine asteroid rotation periods.

A spectroscope may be attached to the back of a telescope to provide a spectrum of the light from the asteroid. If this is done in the infrared it can provide details about the surface composition of the body. The NASA Infrared Telescope Facility (IRTF) (Figure 16.10), on the top of the Mauna Kea volcano on the island of Hawaii, has spectroscopes dedicated to this purpose. The high altitude (14,000 ft) is above most of the Earth's atmosphere that absorbs infrared light and so can be operated in the IR spectrum, which is the part of the spectrum that shows compositional features and thus classification of an asteroid.

Observatories imaging asteroids report their observations to the Minor Planet Center [1] with the following details:

- Universal Time accurate to 0.00001 days (1 second);
- Right ascension to 0.01 second;
- Declination to 0.1 arcsecond;
- Photometric brightness to 0.1 magnitude.

```
Designation ----- Date ------ Right Ascen Declination Mag Obs
PJ94P01d C1994 1014.82517 09 57 25.32 +09 06 28.5 13.3 360
```

The MPC performs orbit determination on the observations to either determine a preliminary orbit or to update the orbit.

The JPL CNEOS website, cneos.jpl.nasa.gov, has a host of useful NEO information for those interested in following NEO statistics, including latest close approaches.

16.7.1 Neo Hazard Scales

Two NEO hazard scales have been developed.

1. The TORINO public scale, developed by Richard Binzel at MIT, has 11 classes from 0 to 10, which are based on NEO kinetic energy and collision probability (Figures 16.11 and 16.12).

(a) (b)

Figure 16.10 (a) NASA IRTF, and (b) SPEX IR spectrograph. (*Source:* NASA images.)

Category	Rating	Description
Events Having No Likely Consequences	0	The likelihood of a collision is zero, or well below the chance that a random object of the same size will strike the Earth within the next few decades. This designation also applies to any small object that, in the event of a collision, is unlikely to reach the Earth's surface intact.
Events Meriting Careful Monitoring	1	The chance of collision is extremely unlikely, about the same as a random object of the same size striking the Earth within the next few decades.
Events Meriting Concern	2	A somewhat close, but not unusual encounter. Collision is very unlikely.
	3	A close encounter, with 1% or greater chance of a collision capable of causing localized destruction.
	4	A close encounter, with 1% or greater chance of a collision capable of causing regional devastation.
Threatening Events	5	A close encounter, with a significant threat of a collision capable of causing regional devastation.
	6	A close encounter, with a significant threat of a collision capable of causing a global catastrophe.
	7	A close encounter, with an extremely significant threat of a collision capable of causing a global catastrophe.
Certain Collisions	8	A collision capable of causing localized destruction. Such events occur somewhere on Earth between once per 50 years and once per 1000 years.
	9	A collision capable of causing regional devastation. Such events occur between once per 1000 years and once per 100,000 years.
	10	A collision capable of causing a global climatic catastrophe. Such events occur once per 100,000 years, or less often.

Figure 16.11 The TORINO scale of NEO collision hazard rating.

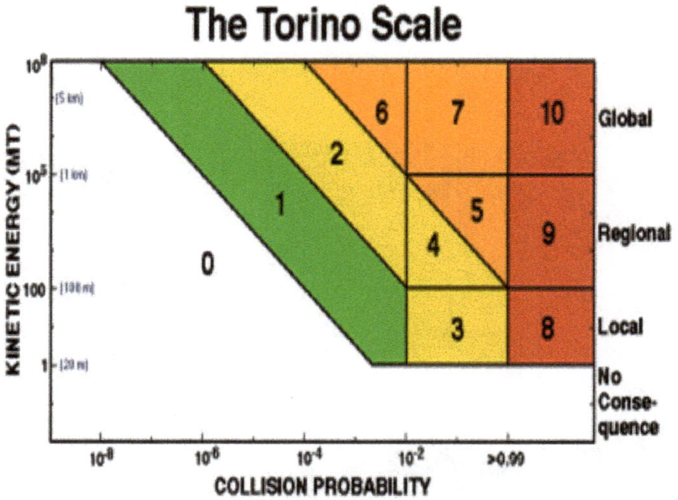

Figure 16.12 The TORINO scale as based on NEO kinetic energy and collision probability.

2. An alternative PALERMO technical hazard scale where values above zero are significant.

The Palermo Scale is the base 10 logarithm of the *relative risk*.

$$PS = \log_{10} R.$$

The relative risk R is given by

$$R = \frac{P_I}{f_B \times DT},$$

where P_I is the impact probability of the event in question, and DT is the time until the potential event, measured in years. The annual background impact frequency,

$$f_B = 0.03 \times E^{-4/5}$$

is the annual probability of an impact event with energy (E, in megatons of TNT) at least as large as the event in question.

The cumulative Palermo scale value reflects the seriousness of the entirety of detected potential collision solutions. It is the base 10 logarithm of the sum of the individual relative risk values.

$$PS_{\text{cum}} = \log_{10}\left(10^{PS1} + 10^{PS2} + 10^{PS3} + \ldots\right)$$

A similar summation can be done to rate the relative hazard posed by a collection of objects.

16.7.2 Neo Observation Programs

There were five initial major NEO observational programs: Spacewatch, Linear, Loneos, Neat, and the Catalina Sky Survey.

Spacewatch was the first dedicated NEO program with two telescopes, 0.9m and 1.8m. It was run by the Spacewatch team from the Lunar and Planetary Laboratory at the University of Arizona.

The LINEAR system was devised by the Lincoln Laboratory at the Massachusetts Institute of Technology. It was based on the US Air Force GEODSS system with two 1m telescopes, and when it came online initially it had the highest NEO discovery rate. It was based near Socorro, New Mexico. The Lincoln Laboratory has since developed a 3m Space Surveillance Telescope, which was initially located at Socorro, but has since been moved to Exmouth on North West Cape in Western Australia. It is capable of making 6,000 observations per night, and is specifically used for orbital space debris observations.

The Lowell Observatory Near-Earth-Object Search (LONEOS) is located near Flagstaff, Arizona. It used a 0.6m f/1.8 telescope with a 16-megapixel CCD camera to provide a field of view of almost 3 × 3 degrees. It ran from 1993 to 2008. It was responsible for recovering the long lost asteroid *1937 UB Hermes*.

The JPL NEAT ran two programs. The NEAT/PALOMAR with a 1.2m Schmidt and the NEAT/MSS with a 1.2m telescope at the US Air Force Maui Space Surveillance site.

The Catalina Sky Survey was run by the Lunar and Planetary Laboratory at the University of Arizona, and consisted of a distributed network of telescopes, including a 0.7m Schmidt at Mount Bigelow, a 1.5m telescope at Mount Lemmon, the 0.6m Uppsala Schmidt at Siding Springs in Australia, and several more recent telescopes.

PanSTARRS is a new, very productive observatory run by the Institute for Astronomy at the University of Hawaii. The first telescope, Pan-STARRS1 (PS1), is a 1.8m-diameter telescope located near the summit of Haleakala on the Island of Maui. It is equipped with the world's largest digital camera, with almost 1.4 billion pixels.

A similar telescope, Pan-STARRS2 (PS2), has been constructed adjacent to PS1. It has a similar, but slightly larger digital camera, with almost 1.5 billion pixels.

The operation of the Pan-STARRS telescopes is mostly funded by the NASA Near-Earth Observation Program. Each night, PS1 observes about 1,000 square degrees of the night sky, using a sequence of four exposures that span a period of about an hour. The images are compared to each other, and objects that move during the one-hour period are identified. Objects that have unusual motions that make them likely to be near-Earth objects are immediately reported to the Minor Planet Center, and a worldwide network of telescopes obtains additional observations of these near-Earth object candidates to determine their orbits and sizes and whether any of them pose a threat to Earth. The positions and brightness of all other moving objects are also reported to the Minor Planet Center, usually within 12 hours of observation.

Pan-STARRS1 is the world's leading near-Earth object discovery telescope. It presently discovers over half of the larger near-Earth objects (diameter > 140m). Pan-STARRS1 is also very efficient at discovering new comets, discovering more than half of all new comets each year since 2014.

There are now many other NEO programs around the world, including Japan (the Bisei Spaceguard Center), France (l'Observatoire de la Cote d'Azure), and the Czech Republic (Klet Observatory). Radar observations are performed in the United States, Australia, and Germany.

16.7.3 Summary

- Over 10 NEAs > 10 km;
- About 1,000 NEAs > 1 km;
- Over 10,000 NEAs > 100m;
- Over 30,000 NEAs discovered;
- Over 2,000 PHAs discovered;
- Several asteroids have now been imaged by spacecraft missions;
- Many asteroids are binary;
- Some asteroids may be rubble piles;
- Comets may pose a larger risk than asteroids because of their lack of detectability until close to the Earth.

16.8 Mitigation

The third and final stage of planetary defense, should it be needed, deals with the mitigation of an identified NEO impact hazard in the near future.

This last phase of planetary defense is the most controversial. Many ideas have been proposed. Some involve giving an asteroid a gentle nudge over a period of time, either with rocket propulsion or using reaction mass from the NEO itself. Others involve giving the asteroid a large single impulse. All schemes require a substantial warning time to allow the relatively small total push to translate to a significant path deflection.

In 1953, Kelly and Dachille [2] foresaw the need for a system to survey "a critical envelope of space" to detect objects that might be on a collision course with Earth, and that if such an object was detected "to this end might be used rocket tug boats sent out to deflect and guide the object from the collision course."

The earliest plan for NEA deflection was published by a group of MIT engineering students in 1968. The detailed plan was published under the name *Project Icarus* (MIT Press).

16.8.1 A Gentle Nudge

A rocket with an ion engine is carefully maneuvered into position against the asteroid, and the engines deliver a very small force for several months. This will cause a very small change in the velocity of the NEO.

16.8.2 A Single Large Impulse

A nuclear explosion is detonated above the target asteroid surface to avoid breaking up the asteroid. This produces a small change in the velocity of the NEA.

16.8.3 Solar Solutions

1. Attach a very large solar sail to the NEO to change its velocity vector;
2. Use a solar collector to heat the NEO material causing vapor to be ejected and the mass reaction (rocket principle) to change the NEO velocity.

16.8.4 Time

All methods require a long time for the small velocity change to translate into a significant orbit change so that the potential impactor is diverted by an adequate margin to miss the Earth, possibly several years (Figure 16.13).

16.8.5 Problems

- An adequate delivery system (Saturn V-type rockets);
- An adequate warning time;

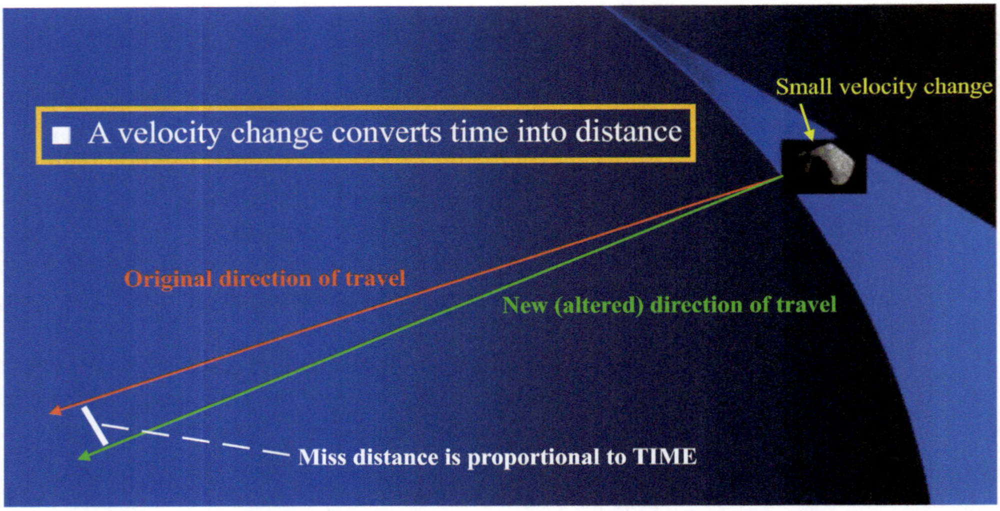

Figure 16.13 Asteroid velocity change becomes a miss distance with adequate time.

- Danger of fragmenting the NEO;
- Rotating asteroid makes some schemes difficult;
- High precision, could make problem worse;
- Socio-political problems.

16.8.6 DART

NASA has already conducted a successful asteroid redirection test. *The Double Asteroid Redirection Test (DART)* [3] project targeted the Didymos-Dimorphos binary asteroid system. The *DART* spacecraft was launched on July 22, 2021, and impacted the smaller component (Dimorphos) on October 7, 2022. A small cubesat (LICIA) was released by *DART* 14 days before impact to observe the changes (Figures 16.14 and 16.15).

An ESA spacecraft *HERA* is expected to pass by Didymos-Dimorphos in 2024 to measure the longer-term effects of the orbit change in the binary asteroid.

16.9 The International Asteroid Warning Network

The International Asteroid Warning Network (IAWN) [4] is a virtual network of institutions that discover, monitor, and characterize near Earth objects and communicate impact risks. It has more than 30 members from observatories and space institutions worldwide. IAWN was established in 2013 as a result of the UN-endorsed recommendations for an international response to a potential NEO impact threat, to create an international group of organizations involved in detecting, tracking, and characterizing NEOs. The IAWN is tasked with developing a strategy using well-defined communication plans and protocols to assist governments in the analysis of asteroid hazards.

Figure 16.14 The NASA DART mission. (*Source:* NASA/DART image.)

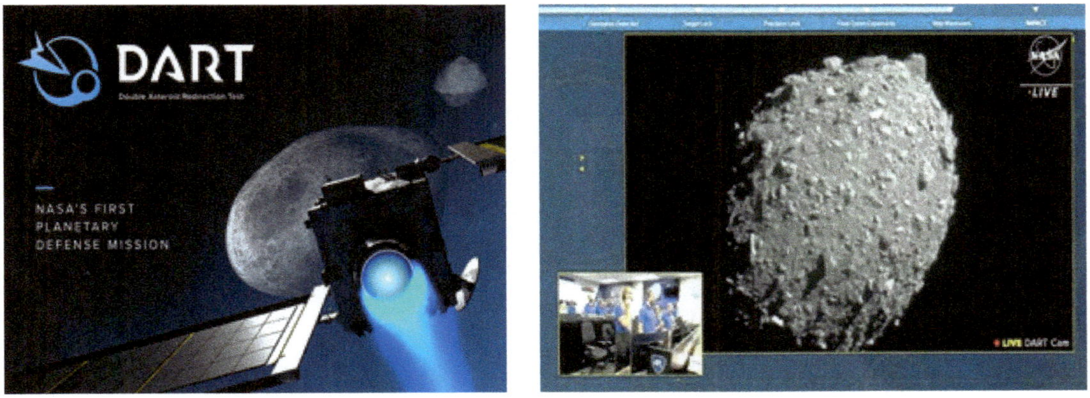

(a) (b)

Figure 16.15 (a) the DART project, and (b) last image of Dimorphos before impact. (*Source:* NASA/DART images.)

References

[1] Minor Planet Center, www.minorplanetcenter.net/iau/TheIndex.html.

[2] Kelly, Allan, and Frank Dachille, *Target Earth*, Pensacola Engraving Company, 1953.

[3] DART Mission, science.nasa.gov/mission/dart.

[4] International Asteroid Warning Network, iawn.net.

Selected Bibliography

Clarke, A. C., *The Hammer of God*, Victor Gollancz, 1993.

Gehrels, T., (ed.), *Hazards Due to Comets and Asteroids*, University of Arizona Press, 1995.

Kelly, A., and F. Dachille, *Target Earth*, Pensacola Engraving Company, 1953.

Lewis, J. S., *Asteroid Mining 101: Wealth for the New Space Economy*, Deep Space Industries, 2014.

Lewis, J. S., *Comet and Asteroid Impact Hazards on a Populated Earth*, Academic Press, 2000.

Lewis, J. S., *Mining the Sky*, Helix Books, 1996.

Lewis, J. S., *Rain of Iron and Ice*, Helix Books. 1996.

Marboe, I. (ed.), *Legal Aspects of Planetary Defence*, Brill, 2021.

MIT Student Project in Systems Engineering, *Project Icarus*, MIT Press, 1968.

Morrison, D., The *Cosmic Impact Hazard (2018)*, Springer Link, 2018.

Schmidt, N. (ed.), *Planetary Defense*, Springer Link, 2019.

Steel, D., *Rogue Asteroids and Doomsday Comets*, Wiley, 1995.

Steel, D., *Target Earth*, Readers Digest, 2000.

Verschuur, G., *Impact: The Threat of Comets and Asteroids*, Oxford University Press, 1996.

Yeomans, D. K., *Near-Earth Objects: Finding Them Before They Find Us*, Princeton University Press, 2012.

CHAPTER 17

Legal Issues in SSA

17.1 The Beginnings of Space Law

Eugene Korovin (1892–1964)

Some see Eugen Korovin, a Russian professor of international law, as the originator of space law with his paper "The Conquest of the Stratosphere and International Law," (English translation) in the journal *Revue Generale de Droit International Public* (*General Review of International Public Law*), in 1934.

Manfred H. Lachs (1914–1993)

A Polish diplomat and International Court of Justice judge, Manfred H. Lachs was heavily involved in international law, and in the early development of space law. After he died, the International Institute of Space Law named their annual space law moot in his honor (i.e., the Manfred Lachs Space Law Moot Court).

International space law is based on international law in general such as the International Law of the Seas and the Antarctic Treaty.

17.2 Issues In Space Law

- Space law is a dynamic and growing area;
- Laws relating to humankind and its current and future activities in outer space;
- Relevant to many everyday activities:
  - Communications, broadcasting, internet (ITU);
  - Remote sensing from space;
  - Global navigation satellite systems (GNSS, e.g., GPS, GLONASS, *Galileo*, *Beidou*).
- Relevant to future activities:
  - Exploration of space and celestial bodies;
  - Exploitation of resources;
  - Settlement;
  - Warfare.

Space law involves states, intergovernmental organizations, and increasingly, private enterprise and individuals.

17.3 Definition and Delineation

Although the United Nations has debated the issue for 40 years, there is no agreed legal definition of the term "outer space" used in the international treaties relating to operations in the volume (somewhere) above the sovereign airspace of member states. One might well ask how laws can be framed on an undefined subject.

Space scientists call the space in which Earth satellites orbit the near-space environment, reserving the term outer space for the interplanetary medium, whereas astronomers reserve the term outer or deep space for the intergalactic medium.

The atmosphere slowly decreases in density with height and so does not offer a guideline as to where space starts. However, air drag does not allow a vehicle to remain in orbit very long once it descends to below about 200 km.

The United States awards astronaut status to anyone who has flown above 80 km in altitude.

Astronautical flight engineers define an "entry interface" at around 120 km as the height at which significant heating occurs to a reentering spacecraft. Natural space objects (meteoroids) also start to ablate or burn up at around this altitude, creating the visible meteors we see in the night sky.

The World Air Sports Federation sets 100 km as the lower boundary of space, and this altitude has been used in recent prizes set in competitions for flying into space. Many groups have adopted this Karman line, named after the aeronautical engineer Theodore von Karman, who predicted the height below which significant lateral thrust would be required to keep a craft flying level.

Figure 17.1 *White Knight* carries *SpaceShipOne* in a private attempt to reach into space, defined in this case at an altitude of 100 km. (Scaled composites image.)

Although many states refuse to legally put an upper limit to their airspace, most do not control their air traffic above 60,000 ft (referred to as Flight Level 600), or about 18 km.

Some states believe that the entire space above their country should belong to them. This makes absolutely no sense whatsoever, and shows a distinct lack of understanding of Earth's place in the universe. It is in fact based on the geocentric cosmology that was espoused until its overthrow by Copernicus. If this principle were to apply, some states might lay claim to a star that is directly above their territory at one point in time, only to find that an hour later the same star belonged to a neighboring state, and so on. Some equatorial states use this reasoning to lay exclusive claim to a portion of the much-prized geosynchronous orbit, and feel that they should derive revenue from allocation of slots in this orbit to others. Fortunately, they are not in a position to enforce these claims. Neither has the absence of a legal delineation of outer space led to any international lawsuits to this end, for it is virtually certain that even with a reluctance to put a lower limit to the term, only a small minority of states would place it above geosynchronous orbit, at an altitude of 36,000 km.

In pragmatic terms, the height at which sovereign airspace above a state becomes outer space is only the height to which

1. The state can defend the airspace below;
2. Other states allow it to so claim.

Anything else is wishful thinking.

A good discussion on the boundary between air space and outer space has been written by Barrett [1].

17.4 Jurisprudence and Jurisdiction

There are essentially four regimes applicable to space law:

1. *State law* (e.g., the issuance of permits for rocket launches);
2. *Bilateral treaties* (e.g., establishing a tracking station in another country);
3. *Multilateral treaties* (e.g., the establishment of the European Space Agency);
4. *International law* (e.g., treaties made at and by the United Nations).

The United Nations Office of Outer Space Affairs has available a compilation of all agreements made available to it by member states [2]. This contains lists of agreements in all of the four space law regimes listed above.

Because of the nature of space operations, international law treaties and agreements are of vital importance. The UN General Assembly has set up a number of bodies to deal with space-related matters (Figure 17.2).

The Committee for the Peaceful Uses of Outer Space reports directly to the United Nations General Assembly. COPUOS has two subcommittees, one legal and the other scientific and technical, which provide advice to it. The Office for

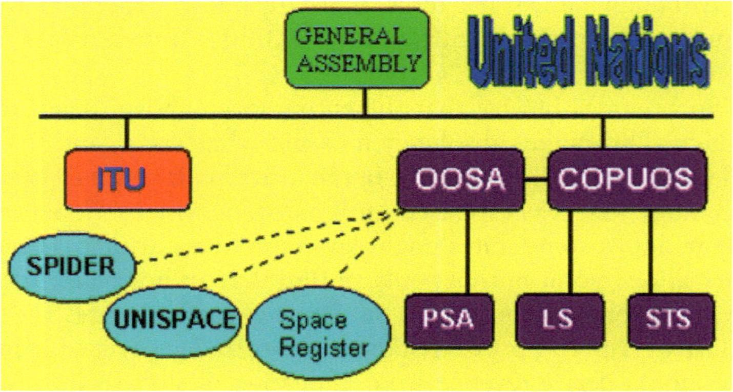

Figure 17.2 United Nations organizations concerned with space law. See the acronym list at the end of this book for an explanation.

Outer Space Affairs serves as the secretariat to COPUOS, and it controls the Program on Space Applications. It also holds the Register of Space Objects, arranges the UNISPACE conferences, and overseas the Space-based Information for Disaster Management and Emergency Response.

Under international law there are now five treaties and five principles governing operations in space. These have been framed to try and ensure that space operations are to the benefit of all states, not just those who can afford to launch vehicles into space.

17.5 The Five United Nations Treaties

17.5.1 1967 OST

"Treaty on Principles Governing the Activities of States in the Exploration and Use of Outer Space, including the Moon and Other Celestial Bodies."

Short title: "The Outer Space Treaty"

> Freedom of space for all. Not subject to national appropriation. International law applies. No military bases. No weapons of mass destruction. Astronauts as envoys—to be rendered assistance. States bear responsibility, ownership and liability for activities carried out under their jurisdiction. Peace, cooperation, and mutual assistance. Opportunities for all. Inspections. Notifications of activities to UN.

This treaty contains elements of the following four treaties, which then reiterate and expand upon individual points declared here.

17.5.2 1968 ARRA

"Agreement on the Rescue of Astronauts, the Return of Astronauts, and the Return of Objects Launched into Outer Space."

Short title: "The Rescue and Return Agreement"

> Astronauts in distress to be given all possible assistance. Astronauts and space objects landing on foreign soil to be returned promptly to the launching state.

17.5.3 1972 LIAB

"Convention on International Liability for Damage Caused by Space Objects."

Short title: "The Liability Treaty"

> States are responsible for any damage their space objects cause and must pay compensation to the damaged state.

17.5.4 1975 REG

"Convention on Registration of Objects Launched into Outer Space."

Short title: "The Registration Treaty"

> All objects launched into space must be notified, by the launching state(s), to the UN Register of Space Objects.

17.5.5 1979 MOON

"Agreement Governing the Activities of States on the Moon and Other Celestial Bodies."

Short title: "The Moon Treaty"

> Strengthens and expands the "Outer Space Treaty" (1967) for space bodies.

Note that the Moon Treaty has never been enacted, because an insufficient number of states ratified it. Initially only 13 states ratified this treaty, and Australia was the only developed country to do so. Not generally known was that they ratified it in the expectation of a (temporary two year) seat on the UN Security Council—which did not happen, at least not at that time.

17.6 The Five United Nations Principles

A. Declaration of Legal Principles Governing the Activities of States in the Exploration and Use of Outer Space

> Exploration and use of outer space should be for the benefit of all mankind. Outer space and any bodies in it are free for all and not subject to national

appropriation. All activities should be peaceful. States bear international responsibility for national activities in space. States shall cooperate with each other in space. States retain control of their objects launched into space, and are liable for any damage such objects might cause. Astronauts shall be regarded as envoys of mankind. All possible assistance shall be rendered them in any emergency or accidental situation.

B. Principles Governing the Use by States of Artificial Satellites for International Direct Television Broadcasting

Access to direct broadcast technology should be available to all states. States should respect the rights and sensitivities of other states (including copyright) in such broadcasts.

C. Principles Relating to Remote Sensing of the Earth from Outer Space

Remote sensing activities to be carried out for the benefit of all states, and be used to protect the environment. Space capable states should make opportunities available to developing states for participation in this area. Data of any sense state should be made available to them.

D. Principles Relevant to the Use of Nuclear Power Sources in Outer Space

In such operations individuals, populations, and the biosphere should be protected against radiological hazards. There should be no significant contamination of outer space with radioactive material. Safety assessments shall be made available, and notifications made of nuclear powered vehicle reentries into the Earth's atmosphere.

E. Declaration on International Cooperation in the Exploration and Use of Outer Space for the Benefit and in the Interest of All States, taking into Particular Account the Needs of Developing Countries

Encouragement of international cooperation in space operations with an onus on space capable states to assist and encourage developing states.

The full documentation on the UN treaties relating to outer space may be found at [3].

17.7 Space Debris

In 1997, with the realization that artificial space debris was becoming a significant issue in outer space operations, the UN convened a committee to discuss the issue. There was bitter argument between many of the parties and no binding resolution could be agreed upon. After 10 years, the only outcome was the issuing of seven nonbinding guidelines on the mitigation of space debris. These were approved by the UN General Assembly in June 2007. These are listed below:

1. *Limit debris released during normal operations.* Space systems should be designed not to release orbital debris during normal operations. If this is not feasible, the effect of any release of debris on the space environment should be minimized.
2. *Minimize the potential for break-ups during operational phases.* Spacecraft and launch vehicle orbital stages should be designed to avoid failure modes that may lead to accidental break-ups. In cases where a condition leading to such a failure is detected, disposal and passivation measures should be planned and executed to avoid breakups.
3. *Limit the probability of accidental collision in orbit.* In developing the design and mission profile of spacecraft and launch vehicle stages, the probability of accidental collision with known objects during the system's launch phase and orbital lifetime should be estimated and limited. If available orbital data indicate a potential collision, adjustment of the launch time or an on-orbit avoidance maneuver should be considered.
4. *Avoid intentional destruction and other harmful activities.* Recognizing that an increased risk of collision could pose a threat to space operations, the intentional destruction of any on-orbit spacecraft and launch vehicle orbital stages or other harmful activities that generate long-lived debris should be avoided. When intentional break-ups are necessary, they should be conducted at sufficiently low altitudes to limit the orbital lifetime of resulting fragments.
5. *Minimize potential for post-mission breakups resulting from stored energy.* In order to limit the risk to other spacecraft and launch vehicle orbital stages from accidental break-ups, all on-board sources of stored energy should be depleted or made safe when they are no longer required for mission operators or post-mission disposal.
6. *Limit the long-term presence of spacecraft and launch vehicle orbital stages in the low Earth orbit region after the end of their mission.* Spacecraft and launch vehicle orbital stages that have terminated their operational phases in orbits that pass through the LEO region should be removed from orbit in a controlled fashion. If this is not possible, they should be disposed of in orbits that limit their long-term presence in the LEO region.
7. *Limit the long-term interference of spacecraft and launch-vehicle orbital stages with the geosynchronous region after the end of their mission.*

The first six guidelines refer to space debris mitigation in low Earth orbit. The seventh guideline is an attempt to limit the creation of space debris in geosynchronous orbit by "leaving objects at the end of their mission in an orbit above the GEO region such that they will not interfere with, or return to the GEO region." This is now generally interpreted as boosting a terminated GEO object into a supersynchronous orbit at least 300 km above geosynchronous orbit. This of course requires that geosats be left with sufficient fuel to perform this end-of-life maneuver.

17.8 Space Law Compliance

The principal judicial organ of the United Nations is the International Court of Justice (ICJ) seated at The Hague. The ICJ is composed of 15 judges elected to 9-year terms of office by the United Nations General Assembly and the Security Council. These organs vote simultaneously but separately. In order to be elected, a candidate must receive an absolute majority of the votes in both bodies.

The ICJ settles legal disputes submitted to it by States in accordance with international law. Compliance issues of space law relate to launch registrations, militarization of space, and interpretation of the space treaties. Launch registration used to be a major issue with some States (e.g., the United States) overlooking this issue. At this time launch registration has increased but with a concomitant increase in the declared function of the launch and satellite.

There have been several interesting cases in space law, but rarely have they made their way to the ICJ, the parties preferring to resolve the issue in other ways. We will discuss three cases here.

Case 1: The Martian Yemenites

After NASA landed the Pathfinder mission on Mars in 1977, the event drew legal ire from a group of Yemenites. According to Arabic news sources, three men, Adam Ismail, Mustafa Khalid, and Abdullah al-Umari, wanted to sue NASA for trespassing. Filing a lawsuit with the Yemeni Prosecutor General, the trio argued that they had inherited the planet from their ancestors 3,000 years ago, according to the mythologies of the ancient Sabaean and Himyaritic civilizations.

The case never made it to the International Court of Justice.

Case 2: Reentering Space Debris

Approximately 200 pieces of space debris make it to the ground each year (Figure 17.3). In some cases these have resulted in damage on the ground. According to the UN treaty (1972 LIAB) this debris is strictly the property of the launching

(a) (b)

Figure 17.3 Both of these objects are part of a rocket booster that launched a satellite into orbit. The object in (a) is from a Delta rocket that landed in the United States (*Source:* NASA), whereas the object in (b) is probably the third stage of an Indian PLSV rocket that landed in the Indian Ocean and finally drifted and was towed onto the beach at Greenhead, north of Perth, Western Australia (*Source:* The Independent).

state, which is also responsible for any reentry damage caused. Only rarely is the treaty invoked.

In some cases where fuel tanks for the lower stages of a rocket are involved, the object may in fact never have made it to outer space (if one accepts the definition of 100-km altitude). In many cases the value of the object to the finder (or eventually a museum) is many times that to the legal owner (the launching authority or state to whom it belongs). In fact, in most cases the latter would be happy to forego ownership.

In a famous case in Western Australia, the US Space Station *Skylab* reentered over southeastern Western Australia in July 1979 with a large number of objects impacting the ground. The Esperance Museum contains the greatest collection of these items, but several other museums have some *Skylab* items on display. And there are still some items to be found. When I visited the Esperance Museum around 2010 and gave a talk on space debris reentries, I was approached by a farmer following the talk. He produced a small stainless steel heat exchanger that I identified as most probably being from *Skylab*. It was a prized possession and he said he had never shown it to anyone before.

The Shire of Esperance, in an essentially advertising gesture, fined NASA $400 for littering. This was paid by a US radio station (Figure 17.4).

(a)

(b)

Figure 17.4 (a) NASA Map showing *Skylab* reentry path, and (b) sign outside Esperance Museum.

Case 3: COSMOS 954-1

Cosmos 954 was a Soviet military radar ocean reconnaissance satellite that carried a nuclear reactor. The reentry footprint occurred over Canada. The main part of the reactor was deposited on a frozen lake, but the containment vessel of the reactor was breached and other radioactive contamination occurred across a large area. Reentry occurred January 24, 1978, over the Canadian Northwest territories in a footprint covering 100,000 square kilometers. The highest radiation levels encountered (~200 R/hr) could cause severe radiation sickness and death from only a few hours exposure. The cleanup operation (Operation Morning Light) (Figures 17.5 and 17.6) lasted about 3 months. Canada made a claim of $6M against the USSR. They paid Canada $3M.

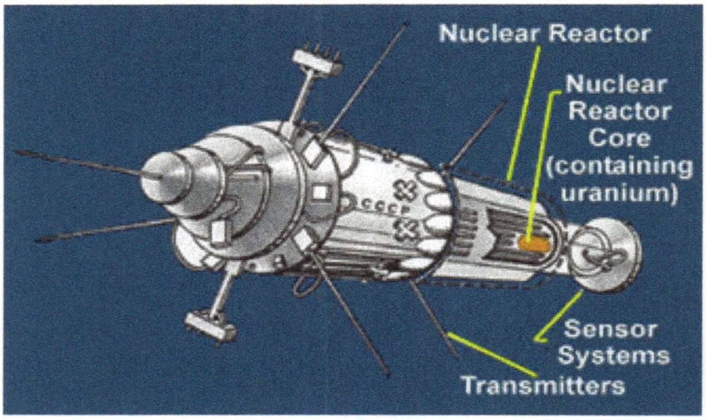

(a)

(b)

Figure 17.5 (a) Schematic of *Cosmos 954*, and (b) Canadian cleanup operation patch.

(a)

(b)

Figure 17.6 Canadians with radiological equipment clean up *Cosmos 954* contamination.

One aftereffect of the *Cosmos 954* reentry was that the Soviets decided that to avoid any further incidents of this kind, they would eject the core of the nuclear reactors and boost them to a higher orbit of around 900 km. However, since then it has been discovered that these cores are now leaking radioactive sodium (a highly reactive metal) droplets into low Earth orbit. No one has filed a lawsuit against the launching party for dangerous pollution of this region of LEO.

Cosmos 954 is not the only spacecraft with radioactive materials onboard to reenter the Earth's atmosphere. On April 21, 1964, the *Transit 5BN-3* along with its radioisotope thermoelectric generator (RTG)—with its fuel of plutonium 238—reentered over the Indian Ocean near Indonesia. It is believed that the spacecraft totally disintegrated in the reentry process.

And then again, in 1996 a failed Russian mission to Mars made an uncontrolled reentry passing over Australia and eventually depositing some of its pieces, including an RTG, somewhere in the Bolivian Alps. It was this event which led Australia to develop procedures for reentry events over the country.

17.9 The Future of International Space Law

With the failure of the Moon Treaty and a subsequent later attempt to pass a treaty on space debris, this does not bode well for any subsequent attempts to use international law to control and remediate the growing problem of space debris. It would appear that state law and bilateral and multilateral treaties are going to be the only way that space operations are controlled in the future. If state law diverges between different nations or blocks of nations, we can only hope that it does not escalate into military action. But with the increasing number of state actors (around 50 as compared to 2 major players in the 1960s), and their increasing belligerence, the prospect of conflict appears likely.

17.10 Multilateral Space Agreements

17.10.1 Establishment of the European Space Agency

The European Space Agency had its foundation in the early 1960s. It was decided to initially create two organizations, the European Launcher Development Organization (ELDO) to develop a launch system, and the European Space Research Organization (ESRO) to develop spacecraft. It was in 1964 that the signed conventions for the two agencies came into force. The signatories were from 10 European nations.

Five specialized space centers were set up in various countries across Europe. The European Space Operations Centre (ESOC) was set up in Darmstadt, Germany in 1967. It has now operated more than 50 satellites in its history. The European Space Research and Technology Centre (ESTEC) was established in Noodwijk in the Netherlands in 1968.

In 1975 ESA was created in its current form, merging ELDO with ESRO. There were 10 founding members: Belgium, Germany, Denmark, France, United Kingdom,

Italy, the Netherlands, Sweden, Switzerland, and Spain. Ireland joined later in the year. ESA is now an international organization with 22 member states.

17.10.2 The ARTEMIS Accords

The Artemis Accords (Figure 17.7) are a multinational treaty initially signed in October 2020 by the United States, Italy, Australia, Canada, Japan, Luxembourg, the United Arab Emirates, and the United Kingdom. They have since been expanded to include 20 nations.

The principles of the Artemis Accords are:

Peaceful uses of space;
Transparency of operations;
Interoperability of equipment between signatories;
Emergency assistance to all;
Registration of space objects;
Release of scientific data obtained through space operations;
Protecting heritage for significant space events and objects;
Space resources to benefit all signatories;
Deconfliction of activities;
Orbital debris mitigation.

The Artemis Accords state that the use of space resources can benefit humanity. NASA plans to exploit lunar resources extensively during the Artemis program, especially the water ice that seems to be plentiful on the permanently shadowed floors of lunar craters. This ice not only can provide life support for astronauts, it

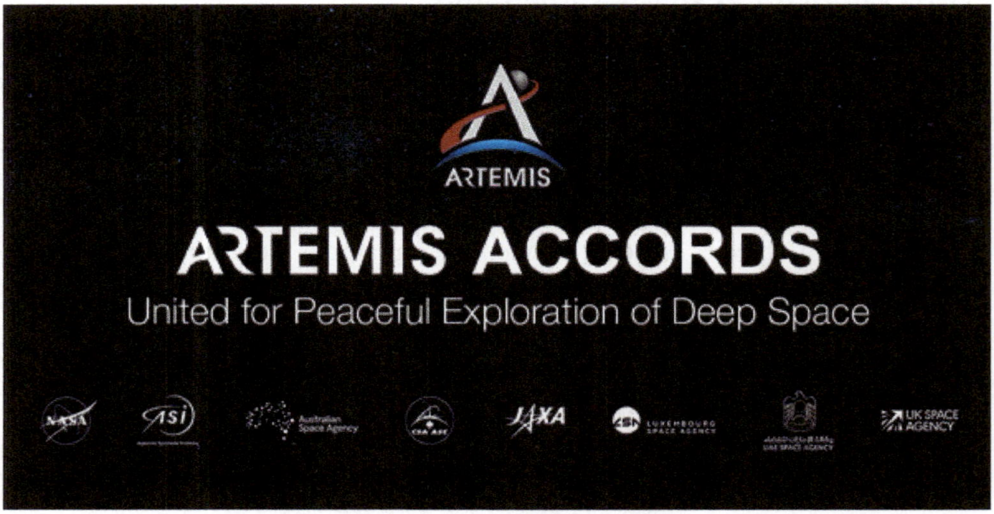

Figure 17.7 The Artemis Accords—Onward to the Moon and Mars. (*Source:* NASA/Artemis.)

can also be split into its constituent hydrogen and oxygen, the chief components of rocket fuel, agency officials have stressed.

Such mining activities will be conducted in full compliance with the OST, the Artemis Accords stress. The OST prohibits any nation from claiming sovereignty over the Moon or any other celestial object. But it does seem to allow the extraction and sale of space resources, many space-law experts say. The US Congress passed a law in 2015 explicitly permitting American companies to mine and sell off-Earth resources.

17.10.3 *The Woomera Manual:* Militarization of Space

The Woomera Manual project is an international research project that is spearheaded by the University of Adelaide, the University of Exeter, the University Of Nebraska, and the University of New South Wales, Canberra. "Our mission is to develop a manual that objectively articulates and clarifies existing international law applicable to military space operations," said founding partner Professor Melissa de Zwart from the University of Adelaide.

It follows on from previous manuals dealing with conflict in the international arena:

San Remo Manual–Conflict at sea;

Harvard Manual–Air and missile warfare;

Tallin Manual–Cyber warfare.

"Conflict in outer space is not a case of if but when," said Zwart. "However, the legal regime that governs the use of force and actual armed conflict in outer space is currently very unclear, which is why *The Woomera Manual* is needed."

17.11 National Space Law

17.11.1 Australia

Australia has enacted many bilateral space agreements with other countries for cooperative space projects. Such agreements have included the establishment of space-tracking stations in Australia. In general, Australia benefits from the data obtained, and now has a policy of bilateral operation of such stations that gives Australian scientists, engineers, technicians, and others job opportunities that they would otherwise have had to travel overseas to acquire. In many cases these agreements have provided benefits that are quite incommensurate with the small amount of funding that the Australian government has committed toward them.

However, it was not until 1998 that the Australian Commonwealth Government enacted specific space legislation, which is termed the Australian Space Activities Act (ASAA). This primarily relates to the regulation of space launches within Australia, and by Australians outside of Australia, and also the return of space objects to Australia. In essence, a space launch license under this act makes the Australian government liable under the UN liability treaty for any damage caused by Australian space objects, and also allows Australia to claim the return of Australian objects under the UN Rescue and Return Treaty. Of course, this all comes at a considerable price to the applicant for such a license. The ASAA was amended by the Space Activities Amendment Bill (SAAB) in 2002.

17.11.2 United States

In August 2006, the United States adopted a new policy document on space [4, 5], in which there is a priority mandate for space programs and activities, together with a commitment to the peaceful use of space for the benefit of all nations and the rejection of claims to national sovereignty in space. However, the same policy includes the rights of the United States to preserve its freedom in space, and to deny access to space to any enemies with hostile intent to US interests.

Current US space policy has also been focused on regulating private space flight. This followed from the success of the scaled composites *SpaceShipOne* in a suborbital flight exceeding 100 km in altitude, and a subsequent proposal to make such flights available to paying passengers. The legislation recognizes the inherent dangers in such operations and recognizes that properly informed passengers should be given the right to undertake such ventures, whilst ensuring the safety of (uninvolved) people and property on the ground, and appropriate legal protection for the space operator, as long as compliance with the regulations is established and maintained.

NASA and other US government agencies have also been responsible for implementing many laws relevant to its own space activities:

- Safety of space operations;
- Rules relating to the minimization of space debris;
- Dissemination of data acquired through space research and activities;
- Space communication and navigation;
- Planetary protection—prevention of biological hazards to and from other worlds.

17.12 Current Issues

17.12.1 Definition and Delineation of Outer Space

Despite the fact that the issue has been discussed without resolution for 40 years, the legal definition and delineation of outer space must still be listed as a current issue urgently in need of resolution. Australia has been proactive in this area, and a 2002 amendment to the 1998 Australian Space Activities Act has taken the boundary to be at an altitude of 100 km.

17.12.2 Definition of an Aerospace Object

A current COPUOS concern is the definition of an aerospace object. Current UN treaties refer to space objects as man-made objects that have been launched from Earth into space. However, some are concerned that the development of vehicles such as space planes, which will be able to fly in the atmosphere as well as use rocket technology to fly in space, may cause problems in traversing sovereign airspace, and that such vehicles may need to be covered by a different definition. To this end, COPUOS has issued a 10-item questionnaire to assess states' beliefs and opinions in this area [6]. A compendium of replies received has been made available [7] and is interesting reading. This question of course relates directly to the definition and delineation of outer space. Some of the problems that may arise in this area are illustrated in the fictional space moot scenario [8].

17.12.3 The Problem of Orbital Space Debris

Since the start of the space age the problem of unwanted material or debris in space has been growing. Each space launch usually leaves considerably more than the desired satellite in orbit. Expended rocket boosters, attachment bolts, shields, solid rocket motor slag, and innumerable other items are placed into Earth's orbit. Some of these decay (lose altitude) and burn up in the atmosphere, while some are large enough to escape complete destruction by ablation and then may pose a potential hazard to life and property on the Earth's surface.

In space, materials degrade and detach from satellites; stored energy in the form of unspent fuel and battery vapors may cause explosive ruptures; and there is fragmentation of space objects. Collisions between space objects at hypervelocity not only causes damage, but also creates thousands of other space objects (i.e., fragments of the original objects), which themselves then pose collision hazards to active spacecraft.

Although collisions between space objects are now very infrequent, the growth of collisional fragments will become exponential in less than a hundred years, and may render the low Earth orbit regime (from about 500- to 1500-km altitude) unusable within a few hundred years. To add to the natural creation of space debris, antisatellite tests will severely exacerbate the problem. On January 11, 2007, China conducted such a test when a ground-launched missile was used to destroy an old Chinese weather satellite. That one test created over 2,000 objects with a diameter 10 cm or larger (that could be tracked from the ground) and an estimated 30,000 objects under 10 cm in size that could still pose a dangerous collision hazard. This event is likely to test the UN liability treaty (which China has ratified) in the next few years. Since 2007 further ASAT tests have been conducted with subsequent debris production.

As mentioned before, COPOUS has discussed this issue via its subcommittees (STC and LC) for the last 10 years, and finally proposed a set of mitigation guidelines that were adopted in June of 2007. These are woefully inadequate and have led to calls for a reexamination of the subject.

17.12.4 Revising and or Enacting New International Space Law

It has been argued that the current UN international space law treaties were enacted many years ago when there were only two major players in space, the USSR and the United States. The situation has now become one where around 50 states have space assets and are involved in space-based activities. It has thus been suggested that it is now time to revise and/or enact new international space laws. This item is discussed further in Chapter 18 on space security.

17.13 International Institute of Space Law

The IISL was established in 1960 and is an autonomous component of the International Astronautical Federation (IAF). It holds annual colloquia on space law. Its aim to promote international cooperation in space law and to foster the development of space law and studies on the legal and sociological aspects of space activities. It is located in France [9].

The IISL sponsors the Manfred Lachs Space Moot competition every year, open to law students around the world. Talk to your law lecturer about entry.

References

[1] www.airpower.au.af.mil/airchronicles/aureview/1973/may-jun/barrett.html.
[2] http://www.unoosa.org/pdf/spacelaw/intlagree.pdf.
[3] www.unoosa.org/pdf/publications/STSPACE11E.pdf.
[4] www.ostp.gov/html/US%20National%20Space%20Policy.pdf.
[5] edocket.access.gpo.gov/2006/pdf/E6-21193.pdf.
[6] www.unoosa.org/pdf/misc/aero/questE.pdf.
[7] www.unoosa.org/docs/misc/aero/aero_compE.doc.
[8] www.spacemoot.org/acrobat/prob2007.pdf.
[9] iislweb.space, the International Institute of Space Law.

Selected Bibliography

Belitsky, B., and A. S. Piradov, *International Space Law*, University Press of the Pacific, 2000.
Lambright, W. H. (ed.), *Space Policy in the Twenty-First Century*, Johns Hopkins University Press, 2002.
Reynolds, G. H., *Outer Space: Problems of Law and Policy*, Westview Press, 1998.

CHAPTER 18

Space Security

Space is very important to many activities undertaken in society every day, many of which go unnoticed or are taken for granted. The security of the space systems that enable these activities is thus very important. However, these systems are also subject to many threats from both natural sources and man-made activities. Space security is thus a balance between opportunities and threats.

18.1 Definitions

Space security relates to guaranteed access to space and the ability to freely exploit space for various purposes.

Counterspace refers to operations, programs, or actions that are against or designed to prevent space operations, programs, or actions.

Space asset refers to either a ground-based or space-based asset that is part of a space operation or space system.

Space-based asset refers to an asset that is resident in the space domain.

Resident space object (also known as an orbital space object) refers to any material object such as a satellite or spacecraft (either active or inactive) or a piece or space debris that is an orbit around the Earth.

18.2 Opportunities

Space opportunities can be classified into three main areas.

18.2.1 Remote Sensing (Earth Observation)

Cartography: the production of more accurate maps in a timely fashion

Damage assessment: monitoring wildfires and flooding

Data gathering: using imagery to gather statistics on infrastructure, crop assessment, for example

Meteorology forecasts: more accurate and timely forecast of weather conditions and extreme events and evacuation warnings

Events: population movements, missile launch warnings

18.2.2 Communication

Personal communication: fixed, mobile, and internet

Business communications: private networks

Government communication: civilian and military

Entertainment: TV distribution and direct broadcast from geosats

Alerts: pagers

Telemetry: remote measurements (e.g., ocean currents from isolated sensors of buoys)

Control: of essential service systems

18.2.3 Positioning/Navigation/Timing

Land, sea, and air travel: minimizing cost, reducing collision threats

Survey: accurate positioning to a global reference

Rescue: saving lives in a timely fashion, SARSAT (search and rescue) in remote areas

Timing: data time stamps, event timing

Transactions: banking and stock market transactions

18.2.4 Opportunities: GNSS

The global navigation satellite system, as typified by the GPS satellite system, has become so ubiquitous that we often don't realize how important it has become to our way of living.

GPS provides both accurate position and time. A GPS receiver is often incorporated in mobile phones. It has taken over from most street directories, it has enabled such applications as Uber, and it is used to track the home delivery of goods and services. It has revolutionized surveying and cartography. The benefits to the military are immeasurable, especially in the area of navigation in the land, sea, and air domains, and in search and rescue operations. GPS also provides an extremely accurate time service that is used in many commercial operations, and enables a plethora of scientific research projects. It is used in financial transactions, such as the stock exchange where timely share transfers can mean the loss or gain of millions of dollars, even down to the microsecond level. Mostly unknown, GPS also monitors for, and can precisely locate, nuclear explosions.

18.2.5 Opportunities: Environmental Satellites

A plethora of Earth observation satellites enables us to keep track of our environment, including weather forecasts, space weather, and other environmental parameters with many civilian and military applications in agriculture, disaster relief, and planning.

18.2.6 Opportunities: Communications Satellites

Today's world relies heavily on communication satellites, both for telecommunication and for broadcast purposes. This reliance has increased heavily with the development and expansion of the internet. Geosynchronous satellites have an unacceptable delay for some communications. Even telephone calls between humans can only use a geosat one way because of the unacceptable delay with a two-way link, undersea cable provides the return.

However, this problem has been addressed by the use of constellations of low Earth orbit satellites.

18.3 Threats

18.3.1 Natural

The natural components of the space environment, which pose a threat to space operations include meteoroids and near-Earth asteroids and comets.

Meteoroids are small pieces of asteroidal and cometary material with sizes from microns (millionths of a meter) to around 10 meters. These exist in large numbers throughout the solar system and pose a collision hazard to orbital satellites. They are produced by outgassing material from comets and collisional impacts between asteroids. The collision hazard was demonstrated through impacts with space shuttle windscreens and recorded by the NASA Long Duration Exposure Facility. Although the mass of meteoroids which hit the Earth each year (~10,000 tons) is considerably larger than man-made space debris in orbit (50 to 5,000 tons depending on how it is counted), man-made space debris is responsible for more OSO impacts because it is resident in orbit, and thus remains in space for a much longer time than natural space debris, which transits low Earth orbit in typically 100 seconds or less.

The upper size limit for meteoroids is purely one of man's devising and they actually merge seamlessly into asteroids with sizes from 10m to 1,000 km. The numbers of objects varies inversely with size according to the square of the size. So as the object size increases their numbers substantially decrease and thus the collision hazard of near Earth objects with Earth-orbiting satellites is very low. We have already shown (Chapter 16) that the collisional hazard to geosats from NEOs is extremely low. Rather, the hazard to ground space infrastructure (and all other infrastructure) is the item of concern. And although the probability of a significant ground impact is very low (once per century to once per millennium), the results can be catastrophic.

Orbital decay due to atmospheric drag can also be listed as a hazard to satellites in the lower part of low Earth orbit (particularly below 500 km). This will reduce a satellite's lifetime below its rated lifetime. This is very significant for the constellations of imaging satellites (Planet Ltd) and the newer networks of LEO communications satellites. In the former case, the low orbit required to provide high resolution. In the latter case, the low orbit is required to provide low communication latency and thus high-speed communications.

On the opposite side, atmospheric drag can be regarded as beneficial to remove space debris from these low orbits.

18.3.2 Artificial

Man-made space debris is caused in a number of ways we have already discussed. At the moment only a small fraction of debris is caused through collisions. Collisions between active or defunct satellites, or more probably between small pieces of debris and larger objects will become more frequent and eventually will predominate the debris-generation process. Many schemes have been proposed to remove debris from orbit, but they are all very expensive and no one really wants to pay the bill. And it is probably more important to remove the many smaller pieces than the large objects, although in the end all orbital sizes will require removal.

The launch of large constellations of communication satellites in low Earth orbit (e.g., Starlink) is beneficial to expand low latency network communications for civilian and military purposes. However, it also increases the risk of collision and production of large quantities of space debris.

The increased collision probability may be controllable in normal conditions (i.e., as long as sufficient tracking and conjunction assessment facilities are available). However, a severe space weather storm, which causes increased and unpredictable atmospheric drag might change the equation drastically.

Although space weather has been rather benign over the last two decades, in the early 1980s such a severe space weather event caused the space-tracking authorities (in the United States) to lose track of one third of the LEO orbital population. Full tracking was not regained for up to three months. Space objects could be detected, but they were so far removed from their original orbits that the tracks could not be correlated with the known objects. If such an event were to occur today it is virtually certain that collisions would result with massive debris creation.

Antisatellite activities may well play a significant role in debris generation in the future, and if space war starts in earnest, we can expect debris numbers to climb exponentially, and limit satellite operations for decades or longer.

Cyber warfare against satellites may not only affect their intended operations, but could possibly cause a loss of control that affects orbital behavior and increases the probability of collision and debris generation.

18.4 Legal Issues

Legal issues pertaining to space security are covered by the United Nations treaties on outer space (see Section 17.5). In particular, the UN legal principles governing outer space read (Section 17.6):

"Exploration and use of outer space should be for the benefit of all mankind. Outer space and any bodies in it are free for all and not subject to national appropriation. All activities should be peaceful. States bear international responsibility for national activities in space. States shall cooperate with each other in space. States retain control of their objects launched into space, and are liable for any damage such objects might cause. Astronauts shall be regarded as envoys of mankind. All possible assistance shall be rendered them in any emergency or accidental situation." Five treaties (although only four have been ratified) provide the legal basis for this space law (see Section 17.5).

The current UN treaties on outer space were negotiated at a time (late 1960s and early 1970s) when the United States and the Soviet Union were almost the only major players involved. Since then the scene has changed dramatically with 50 states involved in an extremely diverse range of space activities. Because of this, it has been suggested that new treaties should be negotiated. However, the bitter 10-year debate over space debris, which resulted in not a treaty but only 7 guidelines, and with the increasing belligerence shown by state actors, it is very unlikely any positive outcome would result. State law, bilateral, and multilateral treaties are likely to be the only new space laws enacted.

18.5 Political Issues

A senior official from Russia's Foreign Ministry has warned that Western commercial satellites could become legitimate targets for Moscow if they were involved in the war in Ukraine.

Konstantin Vorontsov, deputy director of the Russian Foreign Ministry's Department for Nonproliferation and Arms Control, told the United Nations on October 26, 2022 that the use of Western satellites to aid the Ukrainian war effort was "an extremely dangerous trend."

"Quasi-civilian infrastructure may be a legitimate target for a retaliatory strike," Vorontsov told the United Nations First Committee, adding that the West's use of such satellites to support Ukraine was "provocative."

"We are talking about the involvement of components of civilian space infrastructure, including commercial, by the United States and its allies in armed conflicts," Vorontsov was quoted as telling the committee, without naming any specific satellite firms.

18.6 Space Security: The Big Picture

Space security is inevitably a part of, and is linked to, space situational awareness and space sustainability. This triad is illustrated in Figure 18.1. Space safety also plays a considerable role in this. Different people with different ideas about the components of space security have made for a plethora of people who claim a home in this complex mixture. Many of the people involved come from a legal background.

18.7 Space Security Organizations

18.7.1 Space Security Organizations: Private #1

The primary focus of the Secure World Foundation is on space sustainability, the ability of all humanity to continue to use outer space for peaceful purposes and socioeconomic benefit over the long term.

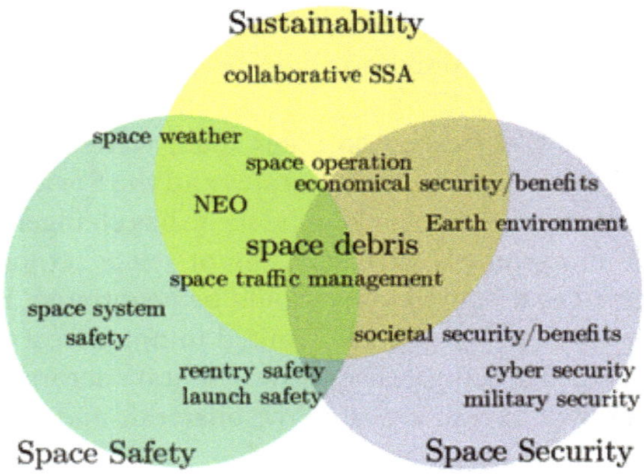

Figure 18.1 An overview of space security.

The Secure World Foundation is a private organization dedicated to promoting space security. They essentially believe that all space activities should be made more transparent, and publish several comprehensive reports to achieve this aim. In particular they try and make public the counterspace capabilities of all nations in the belief that this will reduce the threat from unexpected attacks on space assets. All their reports are freely available on their website, swfound.org.

Our Vision

> Secure World Foundation envisions the secure, sustainable, and peaceful uses of outer space contributing to global stability and benefits on Earth.

Our Mission

> The mission of the Secure World Foundation is to work with governments, industry, international organizations, and civil society to develop and promote ideas and actions to achieve the secure, sustainable, and peaceful uses of outer space benefiting Earth and all its peoples.

18.7.2 Space Security Organizations: Private #2

The Center for Strategic and International Studies (CSIS) is a bipartisan nonprofit policy research organization dedicated to advancing practical ideas to address the world's greatest challenges.

The CSIS addresses many world issues that have no direct involvement with space. The section of CSIS that is relevant to space security is the CSIS Aerospace Security Project (ASP).

The ASP is a leading source of analysis and expertise on the policy and technology issues shaping the future of aerospace security. One recent publication is *Space Threat Assessments 2023*, aerospace.csis.org.

18.7.3 Space Security Organizations: US Government

The US government now carries out space security operations through the US Space Force (USSF). The organization of the USSF is shown in Figure 18.2.

The meaning of space security to an individual country is usually the securing of that country's space assets against threats from any other country. This security may be carried out by diplomatic, political, and military means. Counterspace capabilities (the ability to deny the enemy the "high ground") are almost exclusively carried out by a country's military. In the United States, this is currently assigned to the US Space Force, and mostly carried out by the Space Operations Command.

THE USSF Space Operations Command is built around the Space Mission Deltas (Figure 18.3). These are equivalent to Wing level organizations in the US Air Force. There are nine of these deltas with tasks across the space security spectrum as shown.

18.8 The Counterspace Continuum

Counterspace is defined as those activities carried out to either deny space access to an enemy, or to protect one's own space activities. Figure 18.4 shows the range of operations and actions involved in counterspace activities.

18.8.1 Counterspace Capability Classification

Counterspace capabilities can be used in both defensive and offensive capacities (Table 18.1).

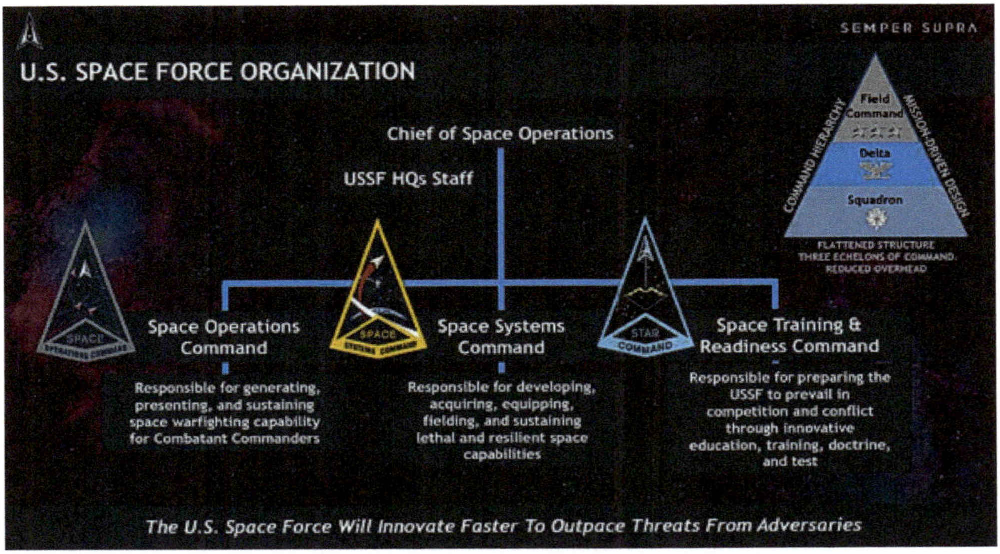

Figure 18.2 Organization of the US Space Force. (*Source:* USSF.)

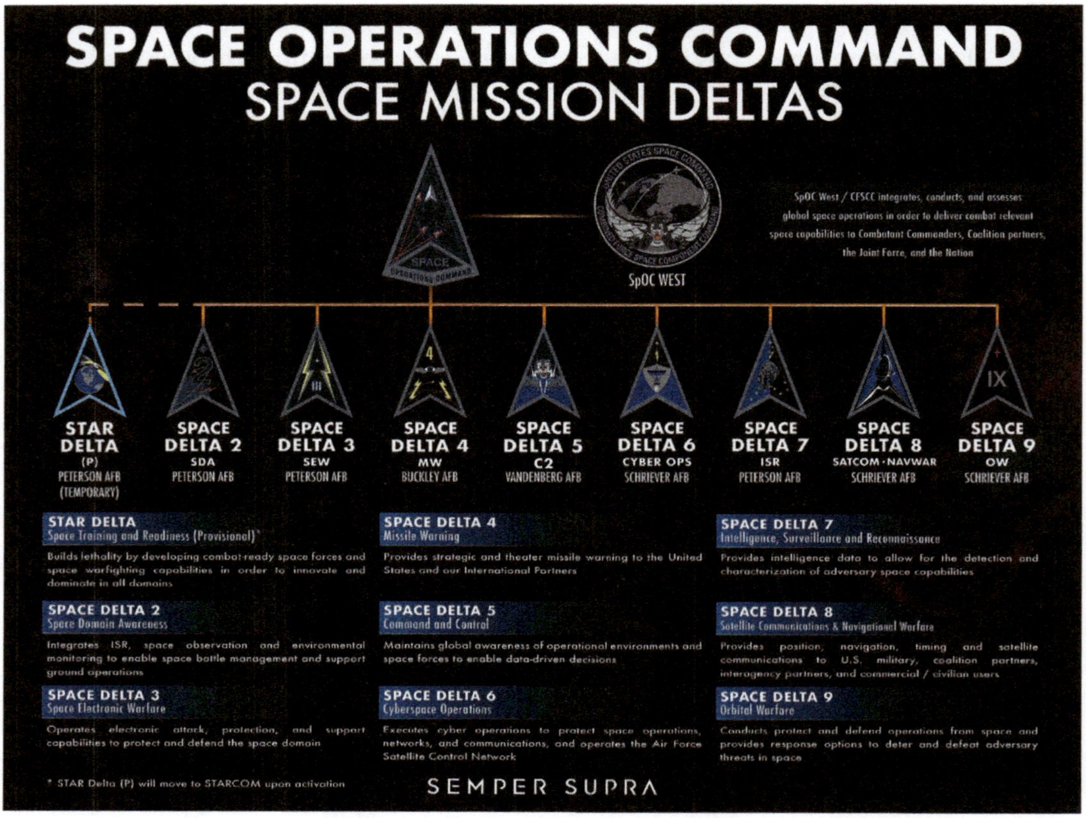

Figure 18.3 US Space Force Deltas and Their Missions. (*Source:* USSF.)

Cyber warfare is involved in the interference with, or destruction of, information technology assets. This may or may not involve the physical destruction of equipment. Cybersecurity is involved with the protection of information technology assets, and usually involves both physical and software security.

Space situational awareness involves a knowledge of the types and orbits of resident space objects, natural space objects, and a knowledge of the space environment. SSA is thus an essential component of space security.

18.8.2 Counterspace Capability Assessment

Figure 18.5 is a global assessment of counterspace capabilities issued by the Secure World Foundation in 2023.

18.8.3 Kinetic Weapons

Direct ascent: A missile is fired from a terrestrial platform (the ground, a ship, or an aircraft), ascends to the required orbit, and is either guided or homes in on the target for a direct collision. The target is destroyed by the hypervelocity kinetic energy of the collision.

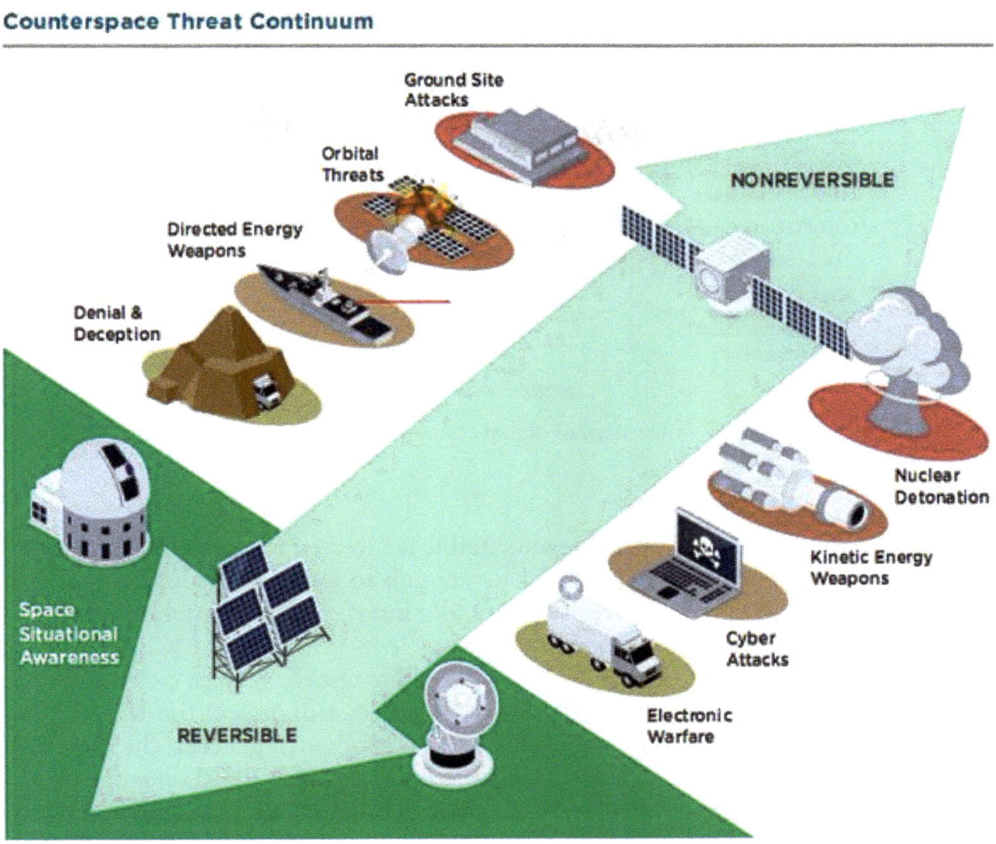

Figure 18.4 The counterspace threat spectrum. (*Source:* SWF.)

Table 18.1 Counterspace Operations

Offensive	
Kinetic	**Non-Kinetic**
Direct ascent	Directed energy
Co-orbital	Lasers
	High power microwave
	Directed energy
	Electronic warfare
	Jamming
	Interference
	Spoofing
	Cyber warfare
Defensive	
Space situational awareness	
Cybersecurity	

2023 Global Assessment

	U.S.	Russia	China	India	Aus.	France	Iran	Japan	North Korea	South Korea	U.K.
LEO Co-Orbital	some	significant	some	none	none	none	none	none	none	none	none
MEO/GEO Co-Orbital	some	significant	some	none	none	none	none	none	none	none	none
LEO Direct Ascent	some	some	significant	some	none	none	none	none	none	none	none
MEO/GEO Direct Ascent	some	some	some	none	none	none	none	none	none	none	none
Directed Energy	some	some	some	none	none	some	none	none	none	none	none
Electronic Warfare	significant	significant	significant	some	some	some	some	none	some	some	none
Space Situational Awareness	significant	significant	significant	some	some	some	some	some	none	some	some

Legend: none some significant

Figure 18.5 SWF assessment of global counterspace capabilities. (*Source:* Secure World Foundation.)

Co-orbital: A counterspace satellite is directed to move to the orbit of the target and slowly approach it until close enough to either explode, damaging the target through debris, or otherwise physically interfere with the target to remove it from orbit or render it inoperative.

Kinetic weapons—ASATS: Antisatellite tests have been carried since the beginning of the space age, initially by the Soviet Union and the United States. Since 2005 they have been joined by China and India. In general, the United States has tended to use direct ascent ASAT methods whereas Russia has favored co-orbital methods. Figure 18.6 is the Secure World Foundation list of antisatellite tests as of 2023. Figure 18.7 is a graph of antisatellite tests by year (Secure World Foundation).

Kinetic weapons—direct ascent: All four countries (United States, Russia, China, and India) involved in antisatellite activities have used a direct ascent missile to destroy their targets (Figures 18.8 and 18.9). In 1985, the United States employed an F-15 aircraft to launch an ASM-135 ASAT missile and by direct ascent destroyed the *P78-1 Solwind* satellite (which was providing space weather data on solar coronal mass ejections!).

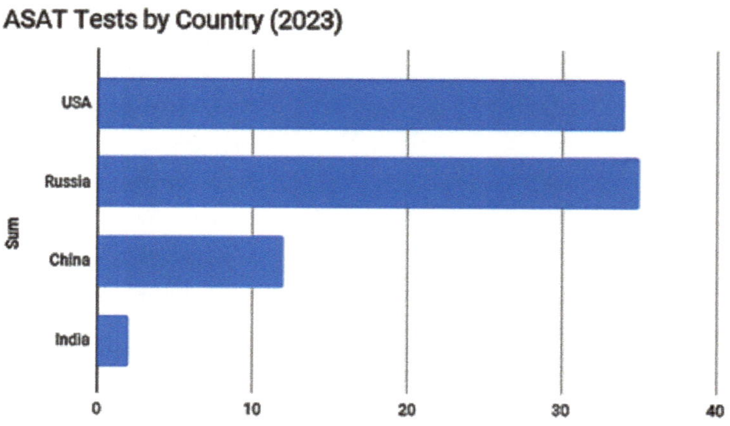

Figure 18.6 Antisatellite test scores. (*Source:* Secure World Foundation.)

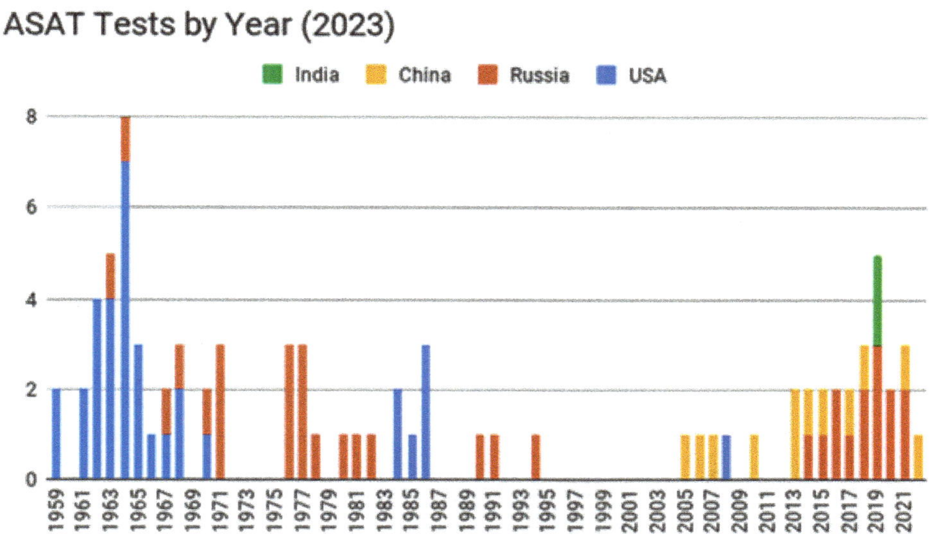

Figure 18.7 Antisatellite tests by year. (*Source:* Secure World Foundation.)

Figure 18.8 The Russian version of a direct ascent antisatellite weapon. (*Source:* US Naval Institute.)

Kinetic weapons—co-orbital: A co-orbital weapon is one previously launched that moves to the target orbit and slowly and (hopefully) stealthily moves close to the target satellite and then uses a kinetic or directed energy weapon to destroy or disable the target.

Figure 18.9 A US antisatellite weapon launched in 1985 from an F-15 aircraft. (*Source:* USAF images.)

The big problem with antisatellite tests is that they always create debris. Figure 18.10, from the Secure World Foundation, gives statistics on this aspect of antisatellite tests.

18.8.4 Nonkinetic: Directed Energy

18.8.4.1 Lasers

A *low- to medium-power laser* on the ground (Figure 18.11) or in space may be used to dazzle or blind a sensor while observing a particular ground or space asset. This is a temporary effect. A *high-power laser* on the ground or in space may be used to permanently damage a satellite or part of a satellite such as a sensor. Usually only a small but vital part of the satellite will be targeted to render it inoperative for its primary function. Satellites most susceptible to this type of attack are imaging satellites. Super power lasers have been mooted to destroy cities from space and totally destroy satellites from the ground (Figure 18.12). So far, these are only in the realm of science fiction.

18.8.4.2 High-Power Microwave

A source of high-power microwave (HPM) energy is directed toward the target to disable or render inoperative its electronics or operations or operators. Because HPM cannot be formed into a laser-like beam, it would require a co-orbital satellite to implement. The effect may be temporary or permanent. HPM might be used by a guard satellite to protect high-risk or high-cost assets, and ground systems have been developed to kill drones and incapacitate humans. HPM weapons are generally short range but some have been used over a few kilometers.

ORBITAL DEBRIS CREATED BY ASAT TESTS IN SPACE

DATE	COUNTRY	ASAT SYSTEM	TARGET	INTERCEPT ALTITUDE	TRACKED DEBRIS	DEBRIS STILL ON ORBIT	TOTAL DEBRIS LIFESPAN
Oct. 20, 1968	Russia	IS	Cosmos 248		252	79	50+ years
Oct. 23, 1970	Russia	IS	Cosmos 373		147	35	50+ years
Feb. 25, 1971	Russia	IS	Cosmos 394		118	45	50+ years
Dec. 3, 1971	Russia	IS	Cosmos 459		29	0	3.3 years
Dec. 17, 1976	Russia	IS	Cosmos 880		127	57	45+ years
May 19, 1978	Russia	IS-M	Cosmos 970		73	64	40+ years
Apr. 18, 1980	Russia	IS-M	Cosmos 1171		48	6	40+ years
Jun. 18, 1982	Russia	IS-M	Cosmos 1375		64	60	35+ years
Sept. 13, 1985	U.S.	ASM-135	Solwind	530 km	287	0	18+ years
Sept. 5, 1986	U.S.	Delta 180 PAS	Delta 2 R/B		17	0	< 1 year
Dec. 26, 1994	Russia	Naryad-V?	Unknown		27	24	25+ years
Jan. 11, 2007	China	SC-19	FengYun 1C	880 km	3536	2786	15+ years
Feb. 20, 2008	U.S.	SM-3	USA 193	220 km	175	0	1+ year
Mar. 27, 2019	India	PDV-MK II	Microsat-R	300 km	130	1	3+ years
Aug.-Dec. 2019	Russia	Cosmos 2535	Cosmos 2536		30	16	3+ years
Nov. 15, 2021	Russia	Nudol	Cosmos 1408	470 km	1790	300	Unknown
				Total	**6850**	**3472**	

Figure 18.10 Statistics on space debris created by antisatellite tests. (*Source:* Secure World Foundation.)

18.8.4.3 Particle Beam

A beam of subatomic particles (protons, electrons, etc.) is used to destroy a satellite through heating, surface charging/discharging, deep dielectric charging/discharging, or radiation damage. Such a weapon would normally be used in space, as the atmosphere poses a large barrier to the propagation of such beams. The target will normally be a particular system on the satellite not the whole satellite. The effect may be permanent or temporary.

18.8.5 Nuclear Explosions

A high-altitude nuclear explosion, either deliberately for the purpose, or from an atmospheric nuclear bomb test or nuclear war, will destroy satellites in orbit over a

Figure 18.11 A medium-power laser to dazzle or blind a satellite. (*Source:* US Navy, John F. Williams.)

(a) (b)

Figure 18.12 Illustrations that imagine (a) a high-power space laser to destroy a city, and (b) a high-power ground laser to destroy a satellite.

period of time due to radiation damage. Unhardened satellites in LEO are particularly susceptible to this. GNSS satellites are hardened to resist the Van Allen belt radiation in which they function and thus are much less susceptible.

18.8.6 Electronic Warfare

Electronic warfare may be passive, such as the fields of signals intelligence (SIGINT), which is subdivided into electronic intelligence (ELINT) whereby information is collected by the interception of telemetry and other command and control signals of an adversary; and communications intelligence (COMINT), whereby verbal or digital communications are intercepted. Electronic warfare may also be active in the interference, jamming, or spoofing of adversarial radio frequency systems.

Electronic warfare involves the reception and transmission of radio frequency energy, and for space systems, whether on the ground or in space, this will normally involve frequencies in the VHF, UHF, EHF, and SHF bands covering from 30 MHz to 300 GHz.

Electronic warfare has been used since the invention of radio in the early twentieth century.

Electronic warfare became widespread in World War II (1939–1946), and has now spread into the space domain.

18.8.6.1 EW: Details

The US military classifies electronic warfare by the acronym MIJI, which stands for: meaconing, intrusion, jamming, and interference. These may overlap. The civilian world normally speaks of interference, jamming, and spoofing.

Interference (Figure 18.13) may be intentional or unintentional (a byproduct of a transmission) and generally refers to degradation of radio signals.

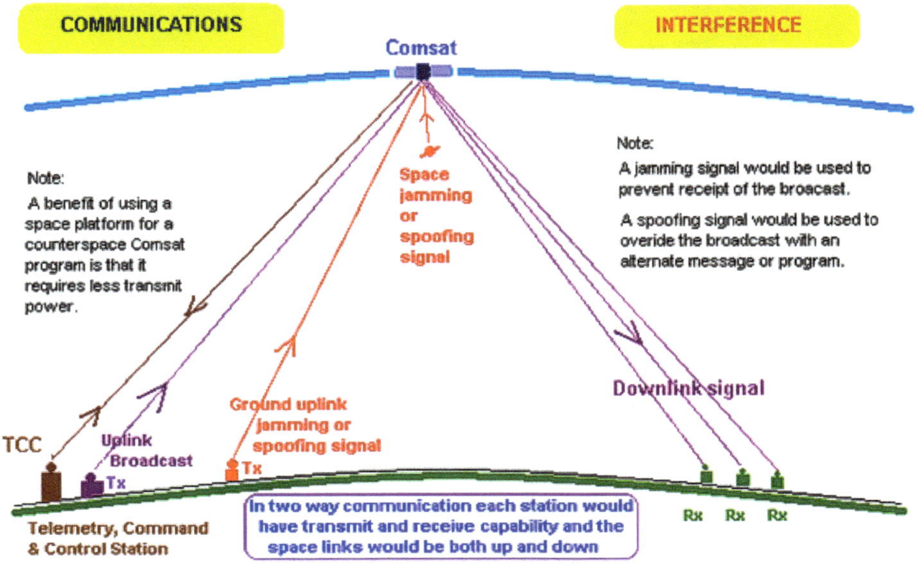

Figure 18.13 Electronic warfare against communication satellites.

Jamming (Figure 18.14) is a where a specific radio transmission is intended to render a communication signal unintelligible.

Spoofing is where a signal similar to the one being attacked is used to alter the message and mislead the enemy or substitute counterfeit information.

Space electronic warfare may be directed toward either a satellite uplink or downlink, and thus be directed toward a ground or a space asset.

The most effective interfering signal is one that uses the same modulation as that being transmitted. For example, an AM signal is most effective at interfering with an AM signal, an FM signal with an FM signal, and a digital signal with a similar type of digital signal. FM receivers are particularly susceptible to takeover by a higher power signal.

Nonkinetic electronic warfare COMSAT interference is depicted in Figure 18.13, and nonkinetic electronic warfare GNSS jamming is depicted in Figure 18.14.

18.8.7 Nonkinetic Cyber Warfare

Nonkinetic cyber warfare involves individuals, teams, or AI that use software and information systems and networks to compromise, control, interfere with, or destroy information and computer systems and networks. The targets may be space systems, whether in space or on the ground (e.g., space command and control systems).

Cyberattacks, not missiles, may be the most worrisome threat to satellites. This is particularly so if satellites are networked. It may then be just as cheap to threaten the network as to threaten a single satellite. Attacks may be ground to ground, ground to space, space to ground, or space to space.

States with limited resources may be just as successful as major players in this field.

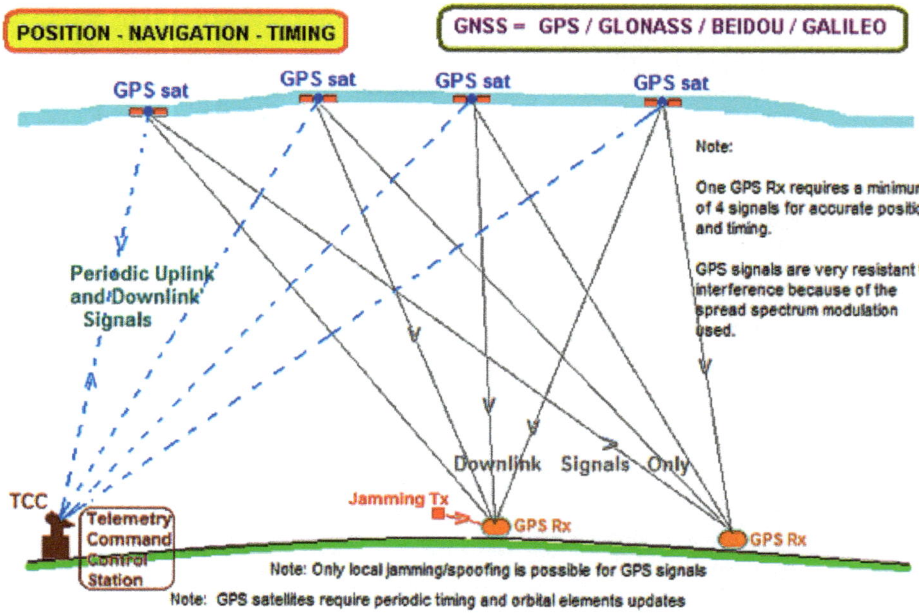

Figure 18.14 Jamming of a GNSS signal.

18.8.6.1 Space Cybersecurity Case Study 1: Weather Satellite

A cyberattack from China breached the US National Weather Service in October 2014, forcing cybersecurity teams to seal off data from weather satellites vital to disaster planning, aviation, shipping, and scores of other crucial uses.

Weather satellites orbit hundreds to thousands of kilometers above Earth and offer continuous views of weather systems, such as hurricanes, thunderstorms, and cold fronts, while also measuring temperature and moisture at different altitudes—all crucial bits that feed prediction models. To get that information to the public, NOAA makes satellite data and imagery publicly available through the web, as well as file-transfer networks for downloads.

18.8.6.2 Space Cybersecurity Case Study 2: Australian Wx Intel

In 2016 a (suspected Chinese) cyberattack was reported by the Australian Bureau of Meteorology. However, it is believed that the real target was the Australian Geospatial-Intelligence Organisation, which supports defense operations through provision of satellite and other imagery.

The Australian Geospatial-Intelligence Organisation is an Australian government intelligence agency that is part of the Department of Defence, responsible for the collection, analysis, and distribution of geospatial intelligence in support of Australia's defense and national interests. The AGO is one of six agencies that form the Australian Intelligence Community.

As an offshoot, an Australian startup firm Newsat had its computer systems so badly corrupted that it went into liquidation. The firm had planned to launch two Australian satellites and build a satellite industry.

18.8.6.3 Space Cybersecurity Case Study 3: Viasat and Ukraine

Just an hour before Russian troops invaded Ukraine, Russian government cyber-agents targeted the American satellite company Viasat. The operation resulted in an immediate and significant loss of communication in the earliest days of the war for the Ukrainian military, which relied on Viasat's services for command and control of the country's armed forces. The attack, on February 24, 2022, launched destructive wiper malware called AcidRain against Viasat modems and routers, quickly erasing all the data on the system. The machines then rebooted and were permanently disabled. Thousands of terminals were effectively destroyed in this way.

18.8.8 Space Situational Awareness

It may seem unusual to include SSA in the list of counterspace capabilities but knowledge of your own space assets, and of the space environment, are essential to protect them and knowledge of anyone else's space assets are essential to perform conjunction assessment and collision avoidance.

A knowledge of enemy space assets is also desirable to avoid a surprise attack (defensive SSA) and to enable you to counter enemy space operations in an offensive capacity if necessary. If your SSA capabilities are extensive, it is also desirable for you to perform CONASS of all space assets in your knowledge base to avoid collisions that will create extensive space debris and thus protect both your space assets and those of friendly states.

18.8.7.1 SSA Activities

- Producing catalogs of RSOs;
- Conjunction assessment (CONASS);
- Collision avoidance (COLA);
- Detecting spacecraft hazards;
- Identifying owner states and liabilities;
- Monitoring spacecraft behavior;
- Diagnosing spacecraft malfunctions.

Tracking spacecraft and space debris is the first essential SSA activity. These observations are used to calculate accurate orbits, so that conjunction assessments can be made, the determination of whether resident space objects may collide.

The USSF maintains the widest and most comprehensive space surveillance network by far of any country in the world (Figure 18.15), and is thus the most able to perform satellite CONASS. Part of their space object catalog is available publicly.

18.9 Space War

On November 6, 2023, an Iranian supplied Ghandar-110 missile was launched from Yemen by the Houthis toward Israel, but was intercepted by the Israeli Arrow-2 system. On November 9, a second Ghandar missile was fired by the Houthis from Yemen toward the city of Eilat in southern Israel. This one was intercepted and destroyed by the newer Israeli Arrow-3 system. While the first intercept and destruction may

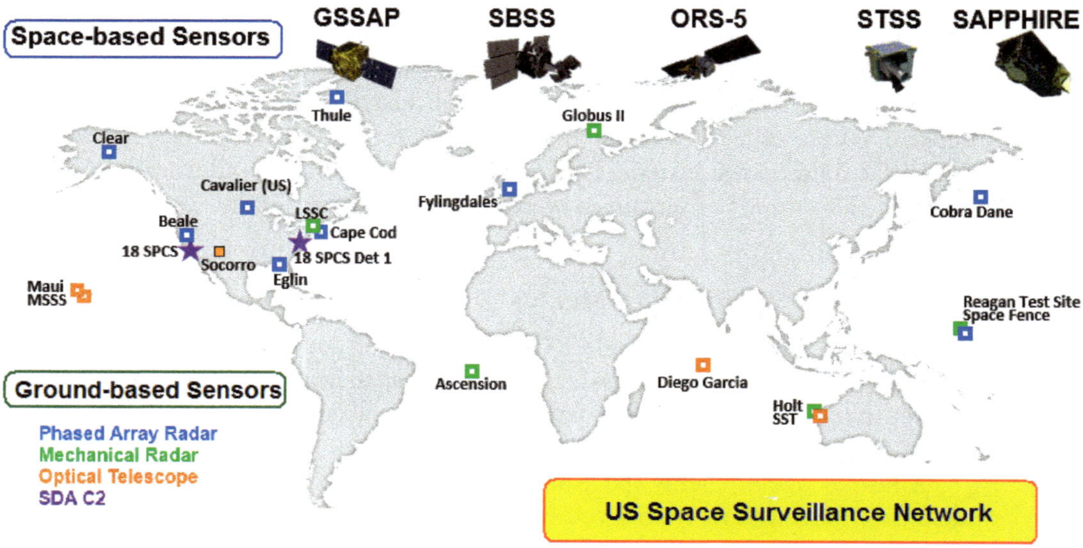

Figure 18.15 The US Space Surveillance Network (ground stations and satellites).

have taken place above a height of 100 km (the unofficial altitude at which outer space starts), the second incident definitely did. Some commentators have defined these incidents as the first rounds on war in space. (Although missiles have previously been destroyed by interceptors in the space environment, both the missile and the interceptor have always been from the same country in a test of such interception.)

The Ghandar-110 missile is specified as a medium-range ballistic missile (MRBM) with a range of 1,800 to 2,000 km and a maximum speed of Mach 9 (~3 km/s). The Arrow 3 or Hetz 3 is an exoatmospheric hypersonic antiballistic missile, jointly funded, developed, and produced by Israel and the United States. Manufactured by the Israel Aerospace Industries and Boeing, it has a specified maximum range of 2,400 km and an interception height of greater than 100 km. A true ballistic missile will in fact ascend to heights well above 100 km to reach its maximum range in the most efficient manner.

Although some may argue that the above scenario falls into ballistic missile defense rather than SSA, the take home message from this table is that any medium-range ballistic missile will reach well into low Earth orbit and will spend around 5 to 10 minutes in this domain. The technology used for these missiles is directly applicable to direct ascent kinetic kill missiles. All it requires is slightly more advance terminal guidance, which is most easily achieved with a camera and ground direction. It would certainly be good to pray for peace and attempt to negotiate a ban on such missiles, particularly with a foresight to limit collisional space debris, but in light of the current world situation it would also be wise to prepare for war. The other take home message is that such technology may also allow a nuclear device (stolen or provided by a third party) to be lofted into LEO and detonated to destroy a large number of satellites over a time period of several months (see Chapter 14 and the section on anthropogenic space weather).

Selected Bibliography

Advancing Space (Australia), Civil Space Strategy.pdf.

Air University Space Primer (US), AU-18.pdf.

Center for Strategic and International Studies (Aerospace), aerospace.csis.org.

Counterspace Capabilities, swf_global_counterspace_capabilities_2023_v2.pdf.

Madi, M., and O. Sokolova (eds), *Space Debris Peril: Pathways to Opportunities*, Boca Raton, FL: CRC Press, 2021.

Sach, R. H., *Astroreality: Avoiding Conflict in Outer Space*, Echo Books, 2015.

Secure World Foundation, swfound.org.

Space Safety Coalition, spacesafety.org.

Space Safety Magazine, spacesafetymagazine.com.

Space Security Index, www.spacesecurityindex.org.

The COPUOS Briefing Book, copuos-briefing-book-digital-1.pdf.

CHAPTER 19

Epilogue

19.1 The Current and Future Scenarios

SSA is a diverse and growing field. The main part of this book has been involved with the principles associated with human use of the near-space environment. This use continues to grow, both quantitatively and qualitatively. To that extent, this chapter is concerned with recent developments and potential future developments in and with the space domain. It thus should not be regarded as a conclusion but as an addition to, after, or above the work presented in the previous chapters.

19.2 Space Traffic Management/Awareness

As discussed in the Chapter 1, space traffic management is a term that is now used widely by a diverse range of actors to indicate a diverse range of actions that might, or should be, undertaken to resolve undesirable outcomes in the use of the space environment.

The term itself is unfortunate in that one cannot "manage" space "traffic" in the sense that the term traffic management is generally used in a terrestrial sense. It is not possible to stop objects in space from moving, or even change their speed from anything other than the speed that gravity defines for the orbit they are in. It is also not possible to stop the dispersion of fragments in a collision into a range of orbits prescribed by the laws of mechanics and gravity. Gravity is the only real manager of orbital motion. An excellent paper by Rebecca Reesman and James Wilson of The Aerospace Corporation, titled *The Physics of Space War: How Orbital Dynamics Constrain Space-to-Space Engagement*, should be required reading for all players in STM.

Although the larger satellites carry fuel to maneuver out of the way of potential impacts, this is not so for the majority of objects (particularly debris) in low Earth orbit. This would be analogous to some vehicles stopping at traffic lights while most ignored them. Not exactly a "managed" situation.

Although engineers and physicists no doubt understand these realities, legal and political professionals may not, and this may lead to unrealistic expectations if not unwarranted legislation. However, because STM has now become a popular bandwagon term and multibillion-dollar industry, even those who understand the unreality of the term are likely to play along to ensure their share in the money available for this fuzzy enterprise.

19.2.1 Definitions of STM

This is not to say that there are no definitions for STM. There are indeed many.

The International Academy of Astronautics defines STM as "the set of technical and regulatory provisions for promoting safe access into outer space, operations in outer space, and return from outer space to Earth free from physical or radio-frequency interference." This is a very interesting definition that few would argue with but it really says nothing about how to achieve the goal.

The US Space Policy Directive-3 (2018) defines STM as "the planning, coordination, and on-orbit synchronization of activities to enhance the safety, stability, and sustainability of operations in the space environment." The big problem with this is that the *on-orbit synchronization* has to be between not just entities in the United States, but between every state that has orbital assets in space! Even entities within the United States cannot agree with the meaning of on-orbit synchronization so there is no way that this concept could be extended to include all space actors.

A shorter definition: Space traffic management refers to the process of monitoring and regulating the movement of objects in outer space to ensure the safe and sustainable use of space resources and prevent collisions. The stumbling block here is the word "regulating." The only regulation possible is the movement of one object, out of the way, to avoid a collision when conjunction analysis of a space-track database indicates one is imminent.

In an article in the journal *Transportation Research Proceedia*, two authors did a linguistic analysis of 10 definitions by other authors and came up with a new definition that stated "Space Traffic Management is a process of control, monitoring and coordination of space traffic with a view to achieve safe operational environment without collisions and interference and with reliable collision avoidance." By this time you might be forgiven for believing that definitions are not worth much, at least in this instance.

19.2.2 The Real World

A talk by US Space Force General DeAnna Burt at the 2021 AMOS Tech Conference in Maui stated that "A lot of the time we talk about space traffic management ... I think we do space traffic awareness." She went on to say one of her organizations (18th Space Control Squadron) track space objects and predict potential collisions, but have little authority to prevent them. This is obviously the case when the two objects in question are pieces of debris with no ability to maneuver. In those instances, "We get to watch the objects collide and then report what happened." Even if one or both objects are active/operational satellites it is not always possible to do anything directly. "If that object is anything other than a US-flagged Department of Defense asset, I am making them aware there is a collision, and it's their choice what they do," she said. "We have no authority or governance to tell them what to do" ("them" in this instance implying the satellite owner).

The possible response depends on the situation. If a collision is flagged between an active satellite and a piece of debris, the satellite owner will usually elect to move their satellite to avoid damage and prevent the creation of further space debris. In the case of two active satellites flagged for collision, the situation is not so clear.

Moving both satellites is not a good option as this could increase the risk of collision, particularly if the coordination in not adequate. Who should move? Moves always involve use of fuel, which will deplete reserves and shorten the satellite lifetime, not to mention disruption of service. Different companies have different criteria that they use to determine if a COLA maneuver is required. A study found that miss-distance criteria varied from 1 to 15 km.

One of the problems is that the SSA data used in CONASS is simply not accurate enough. Typical accuracies from the US military catalog may be several hundred meters or more. Thus a flagged collision will in most instances not occur. Several private companies are now providing CONASS services based on their own space-object tracking data, and may reach better accuracies for a limited set of objects. Laser tracking in particular may provide sub-100m accuracy. However, few services will track objects down to 2 cm, and you only need to miss one, let alone a whole population, to cause significant damage to a satellite.

The USSF Space Command has realized that many satellite operators know the position of their satellites better than they do, and have set up a system to include private data into the DoD space object database. Some companies were, and some still are, reticent about sharing their data, particularly those that provide commercial services. Space Command has now provided a means for users to share data anonymously via a cloud. The problem with this is that error bars need to be accurately determined and specified by the uploader. That is, data quality needs to be closely monitored and not allowed to compromise the better quality data.

It has been stated that satellite dimensions need to be mapped and specified more accurately rather than just assuming a spherical satellite. This might be so if the dimension was even close to the inaccuracies of space-track data, but how many satellites have diameters larger than 100m? Let alone trying to account for satellite orientation. And for a collision with a 10-cm piece of debris all this is really not relevant.

But back to the situation where a possible collision risk exists between two active satellites. Who has the right of way? At the AMOS Tech panel discussion mentioned above, a representative of the European Space Agency said they moved their *Aeolus* Earth observation satellite in September 2019 to avoid a collision with a Starlink satellite. Starlink is a constellation of many communication satellites placed into orbit by SpaceX. SpaceX has been very responsive in addressing these types of issues (as well as those involving radio frequency interference), and have now made an agreement with NASA to move its satellites in the event of any close approaches to NASA satellites.

However, another communication satellite constellation operator, OneWeb, claimed that SpaceX was either unwilling or unable to move one of its satellites to avoid a flagged collision with a OneWeb satellite shortly after its launch from Russia. OneWeb then elected to move its satellite. In a response to the US Federal Communications Commission, SpaceX claimed that its analysis indicated that the risk was below its threshold for a maneuver but OneWeb elected to conduct a COLA maneuver anyway.

Starlink received initial approval for a constellation of 12,000 satellites but now has plans in the United States for a total of 42,000 in orbit. Other US companies, and at least seven other countries, have indicated they plan large LEO satellite constellations as well. It is predicted that at least 100,000 new satellites will be orbited by the end of the decade (2030).

So far the COLA maneuver discussions have mostly been between US entities. There is currently no agreement on standards. Do we think this situation is going to improve when the field is opened up to countries with a dislike for each other? China has made their dislike of the Starlink constellation apparent; and when their chairman announced that China would be orbiting a constellation of 13,000 satellites the language used made me wonder whether these would be used to provide a service to their people, or to attack the Starlink constellation. Countries that like to tightly control their media cannot be pleased with the current situation.

Will we see a concerted effort to jam foreign constellations (either with RF energy or via cyberattacks) when their satellites overfly domestic territory, reminiscent of the use and jamming of HF propaganda broadcasts of the Cold War era. This happened at the start of the Russian war with Ukraine when Viasat terminals were rendered inoperative by cyber means. At this point Elon Musk and Starlink stepped in to provide a communication service that has proved very resistant to attack. Will this continue?

A few other items for thought:

- Space traffic management should be about mitigating risk, not eliminating it.
- Space traffic management terminology is not universally harmonized today. The lack of a shared vocabulary stems from the continued evolution of the STM and space traffic coordination concepts.
- Solutions to the space-traffic problem mostly confront the difficult task of minimizing growth in the number of future objects and maximizing the accuracy of orbital predictions.

Contrast the above statement with the last eight words of the following paragraph.

The US 2018 Space Policy Directive-3, *National Space Traffic Management Policy*, states that "we must develop a new approach to space traffic management" that must "incorporate national security considerations" and "encourage growth of the US commercial space sector."

19.2.3 The Processes of Space Traffic Management

Collision avoidance is usually placed as the prime directive of STM. This is very much dependent upon an accurate and complete database or catalog of all objects in orbit down to a size of 2 cm. Collisions with objects below this size are usually not catastrophic. There is no such catalog in existence today. Until this is rectified we cannot expect CONASS and COLA (see Chapter 6) outputs to produce the most desirable outcomes.

Probably the second STM directive is space debris mitigation. This is a two-part process. The first part is the continuing development of launch procedures that will place a minimal amount of debris into orbit, and which includes the return of all parts of the launch vehicle back to the Earth. The second part is then related to debris removal from orbit. This is normally considered in two parts, the removal of small pieces of space debris and the removal of the large inactive satellites. Totally different techniques are required for the two populations. Removal of large objects that still have maneuvering capability is usually easily accomplished. For large objects in LEO, the satellite is commanded into a lower orbit that will result in a controlled reentry, usually over the southwestern Pacific Ocean devoid of islands. For large objects in GEO, an end-of-life maneuver commands an orbit-raising maneuver to place the object into a supersynchronous (at least 300 km above synchronous) orbit.

If a large debris object or inactive satellite is unable to maneuver it might be grappled by a dedicated disposal satellite and towed into an appropriate disposal orbit. This can take a lot of fuel and could probably only be considered for a limited number of objects.

Small pieces of debris pose a totally different challenge. They cannot maneuver and they cannot effectively be collected by a disposal satellite. The costs would be enormous.

The most promising candidate for removal is a ground laser. The laser would be directed toward a target as it breaches the horizon moving toward the ground-based laser (Figure 19.1). The laser beam would ablate material from the surface of the object producing a jet opposing the orbital motion of the debris. This would result in the object moving into a lower orbit where eventually it would burn up on atmospheric entry. It could not be used on geosynchronous debris. Its use would also have to be carefully timed so that no active satellites in the beam (beyond the intended target) were damaged. This could be a political nightmare as the distinction between a debris disposal laser and a laser weapon would be a difficult call.

Space traffic management is a booming industry. Many private companies (mostly in the United States) have sprung up offering STM services. Some of these provide advice on CONASS and COLA based on the US Space Command Satellite

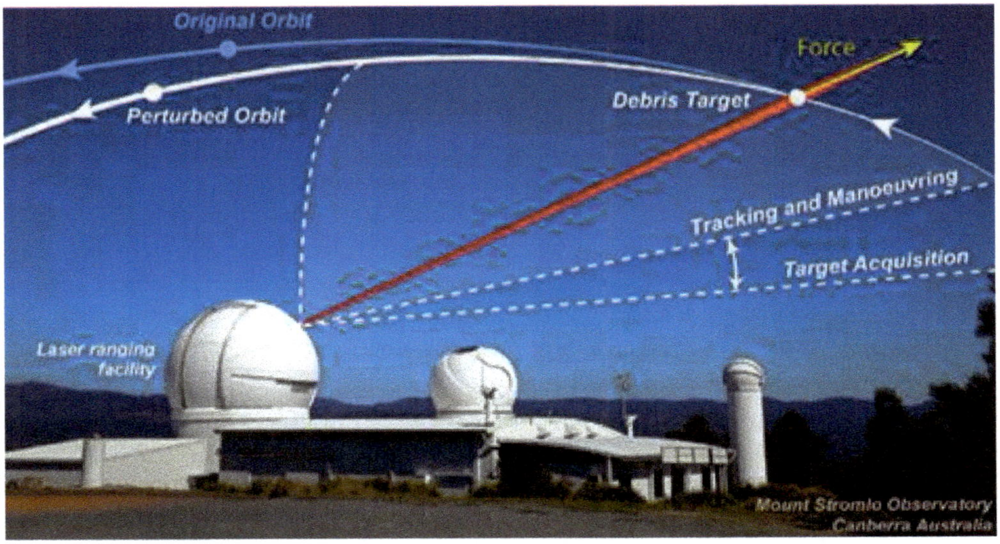

Figure 19.1 Operation of a laser system to remove small space debris from LEO. (*Source:* ElectroOptic Systems, Canberra.)

Catalog, but an increasing number are setting their own global space-tracking stations. These include ExoAnalytics, COMSPoC, Slingshot Aerospace, and LEO-Labs. Some offer STM software services and packages that include advice on the man-made and natural space environment (space weather), all with very attractive graphics. Saber Astronautics is one of these.

Many universities and corporations are developing and testing schemes for space debris removal with prototypes manufactured for actual testing in the space domain.

Astroscale (astroscale.com) is a private orbital-debris removal company headquartered in Tokyo, Japan. The company is developing satellite end-of-life and active debris-removal services to mitigate the growing and hazardous buildup. Two in-space demonstrators have been built, the first launched in 2021. D-Orbit is an Italian company that has developed space tug and in-orbit servicing missions. Both of these companies now have divisions in many other countries including the United Kingdom and the United States. They are only two companies among others that are working on space-debris mitigation services. Some of the others include Clearspace, Airbus, Vyoma, Transastra, NASA, and ESA.

And of course we have a plethora of papers and books with suggested schemes for the regulation of the space industry that will solve the STM problem. Some of these relate to unsolved problems in international space law, some discuss regulation of the new LEO constellations, and a few concentrate on the new era of space tourism.

Figure 19.2 offers a graphical perspective of three principal STM directives.

The three components given in Figure 19.2 are not the only ones, simply those that many consider the most important. In a question and answer report released in February 2022, the European Commission defined five elements in their definition of STM. These were:

1. Space situational awareness activities, including SST;
2. Orbital debris mitigation and remediation;
3. Management of space orbits and radio spectrum;
4. The entire life cycle of space operations including launch phase, in-orbit operations of spacecraft, and end-of-life deorbit operations;
5. Reentry phase of spacecraft (both controlled and uncontrolled).

It is interesting that collision avoidance, which many would regard as the highest priority element of STM, has not been specifically mentioned.

Other aims associated with STM include satellite life extension. This is not a new concept and NASA used the Space Shuttle to refuel high-cost assets such as the Hubble Space Telescope. And not only refuel but repair and upgrade a spacecraft as well. The HST is again a good example of this. After launch, it was discovered that the main mirror had been wrongly configured. It was made with a spherical rather than the required parabolic surface. Engineers developed a corrector lens, which Space Shuttle astronauts were then able to insert into the telescope with great success, as shown by the amazing images we see today. In fact one of the many versatilities of the Space Shuttle touted was its ability to service and repair other satellites. At one point, a failed Intelsat spacecraft was returned to Earth for repair and later relaunched. However, all was not roses. The cost of using the Space Shuttle for

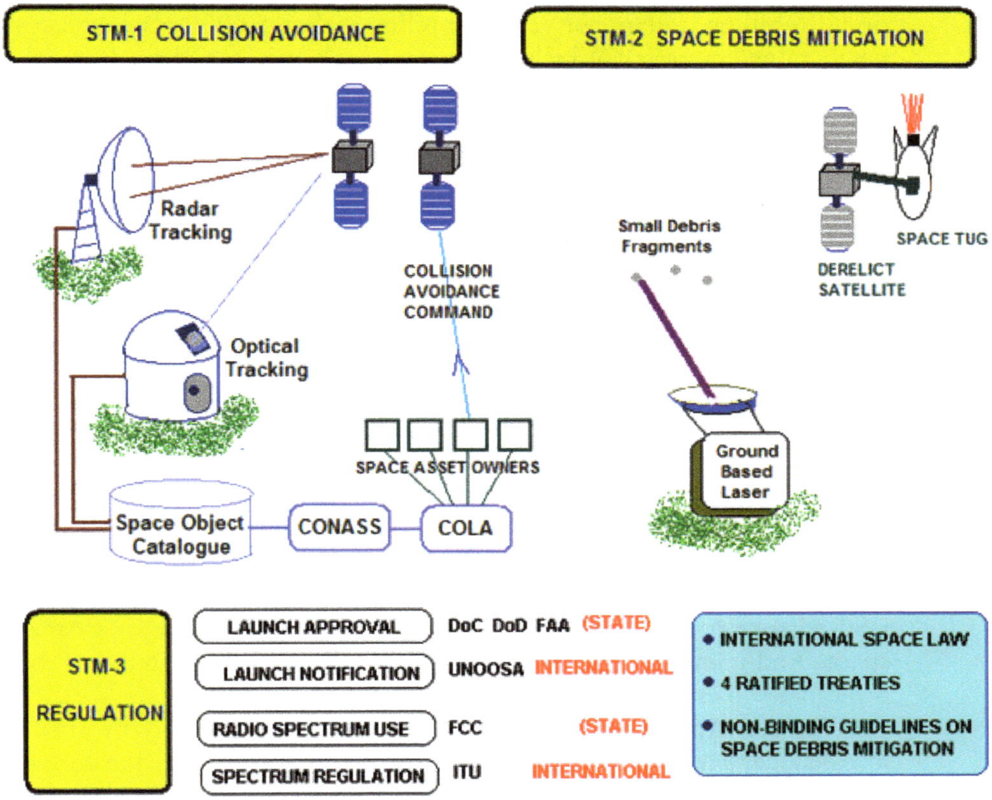

Figure 19.2 Graphical summary of three important STM components. Components one and three are as established at present. Component 2 is still under development.

these missions was greater than the cost of building and launching a new satellite. The cost of "reusable" was so expensive that NASA had to terminate the Shuttle program after 135 launches.

D-Orbit has now developed an unmanned space tug to not only remove defunct satellites from orbit, but also to refuel and repair satellites specifically designed for these operations.

The ultimate STM dream is to develop a circular space economy in which all material sent into orbit can be recycled. However, a quick calculation shows that this does not look to be feasible in the foreseeable future. The total mass of material in orbit has been estimated by ESA to be around 12,000 tons. Much of this will be aluminum (this is what in space debris models the composition of debris particles is taken to be). Currently the amount of aluminum mined each year from the ground is 68 million tons. Thus the total orbital population is only 0.02% of this amount. This is negligible. Assume we launch 1,000 tons (an overestimate) per year, with a cost of $2,000 per ton of aluminum, the total cost of aluminum for the space industry is two million dollars. The cost of collecting and bringing this back to Earth for recycling is likely to be a minimum of $100 million dollars per ton, or \$100 billion dollars per year. Recycling is thus 50,000 times the cost of using newly mined material! Whichever way we look at it, recycling does not look even close to being economically beneficial.

At the moment, regulation of space activities has basically been devolved to the states. Many legal people would like to see this situation changed and move to a system of closer international control and supervision. All current indications are that this is not likely to happen.

19.3 The Australian Connection

Although space situational awareness is most definitely a global endeavor, throughout this book there have been many references and examples from Australian. There are at least two reasons. One is that the author is an Australian and is more aware of space tracking and other space-related activities in this country. The other is that Australia has a long history of space-object tracking, usually through agreements with the United States and to a lesser extent Europe. Australia was also the fifth country to launch its own satellite from its own territory (Woomera), albeit using a US Redstone rocket, the same vehicle-type that launched the first Americans into space. And it enabled the United Kingdom to put its first satellite (*Prospero*) into orbit in 1971 using its own launcher (Black Arrow).

The first Australian satellite-tracking efforts were mainly by amateur groups in Adelaide, Melbourne, Sydney, Perth, and Woomera under the Moontrack program of the Smithsonian Astrophysical Observatory (SAO). This was originally set up and equipment donated to track the US *Vanguard* and *Explorer* satellites. The overall Australian Moonwatch coordinator was Professor Graham Elford from the University of Adelaide Physics Department. However, with the surprise launch of the Soviet *Sputnik-1* in early October the Moonwatch teams were able to observe the satellite on October 8th.

Along with the Moonwatch project, the US Naval Research Lab developed the Minitrack radio interferometer system to give precise positions from satellite beacons transmitting just above 108 MHz. The SAO also developed the Baker-Nunn camera, a high-precision wide-angle field of view optical instrument that could determine highly accurate satellite positions referenced to the background stars, which were then translated into orbits. Both of these instruments were placed near the small town of Woomera that had been built in the middle of nowhere to conduct rocket experiments. The Baker-Nunn camera needed cloud-free skies and these were provided by the arid desert environment in the center of South Australia. It is these same skies in large areas of Western Australia that have more recently attracted numbers of private and government SSA instruments seeking optical viewing facilities.

Since the late 1950s, Australia has been involved in tracking all US manned space flight missions and many unmanned flights. The first manned Mercury missions were tracked from Muchea, just north of Perth, Western Australia. Approximately 1,000 km further north the Carnarvon tracking station was built for the tracking the Gemini manned missions. It also played a critical role in the Apollo missions.

Australia was a founding member of Intelsat, and tracking stations at Sydney, Carnarvon, and Ceduna in South Australia were built to accommodate the needs of satellite tracking, telemetry, and control (TTC) and provide global telecommunication for Australia. Space meteorology services were provided to Australia via satellites orbited by other nations. Because one-third of Australia is located in the tropics, it was often better served by these satellites than was the hosting nation (e.g., Japan). A downlink for Landsat and other environmental observation satellites was provided by a station in Alice Springs.

The US military was also quick to seize upon Australian as a prime location for defense communication and control, and intelligence activities (Woomera, Alice Springs-Pine Gap) and Kojarena (Western Australia). An Australian intelligence analyst once wrote a book entitled *A Suitable Piece of Real Estate*. This was a biased look at the United States' use of Australia for hosting military and intelligence satellite facilities. Australia benefitted enormously from these stations in many ways, and I speak as an Australian on the receiving end of some of these.

Australia also hosts one of the three NASA Deep Space Network stations in the world, the other two being at Goldstone, California, and Madrid in Spain. The Australian DSN station is located at Tidbinbilla in the south of the small area of land known as the Australian Capital Territory. The DSN station is now known as the Canberra Deep Space Communications Complex (CDSCC), and shares in the tracking of space missions that move beyond cislunar space to Mars, Jupiter, and beyond Pluto. It is some distance from the capital city of Canberra to ensure a quiet radio frequency environment.

Although Woomera never achieved its potential as a satellite launch facility, there are now three rocket-launching facilities in Australia:

1. The Arnhem Space Centre (ASC) is Australia's first commercial spaceport, located near Nhulunbuy, in Arnhem Land, in the north of the Northern Territory. It is owned and operated by Equatorial Launch Australia (ELA).
2. Southern Launch is a launch and range service provider, headquartered in Adelaide, South Australia. They own and operate Australia's only rocket launch ranges approved by the Australian Space Agency for launches to space, which includes the Whalers Way Orbital Launch Complex, and the Koonibba Test Range, their suborbital testing facility.
3. Gilmour Space Technologies is developing an indigenous rocket and has a launch site near Bowen on the central Queensland coast.

The current interest in space in general and in placing SSA sensors in Australia was definitely helped by the Australian Space Agency, which was founded in 2018, but was underway before then by agreements between private companies and Australians who had established observatories specifically to host such enterprises. However, since 2018, the number of organizations (including foreign governments) interested in placing their SSA sensors in Western Australia has skyrocketed. When a colleague of mine asked one of these groups why the sudden interest, he was told that "we have interesting neighbors." I elaborated further by pointing out that Beijing is almost due north of Perth, the capital city of Western Australia.

Kerrie Dougherty has done an excellent job of documenting Australia's role in space with the book *Australia in Space* (ATF Press, 2017) for those who wish to learn more.

19.4 SSA Update

The space orbital environment is a rapidly changing domain, and research and development in the area of space situational awareness is changing to meet the challenge. There are a few areas of special interest that we list below.

19.4.1 Megaconstellations and the Paradigm Shift to LEO SSA

Satellite constellations have long been a part of the space domain throughout all orbital regimes. Environmental monitoring, navigation, and communication satellites have all been deployed in constellations of different sizes. What has changed is the initial deployment of megaconstellations in LEO for communication (and environmental monitoring). The definition of the size of a megaconstellation is not clear but it probably has to lie somewhere in the range of 100 to 1,000 satellites. The SpaceX company of Elon Musk really brought the idea of megaconstellations to the fore through the initial simultaneous launch of 60 Starlink communications satellites in 2019, with an intent to launch up to 12,000 more (and a future proposal of up to 42,000). Several other companies have proposed and made initial launches of their own megaconstellations. These include OneWeb (2,000), Boeing (3,000), Amazon (3,000), and China (13,000).

Lists are not very useful at the moment because of the fluidity of the market. Some of the above may drop by the way and others emerge. What is certain (just from Starlink) is that there is now a paradigm shift from GEO to LEO. This is due to the need to reduce communication latency for high-speed data transfers in communication, and the need for higher resolution observations of the terrestrial environment in Earth observation. This order of magnitudes increase in the LEO population has seen a subsequent increase in LEO SSA sensors and services, and the movement of the same from government to private operators. Collision avoidance maneuvers have also seen a substantial rise.

19.4.2 Daytime Optical Observations

One of the latest developments in space-object tracking has been the introduction of daytime optical observations of (mainly) geosats. This requires a number of changes to an optical system. The first is the reduction of scattered skylight into the telescope. The principle comes from the observation that if you are in a deep narrow hole and look up you may see some of the brighter stars. So you place a long cylindrical shield or baffle around the front of your telescope; since generally, refracting telescopes are used as these have less scattered light problems than reflectors. You should also pay attention to and reduce dust particles in the optical train. Next you reduce the bandwidth of your observations and move to a wavelength where the sky light to target light ratio is lower. This generally means infrared wavelengths, but only a band where the Earth's atmosphere allows significant transmission of energy from space. The final technique is to take lots of very short exposures and stack them to produce the final image. In the stacking, images with bad seeing can be eliminated. This might be done looking at the contrast of an image, since images with low contrast indicate lower quality due to atmospheric effects. Contrast can be determined by computing a brightness distribution of the image. A wide distribution indicates a lower quality image. This technique has long been used in automatic camera focusing. A lot of computing, especially with GPUs for parallel processing, is necessary to deal with continuous imaging in real time. Slingshot Aerospace appears to be the leader in this field, but others are working on the problem.

19.4.3 Radar Observations of Geosats

It used to be thought that ground radar observations of space objects could generally be conducted only in low Earth orbit. The reason for this is twofold. Unlike passive optical observations that use reflected sunlight with an object brightness proportional to the inverse square of the object distance, radar observations have a returned signal strength that varies as the inverse fourth power of the object distance. Thus the difference between a LEO object 1,000 km from the sensor and a GEO object 40,000 km away is $(40)^4 = 2.56 \times 10^6$, or a GEO signal over 2.5 million times weaker than the LEO signal. The second reason against a GEO radar is resolution. The resolution at GEO of an S-band radar with a 20m dish antenna is $\theta \sim \lambda/D$ = 10 cm / 20m = 0.005 radians or 0.3 degrees. At a distance of 40,000 km this equates to a distance of 200 km, making it difficult to know what satellite you are pointing at or even if more than one satellite was in the beam.

Despite these formidable obstacles, the Applied Physics Laboratory at John Hopkins University has now developed a geosat radar for the US Department of Defense. Known as the Deep Space Advanced Radar Concept (DARC), the prototype uses around six dishes of 15m diameter in an interferometer. Little more is known about the prototype that has been slated for deployment on North West Cape in Western Australia, where it will join the 3-meter optical Space Surveillance Telescope developed by MIT Lincoln Labs developed optical Space Surveillance Telescope.

Looking at the feed on the antennas, it appears to me that the operating frequency is well above S-band, and may be X-band or higher. The interferometer antenna system should give a resolution comparable to a typical geosat optical system and provide an accurate distance, which the optical system cannot do. It will also allow daytime tracking.

19.4.4 Characterization and Recognition of Unresolved Space Objects

An optical system is generally limited to an angular resolution of one second of arc (~1/206,000 radian) by the optical seeing or turbulence in the atmosphere. At geosynchronous orbit (~40,000 km) this equates to a size or satellite diameter of 200m. There are probably no objects this size in GEO, which implies that all observations of geosats will be unresolved. They will appear as point sources, with no information about their size or shape. Because of this, there is an increasing interest in trying to characterize and even identify a geosat from these limited observations. Object brightness, polarization, and spectral characteristics and how these vary with time can all be used to try and solve this problem. Artificial intelligence techniques are also being used to help in this quest.

19.4.5 Space Observations from Space

Most SSA observations are carried out using instruments on the ground. However, the limitations of these observations with respect to objects in geosynchronous orbit are leading to the development of spacecraft capable of making SSA observations from space. The ability to fly around geosynchronous object and perform close satellite and debris inspections is seen as particularly desirable. Russia, China, and the United States currently have spacecraft in GEO orbits capable of this task.

19.4.6 Cislunar SSA

This is a really hot topic in SSA research at the moment with a large number of papers addressing various aspects of the subject at the 2023 AMOS tech conference in Maui. The term cislunar strictly refers to all the space between the Earth and lunar orbit, but current usage generally means the space around and near the Moon. The interest in the topic stems from an international interest in returning to the Moon and recent space missions from several countries to the surface of the moon and in orbit around it.

The problem is how to track such missions and the debris they will create. Optical observations from the Earth are possible only with the largest telescopes, which are often fully subscribed for astronomical observations and not available for SSA purposes. At a lunar distance of 384,000 km the maximum brightness of a lunar satellite (in magnitudes, a logarithmic scale) is given in Table 19.1.

A 1m telescope has a limiting magnitude of around 21 for a 3-minute exposure, and so could theoretically just detect a 2m object in lunar orbit if the moon were not present. It is the presence of the Moon that makes lunar SSA so difficult, the full Moon has a magnitude of −12, and even when it is only a crescent, the dark side is illuminated by Earthshine and it has a point magnitude below (brighter than) 15 (this depends on many factors). A 10m object is thus very unlikely to be seen above or close to the lunar sphere. A close lunar orbit would be around an altitude of 100 km or less. This distance subtends an angle of nearly one arcminute at the Earth, well above the resolution limit of an optical telescope. It is the object contrast that is the problem.

Radar tracking is out of the question (at least from Earth), but passive radio tracking of a beacon is certainly possible. There are many interesting proposed plans

Table 19.1 Magnitude of a Lunar Satellite Versus Size

Satellite Diameter (m)	*Satellite Magnitude*
0.1	27.2
0.2	25.7
0.5	23.7
1.0	22.2
2.0	20.7
5.0	18.7
10.0	17.2

in this area and the reader is directed to the AMOStech technical papers (amostech.com) for further information.

To add to the tracking we might mention the large volume involved in cislunar space (Figure 19.3). This is about 2.3×10^{17} cubic kilometers (versus 3×10^{14} for the space below GEO). This thousand-fold increase means a lot less potential impacts in the future, but also a much larger search area when an object near the moon becomes lost.

Loss of tracking ability will also occur when a spacecraft is behind the Moon or in front of the Sun.

Orbits near and around the moon are totally unlike orbits around the Earth because the gravitational effects of the Moon, Earth, and the Sun must be considered. This leads to highly nonlinear dynamics and a multibody problem, which makes orbit prediction very difficult and large uncertainties.

Because of significant mass concentrations in the Moon, only a few orbital parameters give rise to stable orbits below 100-km lunar latitude (where the Earth's gravity is substantially less than that of the Moon). Most other orbits will result in a decrease of orbital altitude with subsequent lunar impact within short periods of time. The orbital period of such orbits are around two hours.

Another favored orbit is the NRHO or near rectilinear halo orbit. This is a stable orbit with a period of one week that balances the lunar and terrestrial gravity with the Earth-Moon Lagrange points (zero gravitational forces). Lunar distances range from 3,000 to 70,000 km.

19.4.7 Militarization of Space

The militarization of space is is the most worrisome development in SSA. The University of Adelaide (South Australia) is leading a project to produce the *Woomera*

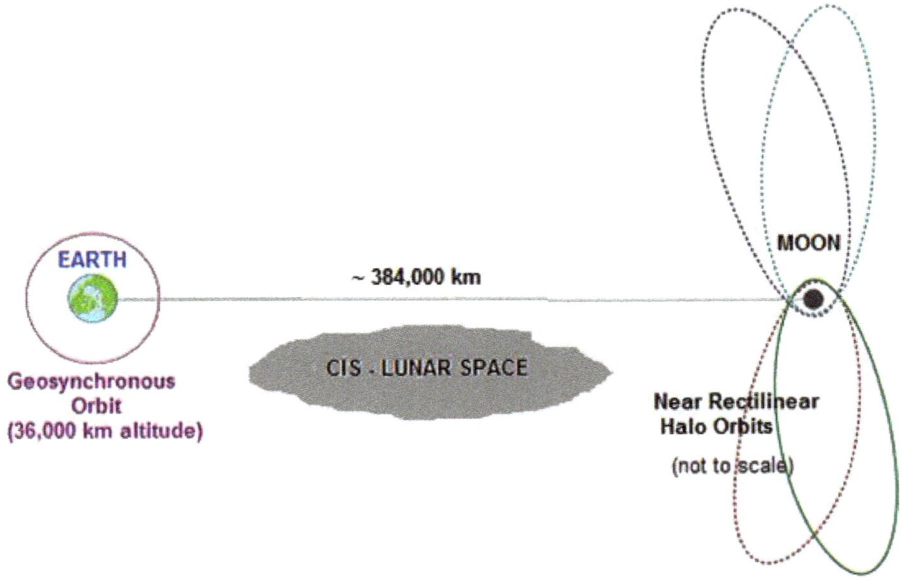

Figure 19.3 Cislunar space, the volume between the Earth and the Moon's orbit.

Manual, a document that will clarify the rules of engagement in space war, for as one of the project founders said, "Conflict in outer space is not a case of *if* but of *when*." Current indications are that the when will be sooner rather than later. Conflict may not always involve kinetic means but may use cyberattacks instead to disable a satellite or a network of satellites. Refer to Chapter 18 for indications that this has already happened.

The US military has always been associated with space since the earliest days, and this has undoubtedly been the case for many other countries. These include military missions for communication and to gain intelligence about activities in other countries. The military has also been at the forefront of new ideas in the use of space. Well before the digital age, the United States was using innovative techniques to obtain high-quality Earth imagery from space. This was done by using film. When the images had been taken, a canister was ejected from the spacecraft, which descended safely through the Earth's atmosphere and deployed a parachute for the final descent. This was then snatched up by a C-130 *Hercules* aircraft configured with a special attachment in the front to pluck up the descending capsule before it reached the ground. Among other things these images were used by President John F. Kennedy to force the Soviets to withdraw their missiles that were secretly placed in Cuba in 1962 (the Cuban Missile Crisis). ASAT testing was also arguably more frequent (although less devastating) several decades ago.

So what is the difference between then and now in the militarization of space? The scale of space operations of all kinds is one factor, the increased number of operators is another, some whom believe that war is essential to bring their savior back to Earth. And the dependence of the major countries on space for their warfighting operations is another. The final factor is the increasing belligerence of countries toward each other in the last few years.

An increasing number of research papers related to space warfare are appearing in various journals. Making war in a GPS-denied environment was on the AMOStech paper list, as were several SSA papers discussing how to determine if specific satellite maneuvers are hostile or otherwise and indications of adversary action. In 2022, five Chinese researchers published a paper in the journal *Modern Defense Technology* titled "The Development Status of Starlink and its Countermeasures." They called for the development of antisatellite capabilities, including a surveillance system with unprecedented scale and sensitivity to track, monitor, and disable each spacecraft in the Starlink constellation. "A combination of soft and hard kill methods should be adopted to make some Starlink satellites lose their functions and destroy the constellation's operating system," said the authors.

The Space World Foundation (swfound.org) has published a very comprehensive report on world counterspace capabilities, which discusses the systems of the major players in space warfare along with some of the minor ones.

Some countries see the incomparable reliance of the United States on space assets for war fighting, not just in space but on the ground, and believe that a space Pearl Harbor is their best means to level the playing field.

Unfriendly rendezvous and proximity operations (RPO) are becoming more common in both LEO and GEO in order for one country to spy on and obtain data about the purpose and activity of another country's satellite(s). This investigation can involve both the visual and radio output of a satellite. It may in the future be the precursor to the destruction of the satellite.

A space war is probably more likely to be started from ground hostilities than in the space domain, and there are many potential flashpoints around the world, including the intense polarization and division within the United States itself, which could lead to a second civil war within the next few years.

List of Acronyms

AGI Analytic Graphics, Inc. (US company)
ASA Australian Space Academy
ASAT Antisatellite
ASM Antisatellite missile
COLA Collision avoidance
CONASS Conjunction assessment
COPUOS Committee for the Peaceful Uses of Outer Space (UN)
DoD Department of Defense (US)
CFE Commercial and Foreign Entities (US JSpOC program)
ECSL European Centre for Space Law
EMP Electromagnetic pulse (produced by a nuclear detonation)
EOS Electro-Optic Systems (Australian company)
ESA European Space Agency
ESTEC European Space Research and Technology Centre
GEO Geosynchronous Earth orbit (36,000-km altitude)
GEODSS Ground-Based Electro-Optical Deep Space Surveillance System (US)
GNSS Global Navigation Satellite System
GPS Global Positioning Satellite (system)
GW Gigawatts (one billion watts)
HANE High-altitude nuclear explosion
HF High frequency (3 to 30 MHz)
HUMINT Human intelligence
IAF International Astronautical Federation
ICG International Committee on Global Navigation Satellite Systems
IISL International Institute of Space Law (part of IAF)
ISS International Space Station
ITU International Telecommunications Union
LS Legal Subcommittee (of COPUOS)
J Joule (SI unit of energy)
JPL Jet Propulsion Laboratory (NASA)

JSpOC Joint Space Operations Center (US)

K Kinetic energy

kHz Kilohertz

LEO Low earth orbit (below 2,000-km altitude)

MHz Megahertz

MIJI Meaconing, Intrusion, Jamming, and Interference (US military)

NASA National Aeronautics and Space Administration (US)

NEA Near-earth asteroids

NEO Near-earth objects (asteroids and comets)

ODPO Orbital Debris Program Office (NASA)

ODQN *Orbital Debris Quarterly Newsletter* (published by ODPO)

OSO Orbital space objects (active satellites and space debris)

OTHR Over the horizon radar

PHA Potentially hazardous

PSA Program on Space Applications (UN)

SAO Smithsonian Astrophysical Observatory

SATCAT STRATCOM Satellite Catalog (US)

SATC Now the preferred term instead of SATCAT

SDA Space domain awareness

SFU Solar flux units (10^{-22} W/m^2/Hz)

SLU Space Law Update (e-publication of UNOOSA)

SMM Solar Maximum Mission (NASA spacecraft, solar observations in the 1980s)

SOT Space-object tracking

SPIDER Space-based Information and Solutions for Disaster Management and Emergency Response (UN)

STS Scientific and Technical Subcommittee (of COPUOS)

SSA Space situational awareness

SSN Space Surveillance Network (US)

SST Space surveillance and tracking

STM Space traffic management

SWE Space weather

SWF Secure World Foundation

SWPC Space Weather Prediction Service (US)

STRATCOM Strategic Command (US)

SWPC Space Weather Prediction Center (US)

SWS Space Weather Services (Australia)

TNT Trinitrotoluene (high-energy explosive)

UN United Nations

UNGA UN General Assembly

UNOOSA United Nations Office of Outer Space Affairs

US United States of America (USA)

USSF US Space Force

UHF Ultra high frequency (300 to 3,000 MHz)

VHF Very high frequency (30 to 300 MHz)

VLF Very low frequency (3 to 30 kHz)

About the Author

John A. Kennewell is a respected Australian physicist and researcher specializing in space situational awareness, space weather, and observational astronomy. With a career spanning both academic and applied research, he has contributed significantly to the understanding and monitoring of near-Earth space environments.

Dr. Kennewell is an adjunct professor at the Curtin University of Technology and holds an adjunct senior lecturer position at the University of Western Australia (UWA) in the School of Physics, Mathematics and Computing. He is also affiliated with the Australian Space Academy, where he has coauthored influential work on SSA, including the widely cited paper, "An Overview of Space Situational Awareness." His research interests include space surveillance and tracking, space weather, near-Earth object monitoring, optical and radio astronomy, and data fusion and visualization for SSA systems.

Dr. Kennewell has collaborated on international projects involving asteroid tracking and transient astronomical events, and has contributed to the development of SSA frameworks that integrate both natural and artificial space object monitoring. He is also known for his work at the Zadko Observatory in Western Australia, contributing to studies of gamma-ray bursts and gravitational wave counterparts.

Index

D

H

I

J

K

L

T

U

V

W

X

Artech House
Intelligence and Information Operations Library

Ed Waltz, Series Editor

Activity-Based Intelligence: Principles and Applications, Patrick Biltgen and Stephen Ryan

Advances in Statistical Multisource-Multitarget Information Fusion, Ronald P. S. Mahler

The Art and Science of Military Deception, Hy Rothstein and Barton Whaley

Aviation Security Engineering: A Holistic Approach, Rainer Kölle, Garik Markarian, and Alex Tarter

Concepts, Models, and Tools for Information Fusion, Éloi Bossé, Jean Roy, and Steve Wark

Counterdeception Principles and Applications for National Security, Michael Bennett

Cyberwarfare: An Introduction to Information-Age Conflict, Isaac R. Porche III

Data Fusion Support to Activity-Based Intelligence, Richard T. Antony

Designing Wireless Sensor Network Solutions for Tactical ISR, Timothy D. Cole

Essential Principles of Signals Collection and Analysis, David A. Garren, Uttam K. Majumder, and Erik P. Blasch

Geospatial Data, Information, and Intelligence, Aaron Jabbour and Renny Babiarz

High-Level Data Fusion, Subrata Das

High-Level Information Fusion Management and Systems Design, Erik Blasch, Éloi Bossé, and Dale A. Lambert, editors

Homeland Security Technology Challenges: From Sensing and Encrypting to Mining and Modeling, Giorgio Franceschetti and Marina Grossi, editors

Human-Centered Information Fusion, David L. Hall and John M. Jordan

Information Fusion and Analytics for Big Data and IoT, Éloi Bossé and Basel Solaiman

Information Warfare and Organizational Decision-Making, Alexander Kott, editor

Information Operations Planning, Patrick D. Allen

Information Warfare and Electronic Warfare Systems, Richard A. Poisel

Information Warfare Principles and Operations, Edward Waltz

Knowledge Management in the Intelligence Enterprise, Edward Waltz

Mathematical Techniques in Multisensor Data Fusion, Second Edition, David L. Hall and Sonya A. H. McMullen

Naval ISR Fusion Principles, Operations, and Technologies, Jim Scrofani, Will Williamson, Jihane Mimih, and Ed Waltz

Principles of Data Fusion Automation, Richard T. Antony

Sensor Management in ISR, Kenneth J. Hintz

SSA: Orbital Debris, Space Weather, and Planetary Defense, John A. Kennewell

Strategem: Deception and Surprise in War, Barton Whaley

Statistical Multisource-Multitarget Information Fusion, Ronald P. S. Mahler